AF560104

DPH MATHEMATICS SERIES

TEXT BOOK OF

3-D

(SPHERE, CONE AND CYLINDER)

By

A.K. Sharma

DISCOVERY PUBLISHING HOUSE

NEW DELHI-110002

First Published – 2005

Reprinted – 2025

ISBN: 978-81-7141-965-4

© Author

Text Book of 3-D Sphere, Cone and Cylinder

Published by:

DISCOVERY PUBLISHING HOUSE

4383/4B, Ansari Road, Darya Ganj
New Delhi-110 002 (India)
Phone: +91-11-23279245; 23253475; 43596065
Mobile: +91 9811179893 / +91 9871656464
E-mail: discoverybooksindia@gmail.com
orderdphbooks@gmail.com
namitwasan9@gmail.com
web: www.discoverypublishinggroup.com

Printed at:

Infinity Imaging Systems
Delhi

Preface

This book Text book of 3-D Sphere, Cone and Cylinder has been written for the use of B.A/B.Sc., students of various Indian Universities. The subject matter has been explained in details and in such a nice manner that the students has no difficulty to understand. All the examples have been completely solved.

I believe that this book will serve specific need of the students appearing in the examination of B.A/B.Sc., and other competitive examinations.

Suggestion and comments for the improvement of the book from reader will be thankfully received and duly incorporated in this subsequent editions.

A.K. Sharma

Preface

CONTENTS

SPHERE

Definition : *A sphere is the locus of a point which moves such that its distance from a fixed point always remains constant.*

The constant distance is called the *radius* and the fixed point is called the *centre* of the sphere.

THE EQUATION OF A SPHERE

(a) When Centre and Radius are Given (Central Form)

Let C(a, b, c) be the centre and r the radius of the sphere. Then if P(x, y, z) be any point on the surface then we have

$$CP = \text{radius of the sphere} = r \text{ i.e., } CP^2 = r^2,$$

and therefore $\mathbf{(x - a)^2 + (y - b)^2 + (z - c)^2 = r^2}$ ..(i)

$$x^2 + y^2 + z^2 - 2ax - 2by - 2cz + (a^2 + b^2 - x^2 - r^2) = 0$$

We note the following characteristics of the equation (A) o the sphere:

(1) It is the second degree in x, y, z.

(2) The coefficient of x^2, y^2, z^2 are all equal.

(3) The product terms xy, yz, zx are absent conversely we consider the equation. $ax^2 + ay^2 + 2ax + avy + 2wz + d = 0$, $a \neq 0$ having the about three characteristics (B) $a_1u_1v_1w_1d$ being given constants and $a \neq 0$. The equation (B) can be rewritten is

$$\left(x+\frac{y}{a}\right)^2 + \left(y+\frac{v}{a}\right)^2 + \left(z+\frac{w}{a}\right)^2 = \frac{u^2+v^2+w^2-ad}{a}$$

this manner of rewriting shows that the distance between the variable points $(x_1y_1z_1)$ and the fixed point.

$$\left(\frac{-y}{a}+\frac{-v}{a}-\frac{w}{a}\right)$$

is $\sqrt{u^2+v^2+w^2}$ ad . $u^2 + v^2 + w^2 - ad \geq 0$ and is therefore constant the locus of the equation B is this sphere if $u^2 + v^2 + w^2 - ad \geq 0$.

Particular Case: Let origin O(0, 0, 0) be the centre and r the radius of the sphere. Let P(x, y, z) be any point on this sphere.

Then OP = radius of the sphere = r (given)

$\Rightarrow$ $OP^2 = r^2$ or $(x - 0)^2 + (y - 0)^2 + (z - 0)^2 = r^2$

$\Rightarrow$ $\mathbf{x^2 + y^2 + z^2 = r^2}$ ...(ii)

which is called the *standard form* of the equation of the sphere.

(b) General Equation of a Sphere

The equation (i) above can be expanded and written as

$$x^2 + y^2 + z^2 - 2ax - 2by - 2cz - (a^2 + b^2 + c^2 - r^2) = 0$$

which is of the form

$$\mathbf{x^2 + y^2 + z^2 + 2ux + 2vy + 2wz = 0.} \qquad \text{...(iii)}$$

and this is known as the *general form* of the equation of a sphere.

Centre and Radius for General Form

The equation (iii) can be rewritten as

$$(x + u)^2 + (y + v)^2 + (z + w)^2 = u^2 + v^2 + w^2 - d.$$

$\Rightarrow$ r $[x - (-u)^2] + [y - (-v)]^2 + [z - (-w)]^2 = u^2 + v^2 + w^2 - d.$

Comparing this with equation (i) above, we find that the centre and radius of the sphere given by (iii) are $(-u, -v, -w)$ and $\sqrt{(u^2 + v^2 + w^2 - d)}$ respectively.

Hence, *centre is $(-u, -v, -w)$ and radius* = $\sqrt{(u^2 + v^2 + w^2 - d)}$.

METHOD OF WRITING THE CENTRE AND RADIUS OF A SPHERE

(i) Write down the coefficients of x^2, y^2 and z^2 as 1, if these are not so, by dividing the equation by the coefficient of x^2. For example if the equation of the sphere is given as $2x^2 + 2y^2 + 2z^2 - 10x - 12y + 16z + 23 = 0$, then we should divide each term by the coefficient 2 of x^2 and write the given equation as

$$x^2 + y^2 + z^2 - 5x - 6y + 8z + (23/2) = 0.$$

(ii) Then the coordinates of the centre of the sphere are given as

$[-\frac{1}{2}$(coefficient of x), $-\frac{1}{2}$(coeff. of y), $-(\frac{1}{2}$ coeff. of z)] and the radius of the sphere = $\sqrt{[(\frac{1}{2}\text{coeff. of x})^2 + (\frac{1}{2}\text{coeff. of y})^2 + (\frac{1}{2}\text{coeff. of z})^2 - (\text{constant term})]}$

Notes : (1) If $u^2 + v^2 + w^2 - d < 0$, then the radius of the sphere (iii) is imaginary whereas the centre is real. Such sphere is called *pseudo-sphere or a virtual sphere.*

(2) The equation (iii) of the sphere contains four unknown constants u, v, w and d and therefore a sphere can be found to satisfy four conditions.

EQUATION OF A SPHERE THROUGH FOUR GIVEN POINTS

Let the co-ordinates of the four given points A, B, C and D be (x_1, y_1, z_1), (x_2, y_2, z_2), (x_3, y_3, z_3) and (x_4, y_4, z_4) respectively.

Let the equation off the sphere passing through these four points be

$$x^2 + y^2 + z^2 + 2ux + 2vy + 2wz + d = 0 \quad \ldots(i)$$

If this sphere passes through the given points A, B, C and D, then we have

$$\left(x_1^2 + y_1^2 + z_1^2\right) + 2ux_1 + 2vy_1 + 2wz_1 + d = 0 \quad \ldots(ii)$$

$$\left(x_2^2 + y_2^2 + z_2^2\right) + 2ux_2 + 2vy_2 + 2wz_2 + d = 0 \quad \ldots(iii)$$

$$\left(x_3^2 + y_3^2 + z_3^2\right) + 2ux_3 + 2vy_3 + 2wz_3 + d = 0 \quad \ldots(iv)$$

$$\left(x_4^2 + y_4^2 + z_4^2\right) + 2ux_4 + 2vy_4 + 2wz_4 + d = 0 \quad \ldots(v)$$

Eliminating u, v, w and d from (i), (ii), (iii), (iv) and (v), we have

$$\begin{vmatrix} x^2 + y^2 + z^2 & x & y & z \\ x_1^2 + y_1^2 + z_1^2 & x_1 & y_1 & z_1 \\ x_2^2 + y_2^2 + z_2^2 & x_2 & y_2 & z_2 \\ x_3^2 + y_3^2 + z_3^2 & x_3 & y_3 & z_3 \\ x_4^2 + y_4^2 + z_4^2 & x_4 & y_4 & z_4 \end{vmatrix} = 0$$

Note: In numerical problems, evaluation of determinant takes much time and so the values of u, v, w and d should be found from (ii), (iii), (iv) and (v) and then these should be substituted in (i), to get the required equation.

EQUATION OF A SPHERE ON THE LINE JOINING TWO GIVEN POINTS AS DIAMETER (DIAMETER FORM)

Let A $(x_1, y_1\ z_1)$ and $B(x_2, y_2, z_2)$ be two given points.

Let P(x, y, z) be any point on the surface of the sphere, drawn on the line joining A and B as diameter. Then AP is perpendicular to BP.

Now the direction ratios of the lines AP and BP are given by

$x - x_1, y - y_1, z - z_1$ and $x - x_2, y - y_2, z - z_2$ respectively.

AS AP is perpendicular to BP, then we have

$$(x - x_1)(x - x_2) + (y - y_1)(y - y_2) + (z - z_1)(z - z_2) = 0, \quad \ldots(i)$$

which is the required equation of the sphere joining the $B(x_2, y_2, z_2)$ as diameter.

Example 1:

OA, OB, OC are three mutually perpendicular lines through the origin and their direction cosines are $l_1, m_1, n_1; l_2, m_2, n_2; l_3, m_3, n_3$. If $OA = a$, $OB = b$, $OC = c$, prove that equation of the sphere through OABC is

$$x^2 + y^2 + z^2 - x(al_1 + bl_2 + cl_3) - y(am_1 + bm_2 + cm_3) - z(an_1 + bn_2 + cn_3) = 0.$$

Solution:

Co-ordinates of points A, B, C respectively are (l_1a, m_1a, n_1a); (l_2b, m_2b, n_2b); (l_3c, m_3c, n_3c).

Let the equation of the sphere be

$$x^2 + y^2 + z^2 + 2xu + 2vy + 2wz + d = 0$$

as it passes through origin, hence, $d = 0$

It passes through A (l_1a, m_1a, n_1a)

$$\therefore \quad l_1^2a^2 + m_1^2a^2 + n_1^2a^2 + 2ul_1a + 2vm_1a + 2wn_1a = 0$$

$$\Rightarrow \quad a + 2ul_1 + 2vm_1 + 2wn_1 = 0. \quad \ldots(i)$$

Similarly for B and C

$$b + 2ul_2 + 2vm_2 + 2wn_2 = 0. \quad \ldots(ii)$$

and
$$c + 2ul_3 + 2vm_3 + 2wn_3 = 0. \quad \ldots(iii)$$

Now, lines OA, OB, OC are mutually perpendicular, hence, l_1, l_2, l_3; m_1, m_2, m_3; n_1, n_2, n_3 are the direction cosines of OX, OY, OZ referred to OA, OB, OC as axes.

Now multiplying (i) by l_1, (ii) by l_2, (iii) by l_3 and adding, we get

$$al_1 + bl_2 + cl_3 + 2u = 0 \Rightarrow u = -\tfrac{1}{2}(al_1 + bl_2 + cl_3)$$

Similarly,

$$v = -\tfrac{1}{2}(am_1 + bm_2 + cm_3)$$

and
$$w = -\tfrac{1}{2}(an_1 + bn_2 + cn_3).$$

Substituting the values of u, v, w, the required sphere is obtained.

Example 2:

A variable plane through a fixed point (a, b, c) cuts the co-ordinate axes in the point A, B, C. Show that the locus of the centres of the sphere OABC is

$$a/x + b/y + c/z = 2.$$

Solution:

Let the sphere OABC be

$$x^2 + y^2 + z^2 + 2ux + 2vy + 2wz + d = 0, \qquad \text{...(i)}$$

so that u, v, w are different for different spheres. The points A, B, C where it cuts the three axes are (–2u, 0, 0), (0, –2v, 0), (0, 0, –2w). The equation of the plane ABC is

$$\frac{x}{-2u} + \frac{y}{-2v} + \frac{z}{-2w} = 1.$$

Since this plane passes through (a, b, c), we have

$$\frac{a}{-2u} + \frac{b}{-2v} + \frac{c}{-2w} = 1. \qquad \text{...(ii)}$$

If x, y, z be the centre of the sphere (i).

$$x = -u,\ y = -v,\ z = -w. \qquad \text{...(iii)}$$

From (ii) and (iii), we obtain

$$\frac{a}{x} + \frac{b}{y} + \frac{c}{z} - 2$$

as the required locus.

Example 3

Prove that the centres of the spheres which touch the lines $y = mx$, $z = c$; $y = -mx$, $z = -c$; lie upon the conicoid $mxy + cz\ (l + m^2) = 0$.

Solution:

Let the equation of the sphere be

$$x^2 + y^2 + z^2 + 2ux + 2vy + 2wz + d = 0.$$

Line $y = mx,\ z = c$ meets it, where

$$x^2 + m^2x^2 + c^2 + 2ux + 2vmx + 2wc + d = 0$$

or $$(l + m^2)\ x^2 + 2(u + vm)\ x + (c^2 + 2wc + d) = 0$$

The line is given to be tangent to the sphere; hence the two values of x given by above equation must be coincident, which gives

$$(u, vm)^2 = (l + m^2)(c^2 + 2wc + d) \quad \ldots(i)$$

In the same way, line $y = -mx$, $z = -c$ will touch the sphere, if

$$(u - vm)^2 = (l + m^2)(c^2 - 2wc + d) \quad \ldots(ii)$$

Subtracting (ii) from (i), we have

$$4uvm = 4wc(l + m^2)$$

or

$$uvm - w(l + m^2) = 0$$

Hence, the locus of centre $(-u, -v, -w)$ will be

$$xym + zc(l + m^2) = 0.$$

Example 4:

A sphere of constant radius 2k passes through the origin and meets the axes in A, B, C. Find the locus of the centroid of the tetrahedron OABC.

Solution:

Let co-ordinates of the points A, B, C be (a, 0, 0), (0, b, 0) and (0, 0, c) respectively.

The equation of the sphere OABC is

$$x^2 + y^2 + z^2 - ax - by - cz = 0.$$

Radius of this sphere is given equal to 2k.

$$\therefore \quad a^2 + b^2 + c^2 = 4(2k)^2 = 16k^2 \quad \ldots(i)$$

Let (x, y, z) be the co-ordinates of the centroid of the tetrahedron OABC; then

$$x = a/4,\ y = b/4,\ z = c/4$$

$$\Rightarrow \quad a = 4x,\ b = 4y,\ c = 4z.$$

Eliminating a, b, c from (i), the required locus is

$$x^2 + y^2 + z^2 = k^2.$$

Example 5:

Find the equation to the sphere through the points (0, 0, 0), (0, 1, –1), (–1, 2, 0) and (1, 2, 3).

Solution:

Let the equation of the sphere by given as

$$x^2 + y^2 + z^2 + 2ux + 2vy + 2wz + d = 0 \quad \ldots(i)$$

If it passes through (0, 0, 0), then $d = 0$ ...(ii)

If it passes through (0, 1, –1), then

$$0 + 1 + 1 + 0 + 2v - 2w + 0 = 0, \qquad \because d = 0$$

$\Rightarrow$ $$v - w + 1 = 0 \qquad \text{...(iii)}$$

If it passed through (–1, 2, 0), then $1 + 4 + 0 - 2u + 4v + 0 + 0 = 0$.

$\Rightarrow$ $$5 - 2u + 4v = 0 \qquad \text{...(iv)}$$

If it passes through (1, 2, 3), then

$$1 + 4 + 9 + 2u + 4v + 6w + 0 = 0, \qquad \because d = 0$$

$\Rightarrow$ $$u + 2v + 3w + 7 = 0 \qquad \text{...(v)}$$

From (iii), we get $w = v + 1$ and from (iv), $u = (5 + 4v)/2$

Substituting these values in (v) we get

$$[(5 + 4v)/2] + 2v + 3(v + 1) + 7 = 0$$

$\Rightarrow$ $5 + 4v + 4v + 6v + 6 + 7 = 0$

$\Rightarrow$ $14v = -18$ or $v = -9/7$

$\therefore$ $w = v + 1 = (-9/7) + 1 = -2/7$

and $u = \frac{1}{2}(5 + 4v) = \frac{1}{2}\left(5 - \frac{36}{7}\right) = -\frac{1}{14}$

$\therefore$ From (i), we get the required equation of the sphere as

$$x^2 + y^2 + z^2 + 2(-1/14)x + 2(-9/7)y + 2(-2/7)z + 0 = 0$$

$\Rightarrow$ $14(x^2 + y^2 + z^2) - 2x - 36y - 8z = 0.$ **Ans.**

Example 6:

A point moves so that sum of squares of its distances from a given number of points is constant. Prove that the locus of this point is a sphere and show that its centre is the centroid of the given points.

Solution:

Let there be n given points (x_1, y_1, z_1), (x_2, y_2, z_2) etc.

Let the moving point be (α, β, γ). Then according to the given problem the sum of the squares of its distances from the given points is constant.

i.e., $\Sigma[(\alpha - x_1)^2 + (\beta - y_1)^2 + (\gamma - z_1)^2]$ = constant = k (say)

$\Rightarrow$ $n\alpha^2 + n\beta^2 + n\gamma^2 - 2\alpha\Sigma x_1 - 2\beta\Sigma y_1 - 2\gamma\Sigma z_1 + \Sigma x_1^2 + \Sigma y_1^2 + \Sigma z_1^2 = k$

$\Rightarrow$ $\alpha^2 + \beta^2 + \gamma^2 - 2\alpha\left(\frac{\Sigma x_1}{n}\right) - 2\beta\left(\frac{\Sigma y_1}{n}\right) - 2\gamma\left(\frac{\Sigma z_1}{n}\right) + (...) = 0$

$\therefore$ Locus of (α, β, γ) is the sphere

$$x^2 + y^2 + z^2 - 2\left(\frac{\Sigma x_1}{n}\right)x - 2\left(\frac{\Sigma y_1}{n}\right)y - 2\left(\frac{\Sigma z_1}{n}\right)z + (\text{constant}) = 0$$

$\therefore$ Its centre is $\left(\frac{\Sigma x_1}{n}, \frac{\Sigma y_1}{n}, \frac{\Sigma z_1}{n}\right)$ which is evidently the centroid of the points (x_1, y_1, z_1), (x_2, y_2, z_2) ...., (x_n, y_n, z_n). **Hence proved.**

Example 7:

Obtain the equation of the sphere having its centre on the line $5y + 2z = 0 = 2x - 3y$ and passing through the points $(0, -2, -4)$ and $(2, -1, -1)$.

Solution:

Let the equation of the sphere be given as

$$x^2 + y^2 + z^2 + 2ux + 2vy + 2wz + d = 0 \quad \text{...(i)}$$

Its centre is $(-u, -v, -w)$. If it lies on the lines $5y + 2z = 0 = 2x - 3y$, then we have $5(-v) + 2(-w) = 0 = 2(-u) - 3(-v)$

i.e., $$5v + 2w = 0 \quad \text{...(ii)}$$

and $$2u - 3v = 0 \quad \text{...(iii)}$$

If the sphere (i) passes through $(0, -2, -4)$ and $(2, -1, -1)$, then we have

$$0 + 4 + 16 - 4v - 8w + d = 0 \quad \text{...(iv)}$$

and $$4 + 1 + 1 + 4u - 2v - 2w + d = 0 \quad \text{...(v)}$$

Solving (ii), (iii) and (v) we obtain

$u = -3;\ v = -2;\ w = 5;\ d = 12$, which satisfy (iv)

$\therefore$ From (i), the required equation of the sphere is

$$x^2 + y^2 + z^2 - 6x - 4y + 10z + 12 = 0$$ **Ans.**

Example 8:

Find the equation to the sphere through the points $(0, 0, 0)$ $(0, 1, -1)$, $(-1, 2, 0)$, $(1, 2, 3)$.

Solution:

Let the equation of the sphere be

$$x^2 + y^2 + z^2 + 2ux + 2vy + 2wz + d = 0 \quad \text{...(i)}$$

As it passes through given points, we have

$$d = 0;$$

$$2 + 2v - 2w + d = 0;$$

$$5 - 2u + 4v + d = 0;$$

$$14 + 2u + 4v + 6w + d = 0$$

yielding $u = -15/14,\ v = -25/14,$

$w = -11/14$ and $d = 0.$

Hence, the equation of sphere becomes

$$x^2 + y^2 + z^2 - \frac{15}{7}x - \frac{25}{7}y - \frac{11}{7}z = 0$$

$$\Rightarrow \quad 7(x^2 + y^2 + z^2) - 15x - 25y - 11z = 0.$$

PLANE SECTION OF A SPHERE

Consider a sphere and a plane. We suppose that the sphere and the plane have points to common *i.e.*, intersect. The set of points common to a sphere and a plane, assuming that the sphere and the plane intersect, is called a plane section of a sphere. We show that the locus of points common to a sphere and a plane is a circle, *i.e., a plane section of a sphere is a circle*

The prove that the section of a sphere

$$x^2 + y^2 + z^2 + 2xu + 2vy + 2wz + d = 0 \qquad \ldots\text{(i)}$$

by a plane $lx + my + nz = p$ ...(ii)

is a circle and to find its radius and centre.

O is the centre $(-u, -v, -w)$ of the sphere (i) and C is the foot of the perpendicular OC from O to the plane (ii). Let A be any point on the section of the sphere (i) by the plane (ii), then CA is any line through C and as such perpendicular to OC.

Now OA = the radius of the sphere = $\sqrt{(u^2 + v^2 + w^2 - d)}$ and OC = length of the perp. from $O(-u, -v, -w)$ to the plane (iii) then we have

$$= \frac{l(-u) + m(-v) + n(-w) - p}{\sqrt{(l^2 + m^2 + n^2)}} = \frac{lu + mv + nw + p}{\sqrt{(l^2 + m^2 + n^2)}}, \text{ numerically}$$

Now from above fig. it is evident the $CA^2 = OA^2 - OC^2$ ...(iii)

i.e., CA^2 = constant, as the radius of the sphere (i) and the distance of its centre from the plane (ii) are constant.

Hence, the distance of any point A on the section of the sphere (i) by the plane (ii) is at a constant distance from the foot C of the perpendicular from centre of (i) to (ii), *i.e.*, the locus of A is a circle with centre C and radius CA given by (iii).

Equations of a Circle

From above we have found that the intersection of the sphere $x^2 + y^2 + z^2 + 2ux + 2vy + 2wz + d = 0$ by the plane $lx + my + nz = p$ is a circle and so the equations of this circle is obtain by

$$x^2 + y^2 + z^2 + 2ux + 2vy + 2wz + d = 0,\ lx + my + nz = p$$

taken together.

Hence, in general, the equations of a circle consist of the equations of a sphere and that of a plane (taken together)

Notes:

(1) From above we conclude that:

(i) The foot of perpendicular from the centre of the sphere on the plane of the circle is the centre of the circle, and

(ii) and radius of the circle = CA.

$= \sqrt{(OA^2 - OC^2)} = \sqrt{[(\text{radius of the sphere})^2 - (\text{length of perpendicular from the centre of the sphere on the plane of the circle})^2]}$

(2) The section of a sphere by a plane passing through the centre of the sphere is called a great circle. Its centre and radius is the same at that of the sphere.

INTERSECTION OF TWO SPHERES

We now consider two spheres and assume that the given spheres have points in common *i.e.*, intersect. Assuming that two given spheres instersect, we show that the *locus of the points of intersection o two spheres is a circle.*

The coordinate of points, iff any common of the two spheres

$$S_1 \equiv x^2 + y^2 + z^2 + 2u_1x + 2v_1y + 2w_1z + d_1 = 0$$

and $$S_2 \equiv x^2 + y^2 + z^2 + 2u_2x + 2v_2y + 2w_2z + d_2 = 0.$$

The points common to these two circles satisfy both these equations $S_1 = 0$, $S_2 = 0$ and hence satisfy $S_1 - S_2 = 0$.

i.e., $2(u_1 - u_2)\,x + 2(v_1 - v_2)\,y + 2(w_1 - w_2)\,z + 2(d_1 - d_2) = 0$

which being a linear equation in x, y, z represents a plane.

Thus, the curve of intersection of the two spheres $S_1 = 0$, $S_2 = 0$ is the same as the curve of intersection of any one of these spheres and the plane $S_1 - S_2 = 0$ and so it is a circle.

SPHERE THROUGH A GIVEN CIRCLE

Let the equations of the circle be

$$S \equiv x^2 + y^2 + z^2 + 2ux + 2vy + 2wz + d = 0 \qquad \text{...(i)}$$

and $$P \equiv lx + my + nz - p = 0. \qquad \text{...(ii)}$$

Then the equation $S + \lambda P = 0,$ ...(iii)

in which λ is a constant represents a sphere as in this equation the coefficient of x^2, y^2, z^2 are equal and the terms containing xy, yz, zx are absent.

Also the equation $S + \lambda P = 0$ is satisfied by all points which satisfy both $S = 0$ and $P = 0$ *i.e.*, which lie on the circle $S = 0$, $P = 0$.

Hence $S + \lambda P = 0$ represents a sphere through circle $S = 0$, $P = 0$.

In a similar manner we can show that the equation $S_1 + \lambda S_2 = 0$ represents a sphere through the circle of intersection of the spheres $S_1 = 0$ and $S_2 = 0$.

Example 1:

A variable plane is parallel to the given plane $x/a + y/b + z/c = 0$ and meets the axes in A, B, C. Prove that the circle ABC lies on the cone

$$yz(b/c + c/b) + zx\,(c/a + a/c) + xy(a/b + b/a) = 0).$$

Solution:

Let the variable plane, which is parallel to the given plane, be

$$x/a + y/b + z/c = k \qquad \ldots(i)$$

This meets the axes in A(ak, 0, 0), B(0, bk, 0) and C(0, 0, ck). Hence, equation of the sphere OABC is

$$x^2 + y^2 + z^2 - akx - bky - ckz = 0$$

or

$$x^2 + y^2 + z^2 - k(ax + by + cz) = 0. \qquad \ldots(ii)$$

The circle ABC lies on both, the plane (i) and the sphere (ii). Hence, (i) and (ii) together represent the circle ABC and the locus of the circle ABC will be obtained by eliminating k from (i) and (ii). Thus, the locus of circle ABC is

$$(x^2 + y^2 + z^2) - (x/a + y/b + z/c)\,(ax + by + cz) = 0$$

or

$$yz\,(b/c + c/b) + zx\,(c/a + a/c) + xy\,(a/b + b/a) = 0.$$

Example 2:

Show that the sphere

$$S_1 \equiv x^2 + y^2 + z^2 + 2u_1x + 2v_1y + 2w_1z + d_1 = 0$$

cuts

$$S_2 \equiv x^2 + y^2 + z^2 + 2u_2x + 2v_2y + 2w_2z + d_2 = 0$$

in a great circle if

$$2(u_2^2 + v_2^2 + w_2^2) - d_2 = 2\,(u_1u_2 + v_1v_2 + w_1w_2) - d_1$$

or

$$2(u_1u_2 + v_1v_2 + w_1w_2) = 2r_2^2 + d_1 + d_2,$$

where r_2 is the radius of the second sphere.

Solution:

The plane of the circle, *i.e.*, the plane in which their circle of intersection lies, is $S_1 - S_2 = 0$,

$\Rightarrow \quad 2(u_1 - u_2)x + 2(v_1 - v_2)y + 2(w_1 - w_2)z + d_1 - d_2 = 0. \qquad ...(i)$

The circle of intersection will be the great circle of the sphere S_2 only when the above plane passes through the centre of the sphere S_2, *i.e.*, passes through the point $(-u_2, -v_2, -w_2)$. Hence,

$$2(u_1 - u_2)(-u_2) + 2(v_1 - v_2)(-v_2) + 2(w_1 - w_2)(-w_2) + d_1 - d_2 = 0$$

$$\Rightarrow \quad 2(u_2^2 + v_2^2 + w_2^2) - d_2 = 2(u_1u_2 + v_1v_2 + w_1w_2) - d_1$$

$$\Rightarrow \quad 2(u_1u_2 + v_1v_2 + w_1w_2) = 2r_2^2 + d_1 + d_2,$$

Example 3:

A is a point on OX and B on OY so that the angle OAB is constant (= α). On AB as diameter a circle is described whose plane is parallel to OZ. Prove that as AB varies, the circle generates the cone

$$2xy - z^2 \sin 2\alpha = 0.$$

Solution:

Let A be the point (a, 0, 0) and B (0, b, 0).

Then since ∠ OAB = α, we have

$$\tan \alpha = b/a \qquad ...(i)$$

Now, a sphere on AB as diameter is

$$(x - a)x + (y - b)y + z^2 = 0$$

$$\Rightarrow \quad x^2 + y^2 + z^2 = ax + by \qquad ...(ii)$$

A plane through AB parallel to OZ is

$$x/a + y/b = 1 \qquad ...(iii)$$

The required circle is given by intersection of (ii) and (iii). Now to determine the locus of this circle, we will eliminate a and b from (ii), with the help of (i) and (iii). So, we have

$$x^2 + y^2 + z^2 = (ax + by)(x/a + y/b)$$

$$\Rightarrow \quad x^2 + y^2 + z^2 = x^2 + y^2 + xy(a/b + b/a)$$

$$\Rightarrow \quad z^2 = xy(\tan \alpha + \cot \alpha)$$

$$\Rightarrow \quad z^2 = 2xy/\sin \alpha$$

$$\Rightarrow \quad 2xy - z^2 \sin 2\alpha = 0.$$

Example 1:

POP' is a variable diameter of the ellipse $z = 0$, $x^2/a^2 + y^2/b^2 = 1$, and a circle is described in the plane PP'ZZ' on PP' as diameter. Prove that as PP' varies, the circle generates the surface

$$(x^2 + y^2 + z^2)(x^2/a^2 + y^2/b^2) = x^2 + b^2.$$

Solution:

Let P be the point (a, cos ϕ, b sin ϕ, 0); then P' is the point ($-$a cosϕ, $-$b sin ϕ, 0) where ϕ varies as POP' varies.

Equation of the sphere on PP' as diameter is

$$(x - a\cos\phi)(x + a\cos\phi) + (y - b\sin\phi)(y + b\sin\phi) + z^2 = 0$$

or $$x^2 + y^2 + z^2 = a^2\cos^2\phi + b^2\sin^2\phi \qquad \text{...(i)}$$

The equation PP'ZZ', *i.e.*, the plane through PP' parallel to z-axis, is

$$x/a\cos\phi = y/b\sin\phi \qquad \text{...(ii)}$$

The required circle is given by the intersection of (i) and (ii). The locus of the circle will be obtained by eliminating ϕ from (i) and (ii).

From (ii), we have

$$\frac{x/a}{\cos\phi} = \frac{y/a}{\sin\phi} = \frac{\sqrt{(x^2/a^2 + y^2/b^2)}}{1}$$

$\therefore$ $a\cos\phi = x/\sqrt{(x^2/a^2 + y^2/b^2)}$ and $b\sin\phi = y/\sqrt{(x^2/a^2 + y^2/b^2)}$

Putting these values in (i), we get

$$x^2 + y^2 + z^2 = \frac{x^2}{(x^2/a^2 + y^2/b^2)} - \frac{y2}{(x^2/a^2 + y^2/b^2)}$$

$$\Rightarrow (x^2 + y^2 + z^2)(x^2/a^2 + y^2/b^2) = x^2 + y^2.$$

Example 2:

Prove that the circles $x^2 + y^2 + z^2 - 2x + 3y + 4z - 5 = 0$, $5y + 6z + 1 = 0$ and $x^2 + y^2 + z^2 - 3x - 4y + 5z - 6 = 0$, $x + 2y - 7z = 0$ lies on the same sphere and find its equation. Also find the value of a for which $x + y + z = a\sqrt{3}$ touches the sphere.

Solution:

The equation of any sphere through the first circle is given as

$$(x^2 + y^2 + z^2 - 2x + 3y + 4z - 5) + \lambda(5y + 6z + 1) = 0 \qquad \text{...(i)}$$

Similarly the equation of any sphere through the second circle is given as

$$(x^2 + y^2 + z^2 - 3x - 4y + 5z - 6) + \mu(x + 2y - 7z) = 0 \quad \text{...(ii)}$$

If the given circles lies on the same sphere then (i) and (ii) should represent the same sphere, so comparing the coefficients of x, y, z and constant terms in (i) and (ii) we have

$$-2 = -3 + \mu \quad \text{...(iii)}$$

$$3 + 5\lambda = -4 + 2\mu \quad \text{...(iv)}$$

$$4 + 6\lambda = 5 - 7\mu \quad \text{...(v)}$$

and $$-5 + \lambda = -6 \quad \text{...(vi)}$$

From (iii) and (vi) we get $\mu = -1$, $\lambda = -1$.

These values of λ and μ satisfy (iv) and (v), hence the given circles lie on the same sphere. Putting $\lambda = -1$ in (i) we have the required equation of the sphere as $x^2 + y^2 + z^2 - 2x - 2y - 2z - 6 = 0$ ...(viii)

Its centre in (1, 1, 1) and radius $= \sqrt{(1^2 + 1^2 + 1^2 + 6)} = 3$. If the plane $x + y + z = a\sqrt{3}$ ...(viii) touches the sphere (vii), then the length of the perpendicular from the centre (1, 1, 1) of the sphere to the plane (viii) must be equal to the radius of the sphere.

i.e., $$\frac{1 + 1 + 1 - a\sqrt{3}}{\sqrt{(1^2 + 1^2 + 1^2)}} = 3 \quad \text{or} \quad 3 - a\sqrt{3} = \pm 3\sqrt{3}$$

or $$a\sqrt{3} = 3 \pm 3\sqrt{3} \quad \text{or} \quad a = \sqrt{3} \pm 3.$$ **Ans.**

Example 3:

Find the equation of the sphere whose centre is the point (1, 2, 3) and which touches the plane $3x + 2y + z + 4 = 0$. Find also the radius of the circle in which the sphere is cut by the plane $x + y + z = 0$.

Solution:

Since the sphere touches the plane $3x + 2y + z + 4 = 0$ ...(i)

so its radius = length of perpendicular from its centre (1, 2, 3) on (i)

$$= \frac{3.1 + 2.2 + 1.3 + 4}{\sqrt{(3^2 + 2^2 + 1^2)}} = \frac{14}{\sqrt{(14)}} = \sqrt{14}$$

$\therefore$ The required equation of the sphere is

$$(x - 1)^2 + (y - 2)^2 + (z - 3)^2 = [\sqrt{(14)}^2],$$

$$\Rightarrow \quad x^2 + y^2 + z^2 - 2x - 4y - 6z = 0$$

Now for the radius of the circle refer.

The coordinates of O, the centre of the sphere are given as (1, 2, 3) and OC being the perpendicular to the plane $x + y + z = 0$ its direction ratios are the same as those of the normal to this plane *i.e.*, 1, 1, 1 the coefficients of x, y, z in the equation $x + y + z = 0$.

$\therefore$ The equation of the line OC are

$$\frac{x-1}{1} = \frac{y-2}{1} = \frac{z-3}{1} = r \text{ (say)}$$

Let OC = r, then the coordinates of C are (r + 1, r + 2, r + 3) and C lies on the plane $x + y + z = 0$, so we have

$(r + 1) + (r + 2) + (r + 3) = 0$ or $3r + 6 = 0$ or $r = -2$

$\therefore$ The coordinates of C are (–2 + 1, –2 + 2, –2 + 3) or **(–1, 0, 1)** **Ans.**

Example 4:

Find the equations of the circle circumscribing the triangle formed by the three points (a, 0, 0), (0, b, 0), (0, 0, c).

Obtain also the co-ordinates of the centre of this circle.

Solution:

The equation of the plane passing through these three points is

$$x/a + y/b + z/c = 1.$$

The required circle is the curve of intersection of this plane with any sphere through the three points.

To find the equation of this sphere, a fourth point is necessary which for the sake of convenience, we take as origin

If

$$x^2 + y^2 + z^2 + 2ux + 2vy + 2wz + d = 0$$

be the sphere through these four points, we have

$$a^2 + 2ua + d = 0; \ b^2 + 2vb + d = 0' \ c^2 + 2wc + d = 0;$$

$$d = 0$$

These give $d = 0,\ u = -\frac{1}{2}a,\ v = -\frac{1}{2}b,\ w = -\frac{1}{2}c.$

Thus, the equation of the sphere is

$$x^2 + y^2 + z^2 - ax - by - cz = 0$$

Hence, $\quad x^2 + y^2 + z^2 - ax - by - cz = 0,\ \frac{x}{a} + \frac{y}{b} + \frac{z}{c} = 1$

are the equations of the circle.

The centre of this circle, is the foot of the perpendicular from the centre $\left(\frac{1}{2}a, \frac{1}{2}b, \frac{1}{2}c\right)$ of the sphere to the plane

$$\frac{x}{a} + \frac{y}{b} + \frac{z}{c} = 1.$$

The equation of the perpendicular are

$$\frac{x - \frac{1}{2}a}{1/a} = \frac{y - \frac{1}{2}b}{1/b} = \frac{z - \frac{1}{2}c}{1/c} = r, \text{ say}$$

$$\Rightarrow \qquad \left(\frac{r}{a} + \frac{a}{2}, \frac{r}{b} + \frac{b}{2}, \frac{r}{c} + \frac{c}{2}\right),$$

is any point on the line. Its intersection with the plane is given by

$$r\left(\frac{1}{a^2} + \frac{1}{b^2} + \frac{1}{c^2}\right) + \frac{1}{2} = 0 \Rightarrow r = -\frac{1}{(2\,\Sigma a^{-2})}.$$

Thus, the centre is

$$\left[\frac{a(b^{-2} + c^{-2})}{2\,\Sigma a^{-2}}, \frac{b(c^{-2} + a^{-2})}{2\,\Sigma a^{-2}}, \frac{c(a^{-2} + b^{-2})}{2\,\Sigma a^{-2}}\right].$$

Example 5:

Show that the centres of all sections of the sphere

$$x^2 + y^2 + z^2 = r^2$$

by planes through a point $(x^2 + y^2 + z^2)$ lie on the sphere

$$x(x - x') + y(y - y') + z(z - z') = 0.$$

Solution:

The plane which cuts the sphere in a circle with centre (f, g, h) is

$$f(x - f) + g(y - g) + h(z - h) = 0.$$

It will pass through (x', y', z'), if

$$f(x' - f) + g(y' - g) + h(z' - h) = 0,$$

and accordingly the locus of (f, g, h) is the sphere

$$x(x' - x) + y(y' - y) + z(z' - z) = 0.$$

Example 6:

If r be the radius of the circle

$$x^2 + y^2 + z^2 + 2ux + 2vy + 2wz + d = 0,\ lx + my + nz = 0,$$

prove that

$$(r^2 + d^2)\ (l^2 + m^2 + n^2) = (mw - nv)^2 + (nu - lw)^2 + (lv - mu)^2.$$

Solution:

The equation of the given sphere is

$$x^2 + y^2 + z^2 + 2ux + 2vy + 2wz + d = 0 \qquad \text{...(i)}$$

having centre at (–u, –v, –w) and radius CP = $\sqrt{(u^2 + v^2 + w^2 - d)}$.

Now distance CN of centre of sphere from the plane is length of perpendicular from centre of sphere on the plane

$$lx + my + nz = 0$$

$$CN = \frac{lu + mv + nw}{\sqrt{(l^2 + m^2 + n^2)}}$$

$$CP = \sqrt{(u^2 + v^2 + w^2 - d)},\ NP = r$$

Then from Δ CPN

$$r^2 = CP^2 - CN^2 = u^2 + v^2 + w^2 - d - \frac{(lu + mv + nw)^2}{l^2 + m^2 + n^2}$$

$$\Rightarrow (r^2 + d)\ (l^2 + m^2 + n^2) = (u^2 + v^2 + w^2)\ (l^2 + m^2 + n^2) - (lu + mv + nw)^2$$

$$\Rightarrow (r^2 + d)\ (l^2 + m^2 + n^2) = (mw - nv)^2 + (nu - lw)^2 + (lv - mu)^2,$$

by Lagrange's identity.

INTERSECTION OF A STRAIGHT LINE AND A SPHERE

Let the equations of the sphere and the straight line be

$$x^2 + y^2 + z^2 + 2ux + 2vy + 2wz + d = 0 \qquad \text{...(i)}$$

and $$\frac{x - \alpha}{l} = \frac{y - \beta}{m} = \frac{z - \gamma}{n} = r \text{ (say)} \qquad \text{...(ii)}$$

Any point on the line (ii) is $(\alpha + lr, \beta + mr, \gamma + nr)$.

If this point lies on the sphere (i), then we have

$$(\alpha + lr)^2 + (\beta + mr)^2 + (\gamma + nr)^2 + 2u\ (\Rightarrow\alpha + lr) + 2v\ (\beta + mr) + (2w\ (\gamma + nr) + d = 0$$

$$\Rightarrow r^2\ (l^2 + m^2 + n^2) + 2r\ [l(u + \alpha) + m(v + \beta) + n\ (w + \gamma)] + (\alpha^2 + \beta^2 + \gamma^2 + 2u\alpha + 2v\beta + 2w\gamma + d) = 0 \quad \text{...(iii)}$$

This is a quadratic equation in r and so gives two values of r and therefore the line (ii) meets the sphere (i) in two points which may be real coincident or imaginary according as roots of (iii) are so.

Note: If l, m, n are the actual direction cosines of the line (iii), then $l^2 + m^2 + n^2 = 1$ and then the equation (iii) can be simplified.

EQUATION OF THE TANGENT PLANE

(a) Let us find the equation of the tangent plane to the sphere

$$x^2 + y^2 + z^2 + 2ux + 2vy + 2wz + d = 0 \quad \ldots(i)$$

at the point (α, β, γ).

As (α, β, γ) is the point on the sphere (i),

$$\alpha^2 + \beta^2 + \gamma^2 + 2u\alpha + 2v\beta + 2w\gamma + d = 0. \quad \ldots(ii)$$

The equations of the line through the point (α, β, γ) are given

$$\frac{x - \alpha}{l} = \frac{y - \beta}{m} = \frac{z - \gamma}{n} \quad \ldots(iii)$$

The points of intersection of the line (iii) and the sphere (i) as in 5.9 above are given by

$$r^2 (l^2 + m^2 + n^2) + 2r\{l(\alpha + u) + m(\beta + v) + n(\gamma + w)\}$$
$$+ (\alpha^2 + \beta^2 + \gamma^2 + 2u\alpha + 2v\beta + 2w\gamma + d) = 0$$

$$\Rightarrow \quad r^2 (l^2 + m^2 + n^2) + 2r \{l (\alpha + u) + m(\beta + v)$$
$$+ n(\gamma + w)\} = 0, \text{ from (ii)}$$

As one of the roots of (iv) is zero, so one of the points of intersection of the sphere (i) and the line (iii) coincides with the point (α, β, γ). If the line (iii) is a tangent line to the sphere (i) at (α, β, γ), then the other point of intersection $\gamma + w$. But these are also the d.r.'s of the radius through α, β, γ of the sphere as shown in cor. 1, above. Hence the normal to the tangent plane at (α, β, γ) is parallel to radius of the sphere at the point *i.e.*, the tangent at (α, β, γ) is perpendicular to the radius through that point.

(b) Let us find the equation of the tangent plane to the sphere

$$x^2 + y^2 + z^2 = a^2 \quad \ldots(i)$$

at the point (α, β, γ).

As (α, β, γ) is a point on the sphere (i), so we get $\alpha^2 + \beta^2 + \gamma^2 = a^2$ ...(ii)

The equations of any line through (α, β, γ) are

$$\frac{x - \alpha}{l} = \frac{y - \beta}{m} = \frac{z - \gamma}{n} = r \text{ (say)} \quad \ldots(iii)$$

The point of intersection of the line (iii) and the sphere (i) are given by

$$(\alpha + lr)^2 + (\beta + mr)^2 + (\gamma + nr)^2 = a^2$$

$\Rightarrow$ $r^2(l^2 + m^2 + n^2) + 2r(l\alpha + m\beta + n\gamma) + (\alpha^2 + \beta^2 + \gamma^2 - a^2) = 0$

$\Rightarrow$ $r^2 (l^2 + m^2 + n^2) + 2r (l\alpha + m\beta + n\gamma) = 0$, from (ii) ...(iv)

As one of the roots of (iv) is zero, so one of the points of intersection of (i) and (ii) coincides with the point (α, β, γ). If the line (iii) is a tangent line to (i) at (α, β, γ) then the other point of intersection should also coincide with (α, β, γ), *i.e.,* the second root of (iv) should also vanish and so from (v) we have

$$l\alpha + m\beta + n\gamma = 0 \qquad ...(v)$$

$\therefore$ The line (iii) is tangent line to (i) if d.c's of the line (iii) viz. l, m, n should satisfy the condition (v).

The tangent plane at (α, β, γ) to the sphere (i) is the locus of all such tangents and its equation is obtained by eliminating l, m, n between (iii) and (v).

$\therefore$ The required equation of the tangent plane is

$$(x - \alpha)\,\alpha + (y - \beta)\,\beta + (z - \gamma)\,\gamma = 0$$

$$\Rightarrow \quad \alpha x + \beta y + \gamma z = \alpha^2 + \beta^2 + \gamma^2 = a^2, \text{ from (iii)}$$

or $\alpha x + \beta y + \gamma z = a^2$, which is the required equation of the tangent plane to the sphere (i) at (α, β, γ).

CONDITION FOR THE PLANE $lx + my + nz = p$ TO TOUCH THE SPHERE $x^2 + y^2 + z^2 + 2ux + 2vy + 2wz + d = 0$

The centre of the given sphere is $(-u, -v, -w)$ and its radius = $\sqrt{(u^2 + v^2 + w^2 - d)}$. If the given plane touches the given sphere, then the length of the perpendicular from the centre $(-u, -v, -w)$ of the sphere to the given plane must be equal to the radius of the sphere.

i.e.,
$$\frac{l(-u) + m(-v) + n(-w)}{\sqrt{(l^2 + m^2 + n^2)}} = \sqrt{(u^2 + v^2 + w^2 - d)}$$

should also coincide with (α, β, γ), *i.e.,* the second root of (iv) should also vanish and for this form (iv), we have

$$l\,(\alpha, u) + m(\beta + v) + n(\gamma + w) = 0. \qquad ...(v)$$

$\therefore$ The line (iii) is a tangent line to the sphere (i) if the d.c's of the line (iii) viz. l, m, n should satisfy the condition (v).

The tangent plane at (α, β, γ) to the sphere (i) is the locus of all such tangents and its equation is obtained by eliminating l, m, n, between (v) and the equation (iii) of the line.

$\therefore$ The required equation of the tangent plane is

$$(x - \alpha)(\alpha - u) + (y - \beta)(\beta + v) + (z - \gamma)(\gamma + w) = 0$$

$\Rightarrow$ $x(\alpha + u) + y(\beta + v) + z(\gamma + w) - (\alpha^2 + \beta^2 + \gamma^2 + u\alpha + v\beta + w\gamma) = 0$

$\Rightarrow$ $x(\alpha + u) + y(\beta + v) + z(\gamma + w) + u\alpha + v\beta + w\gamma + d) = 0,$

with the help of (ii)

$\Rightarrow$ $x\alpha + y\beta + z\gamma + u(x + \alpha) + v(y + \beta) + w(z + \gamma) + d = 0.$ ...(vi)

The equation (vi) is the required equation of the tangent plane to the sphere (i) at the point (α, β, γ).

Corollary :

Target line at any point is perpendicular to the radius through that point.

We know that if $\frac{x - \alpha}{l} = \frac{y - \beta}{m} = \frac{z - \gamma}{n}$ is a tangent line to the sphere $x^2 + y^2 + z^2 + 2ux + 2vy + 2wz + d = 0$ at (α, β, γ), then

$$l(u + \alpha) + m(v + \beta) + n(w + \gamma) = 0. \quad \text{...(i)}$$

Also the direction ratios of the radius through (α, β, γ), are proportional to $\alpha + u$, $\beta + v$, $\gamma + w$ as $(-u, -v, -w)$ are the coordinates of the centre of the sphere.

Hence the result (i) shows that the line with d.r.'s $\alpha + u$, $\beta + v$, $\gamma + w$ is at right angles to the tangent line $\frac{x - \alpha}{l} = \frac{y - \beta}{m} = \frac{z - \gamma}{n}$ at (α, β, γ).

Example 1:

Prove that the plane $x + 2y - z = 4$ cuts the sphere $x^2 + y^2 + z^2 - x + z - 2 = 0$ in circle of radius unity and find the equation of sphere which has this circle for one of its great circle.

Solution:

The centre of the given sphere is $(1/2, 0, -1/2)$ and its radius

$$= \sqrt{(1/2)^2 + 0^2 + (-1/2)^2 - (-1)} = \sqrt{(5/2)} = r.$$

Length of perpendicular from $(1/2, 0, -1/2)$ to the plane is

$$\frac{1}{2}\sqrt{6} = p \text{ (say)}$$

$\therefore$ Radius of circle $= \sqrt{(r^2 - p^2)}$

$$= \sqrt{(5/2 - 6/4)} = 1.$$

Now, equation of a sphere through given circle is

$$x^2 + y^2 + z^2 - x + z - 2 + \lambda (x + 2y - z - 4) = 0$$

$$\Rightarrow \quad x^2 + y^2 + z^2 + (\lambda - 1)x + 2\lambda y + (1 - \lambda)z - (2 + 4\lambda) = 0 \quad \text{...(i)}$$

Its centre is $[-(\lambda - 1)/2, \ -\lambda, -(1 - \lambda)/2]$

If the circle is a great circle of the sphere (i), then its centre should lie on the plane $x + 2y - z - 4 = 0$ of the circle.

$$\therefore \quad -\tfrac{1}{2}(\lambda - 1) + 2(-\lambda) + \tfrac{1}{2}(1 - \lambda) - 4 = 0$$

$$\Rightarrow \quad -3\lambda - 3 = 0 \text{ or } \lambda = -1.$$

From (i), the equation of required sphere is

$$x^2 + y^2 + z^2 - 2x - 2y + 2z + 2 = 0.$$

TOUCHING SPHERES

(i) Two spheres are said to touch *externally,* if the distance between their centres is equal to the sum of their radii.

In this case the point of contact of the sphere divides the line joining the centres internally in the ratio of their radii. [In adjoining in Fig. the point of contact A divides the line joining centres C_1 and C_2 internally in the ratio $r_1 : r_2$]

(ii) Two spheres are said to touch *Internally,* if the distance between their centres is equal to the difference of their radii.

In this case the point of contact of the spheres divides the line joining the centres externally in the ratio of their radii. The point of contact A divides line joining centres C_1 and C_2 externally in ratio $r_1 : r_2$].

Example 2:

Show that the spheres $x^2 + y^2 + x^2 = 25$ and $x^2 + y^2 + z^2 - 24x - 40y - 18z + 225 = 0$ touch externally and find the point of contact.

Solution:

Let the given spheres touch each other at (α, β, γ). The equations of the tangent planes of the spheres at (α, β, γ) are $x\alpha + y\beta + z\gamma = 25$.

and $x\alpha + y\beta + z\gamma - 12(x + \alpha) - 20(y + \beta) - 9(z + \gamma) + 225 = 0$

$$\Rightarrow \quad x\alpha + y\beta + z\gamma = 25 \quad \text{...(i)}$$

and $(\alpha - 12)x + (\beta - 20)y + (\gamma - 9)z = 12\alpha + 20\beta + 9\gamma - 225$ ...(ii)

If the two spheres touch each other, then (i) and (ii) represent the same plane, hence comparing the coefficients of x, y, z and constant terms in these equations, we get

$$\frac{\alpha - 12}{\alpha} = \frac{\beta - 20}{\beta} = \frac{\gamma - 9}{\gamma} = \frac{12\alpha + 20\beta + 9\gamma - 225}{25} = k \text{ (say)}$$

Then $\quad a - 12 = k\alpha$ or $\alpha = 12/(1 - k)$

Similarly $\quad \beta = 20/1 - k)$ and $\gamma = 9/(1 - k)$

Also $\quad 12\alpha + 20\beta + 9\gamma - 225 = 25k$

$$\Rightarrow \quad 12\left(\frac{12}{1-k}\right) + 20\left(\frac{20}{1-k}\right) + 9\left(\frac{9}{1-k}\right) - 225 = 25k$$

$$\Rightarrow \quad 144 + 400 + 81 - 225\ (1 - k) = 25k\ (1 - k)$$

$$\Rightarrow \quad k^2 + 8k + 16 = 0 \quad \text{or} \quad (k + 4)^2 = 0 \quad \text{or} \quad k = -4$$

$$\therefore \quad \alpha = \frac{12}{1+4} = \frac{12}{5};\ \beta = \frac{20}{1+4} = \frac{20}{5} = 4;\ \gamma = \frac{9}{1+4} = \frac{9}{5}$$

$\therefore$ The required point of contact is (12/.5, 4, 9/5). **Ans.**

Also the radii of the spheres are

5, $\sqrt{[(12)^2 + (20)^2 + (9)^2 - 225]}$ *i.e.*, 20.

$\therefore$ Sum of radii = 5 + 20 = 25 ...(iii)

Distance between the centres (0, 0, 0) and (12, 20, 9) of these spheres

$= \sqrt{[(12)^2 + (20)^2 + (9)^2]} = \sqrt{(625)} = 25$

= sum of radii of the spheres, from (iii).

Hence, spheres touch externally.

Example 3:

Show that the spheres $x^2 + y^2 + z^2 - 2x - 3 = 0$ and $x^2 + y^2 + z^2 + 6x + 6y + 9 = 0$ touch externally.

Solution:

The centre and radius of sphere $x^2 + y^2 + z^2 - 2x - 3 = 0$ are (1, 1, 0) and $\sqrt{(1 + 3)}$ *i.e.*, 2.

The centre and radius of the sphere $x^2 + y^2 + z^2 + 6x + 6y + 9 = 0$ are (–3, –3, 0) and $\sqrt{(9 + 9 - 9)}$ *i.e.*, 3.

Now the distance between the centres

$= \sqrt{(1 + 3)^2 + (0 + 3)^2 + (0 - 0)^2]} = \sqrt{(16 + 9)} = 5.$

And the sum of radii of the spheres = 2 + 3 = 5

$\because$ The sum of the radii = distance between the centres so two spheres touch externally

Example 4:

If any tangent plane to the sphere $x^2 + y^2 + z^2 = r^2$ makes intercepts a, b and c on the coordinates axes prove that

$$a^2 + b^2 + c^2 = r^2.$$

Solution:

The equation of the tangent plane at (α, β, γ) to the given sphere is

$$x\alpha + y\beta + z\gamma = r^2. \qquad \text{...(i)}$$

Given the a is the intercept made by the plane (i) on x-axis.

So we have $a\alpha + 0 + 0 = r^2$ or $\alpha = r^2/a$.

Similarly $\beta = r^2/b$ and $\gamma = r^2/c$, as b and c are the intercepts made by the plane (i) on y and z axes respectively.

Also as (α, β, γ) is a point on the sphere, so we have

$$\alpha^2 + \beta^2 + \gamma^2 = r^2 \quad \text{or} \quad (r^2/a)^2 + (r^2/b)^2 + (r^2/c)^2 = r^2$$

$$\Rightarrow \quad a^{-2} + b^{-2} + c^{-2} = r^{-2}$$ **Hence proved.**

$\Rightarrow (lu + mv + nw + p)^2 = (l^2 + m^2 + n^2)(u^2 + v^2 + w^2 - d)$,

which is the required condition.

Example 5:

Find the equation of the tangent plane at the point (1, 2, 3) to the sphere $3(x^2 + y^2 + z^2) - 2x - 3y - 4z - 22 = 0$.

Solution:

Equation of the sphere is

$$x^2 + y^2 + z^2 - (2/3)\,x - y - (4/3)\,z - (22/3) = 0.$$

The equation of the tangent plane to this sphere at (1, 2, 3) is

$$x.1 + y.2 + z.3 - (1/3)(x + 1) - (1/2)(y + 2) - (2/3)(z + 3) - (22/3) = 0$$

$$\Rightarrow x + 2y + 3z - (1/3)\,x - (1/3) - (1/2)\,y - 1 - (2/3)\,z - 2 - (22/3) = 0$$

$$\Rightarrow \quad (2/3)\,x + (3/2)\,y + (7/3)z - (32/3) = 0$$

$$\Rightarrow \quad 4x + 9y + 14z - 64 = 0.$$ **Ans.**

Example 6:

Find the equation of the share through the circle

$$x^2 + y^2 + z^2 = 9,\ 2x + 3y + 4z = 5$$

and the point (1, 2, 3).

Solution:

The sphere

$$x^2 + y^2 + z^2 - 9 + k(2x + 3y + 4z - 5) = 0$$

passes through the given circle for all values of k.

It will pass through (1, 2, 3) if

$$5 + 15k = 0 \Rightarrow k = -\frac{1}{3}.$$

The required equation of the sphere, therefore, is

$$3(x^2 + y^2 + z^2) - 2x - 3y - 4z - 22 = 0.$$

Example 7:

Show that the two circles

$$x^2 + y^2 + z^2 - y + 2z = 0,\ x - y + z - 2 = 0;$$

$$x^2 + y^2 + z^2 + x - 3y + z - 5 = 0,\ 2x - y + 4z - 1 = 0;$$

lie on the same sphere and find its equation.

Solution:

The equations of any sphere through the first circle is

$$x^2 + y^2 + z^2 - y + k(x - y + z - 2) = 0 \quad ...(i)$$

and that of any sphere through the second circle is

$$x^2 + y^2 + z^2 + x - 3y + z - 5 + k'(2x - y + 4z - 1) = 0 \quad ...(ii)$$

The equations (i) and (ii) will represent the same sphere, if k, k' can be chosen so as to satisfy the four linear equations

$$k = 2k' + 1,\ -1 - k = -k' - 3,$$

$$2 + k = 4k' + 1,\ -2k = -k' - 5.$$

The first two of these equations give k = 3, k' = 1, and these values clearly satisfy the remaining two equations also. These four equations in k, k' being consistent, the two circles lie on the same sphere viz.,

$$x^2 + y^2 + z^2 - y + 2z + 3(x - y + z - 2) = 0$$

$$\Rightarrow \quad x^2 + y^2 + z^2 + 3x - 4y + 5z - 6 = 0.$$

PLANE OF CONTACT

Let the tangent plane at (α, β, γ) to the sphere

$$x^2 + y^2 + z^2 + 2ux + 2vy + 2wz + d = 0 \quad ...(i)$$

pass through a point (x_1, y_1, z_1) external to this sphere.

The equation of the tangent plane at (α, β, γ) to the sphere (i) is

$$x\alpha + y\beta + z\gamma + u(x + \alpha) + v(y + \beta) + w(z + \gamma) + d = 0$$

$$\Rightarrow \quad x(u + \alpha) + y(v + \beta) + z(w + \gamma) + (u\alpha + v\beta + w\gamma + d) = 0$$

If it passes through (x_1, y_1, z_1), then

$$x_1(u + \alpha) + y_1(v + \beta) + z_1(w + \gamma) + (u\alpha + v\beta + w\gamma + d) = 0$$

$$\Rightarrow \quad \alpha(x_1 + u) + \beta(y_1 + v) + \gamma(z_1 + w) + (ux_1 + vy_1 + wz_1 + d) = 0$$

This shows that (α, β, γ) lies on the plane

$$x(x_1 + u) + y(y_1 + v) + z(z_1 + w) + (ux_1 + vy_1 + wz_1 + d) = 0$$

$$\Rightarrow \quad xx_1 + yy_1 + zz_1 + u(x + x_1) + v(y - y_1) + w(z + z_1) + d = 0.$$

which is known as the *plane of contact* of the point (α, β, γ).

And the locus of (α, β, γ) is the circle of intersection of this plane and the sphere.

THE POLAR PLANE

Definition : *If the line* $\frac{x - \alpha}{l} = \frac{y - \beta}{m} = \frac{z - \gamma}{n}$ *(where l, m, n are its actual direction cosines) drawn through the point A(α, β, γ) meets the sphere* $x^2 + y^2 + z^2 + 2ux + 2vy + 2wz + d = 0$ *in points P and Q and a point R lies on this such that* $\frac{1}{AP} = \frac{1}{AQ} = \frac{2}{AR}$

(i.e., AR is the harmonic mean of AP and AQ) then the locus of R is defined as the polar plane of (α, β, γ) with respect of the sphere.

The coordinates of any point on the given line are $(\alpha + lr, \beta + mr, \gamma + nr)$.

The distances of the points of intersection of the line and the sphere are given by the equation.

$$(\alpha + lr)^2 + (\beta + mr)^2 + (\gamma + nr)^2 + 2u(\alpha + lr) + 2v(\beta + mr) + 2w(\gamma + nr) + d = 0$$

$$\Rightarrow \quad r^2 + 2r[l(\alpha + u) + m(\beta + v) + n(\gamma + w)] + (\alpha^2 + \beta^2 + \gamma^2 + 2u\alpha + 2v\beta + 2w\gamma + d) = 0, \quad \text{...(i)}$$

remembering that $\quad l^2 + m^2 + n^2 = 1$.

As this line meets the sphere in P and Q so the roots of the equation (i) are AP and AQ

$$\text{Also } \frac{2}{AR} = \frac{1}{AP} + \frac{1}{AQ} = \frac{AP + AQ}{AP \cdot AQ} = \frac{\text{sum of the roots of (i)}}{\text{product of the roots of (i)}}$$

$$= \frac{-2\,[l\,(\alpha + u) + m(\beta + v) + n(\gamma + w)]}{\alpha^2 + \beta^2 + \gamma^2 + 2u\alpha + 2v\beta + 2w\gamma + d}$$

$\Rightarrow$ $\alpha^2 + \beta^2 + 2u\alpha + 2v\beta + 2w\gamma + d$

$$= -\,[l\,.\,AR\,(\alpha + u) + m.AR\,(\beta + v) + n.AR\,(\gamma + w)] \quad ...(ii)$$

Now let R be (x, y, z) and its distance from A (α, β, γ) is AR.

Then from the equation of the line, we have

$$x - \alpha = l\,.\,AR,\; y - \beta = m\,.\,AR \text{ and } z - \gamma = n\,.\,AR$$

$\therefore$ From (ii), we get $\alpha^2 + \beta^2 + \gamma^2 + 2u\alpha + 2v\beta + 2w\gamma + d$

$$= -[(x - \alpha)(\alpha + u) + (y - \beta)(\beta + v) + (z - \gamma)(\gamma + w)]$$

$\Rightarrow$ $$x\alpha + y\beta + z\gamma + u(x + \alpha) + v\,(y + \beta) + w\,(z + \gamma) + d = 0, \quad ...(iii)$$

which is the equation of the locus of R *i.e.,* from definition the equation of the polar of A (α, β, γ) with respect to the sphere.

The point A (α, β, γ) is called the *pole of the plane* (iii) which respect to the sphere.

To find the pole of a given plane with respect to a given sphere : Let the pole be (x_1, y_1, z_1). Then compare the given plane with the polar plane of (x_1, y_1, z_1) with respect to the given sphere and evaluate x_1, y_1 and z_1.

PROPERTIES OF POLE AND POLAR

Property I

The polar plane of a point with respect to a sphere is perpendicular to the line joining the point to the centre of the sphere.

Let the point be P(x_1, y_1, z_1) and then the equation of its polar with respect to the sphere $x^2 + y^2 + z^2 = r^2$ is $xx_1 + yy_1 + zz_1 = r^2$.

$\therefore$ The direction ratios of the normal to this polar plane are x_1, y_1, z_1 which are also the direction ratios of the line joining P (x_1, y_1, z_1) and the centre O (0, 0, 0) of the sphere. Hence OP is perpendicular to the polar plane of P with respect to the sphere.

Property II

If the line joining the centre of the sphere and point P meets the polar plane of P in Q, then OP.OQ = (radius)².

Let P be (x_1, y_1, z_1) and the equation of the sphere be

$$x^2 + y^2 + z^2 = r^2 \quad ...(i)$$

Then the radius of the sphere = r and its centre is (0, 0, 0).

$\therefore$ OP $= \sqrt{(x_1^2 + y_1^2 + z_1^2)}$ and the equations of polar plane of P with respect to the sphere (i) is $xx_1 + yy_1 + zz_1 = r^2$ ...(ii)

$\therefore$ OQ = perp. distance of O(0, 0, 0) from the plane (ii)

$$= \frac{r^2}{\sqrt{(x_1^2 + y_1^2 + z_1^2)}} = \frac{r^2}{OP}.$$

Hence $\quad$ $OP.OQ = r^2$.

Property III—(*Polar Lines*)

Definition : *Two lines which are such that the polar of any point on any one passes through the other are known as polar lines.*

Let AB and CD be two polar lines, with respect to the sphere

$$x^2 + y^2 + z^2 = a^2 \qquad \text{...(i)}$$

Let the line AB be $\dfrac{x-\alpha}{l} = \dfrac{y-\beta}{m} = \dfrac{z-\gamma}{n}$.

Then any point on the line AB is $(\alpha + lr, \beta + mr, \gamma + nr)$. Its polar with respect to the sphere (i) is $x(\alpha + lr) + y(\beta + mr) + z(\gamma + nr) = a^2$

$\Rightarrow$ $$(\alpha x + \beta y + \gamma z - a^2) + r(lx + my + nz) = 0$$

$\Rightarrow$ P + rQ = 0, which for all values of r passes through the line P = 0, Q = 0 *i.e.*, the line $\alpha x + \beta y + \gamma z - a^2 = 0$, $lx + my + nz = 0$, which are therefore the equations of the polar line CD of the line AB.

Also the normal to the plane (i) must be perpendicular to this line with direction ratios 13, –6, –2.

i.e., $$(2r)(13) + (1 + 3r)(-6) + (-3 + 4r)(-2) = 0$$

which is also satisfied for all values of r.

Hence, the polar plane (i) passes through the given line (ii).

Property IV

The distance of the points from the centre of a sphere and proportional to the distance of each from the polar plane of the other. **(Salman's Theorem)**

Let the points be P (x_1, y_1, z_1) and Q (x_2, y_2, z_2) whose polar planes with respect to the sphere $x^2 + y^2 + z^2 = a^2$ are

$$xx_1 + yy_1 + zz_1 = a^2 \qquad \text{...(i)}$$

and $$xx_2 + yy_2 + zz_2 = a^2 \qquad \text{...(ii)}$$

$$\therefore \quad \frac{\text{Distance of P from the polar plane of Q}}{\text{Distance of Q from the polar plane of P}}$$

$$= \frac{(x_1x_2 + y_1y_2 + z_1z_2 - a^2)/\sqrt{(x_2^2 + y_2^2 + z_2^2)}}{(x_2x_1 + y_2y_1 + z_2z_1 - a^2)/\sqrt{(x_1^2 + y_1^2 + z_1^2)}} \quad \textbf{(Note)}$$

$$= \frac{\sqrt{(x_1^2 + y_1^2 + z_1^2)}}{\sqrt{(x_2^2 + y_2^2 + z_2^2)}} = \frac{OP}{OQ},$$

where O(0, 0, 0) is the centre of the sphere.

Property V

If the polar plane of a point P (x_1, y_1, z_1) passes through another point Q (x_2, y_2, z_2), then the polar plane of Q will pass through P and such points are known as conjugate points.

The proof of this and property III below are left as exercises for the reader.

Property VI

If the pole of a given plane α lies on another plane β, then the pole of the plane β will lie on the plane α and such planes are known as *conjugate planes.*

ANGLE OF INTERSECTION OF TWO SPHERES

The angle of intersection of two spheres in the angle between the tangent planes to them at their common point of intersection. As the radii of the spheres at this common point are normal to the tangent planes so this angle is also equal to the angle between the radii of the spheres at their common point of intersection.

If the angle of intersection of two spheres is a right angle, the spheres are said to be *orthogonal.*

CONDITION FOR ORTHOGONALITY OF THE SPHERES

Let the equations of the two spheres be

$$x^2 + y^2 + z^2 + 2ux + 2vy + 2wz + d = 0 \qquad \text{...(i)}$$

and

$$x^2 + y^2 + z^2 + 2u'x + 2v'y + 2w'z + d' = 0 \qquad \text{...(ii)}$$

Let P be a common point of intersection of these spheres and C and C' be their centres.

Then $\quad CP = \sqrt{(u^2 + v^2 + w^2 - d)}$

and $\quad C'P = \sqrt{(u'^2 + v'^2 + w'^2 - d')}$,

Also CC' = distance between C $(-u, -v, -w)$ and C' $(-u', -v', -w')$

$$= \sqrt{[(u - u')^2 + (v - v')^2 + (w - w')^2]}$$

If the two spheres (i) and (ii) cut orthogonally, then

$$\angle CPC' = \tfrac{1}{2}\pi \text{ or } C'C^2 = CP^2 + C'P^2$$

$$\Rightarrow \quad (u - u')^2 + (v - v')^2 + (w - w')^2 = (u^2 + v^2 + w^2 - d) + (u'^2 + v'^2 + w'^2 - d')$$

$\Rightarrow \quad 2uu' + 2vv' + 2ww' = d + d'$, which is the required condition.

Example 1:

Find the equation of the sphere which cuts orthogonally each of the four spheres $x^2 + y^2 + z^2 + 2ax = a^2$; $x^2 + y^2 + z^2 + 2by = b^2$; $x^2 + y^2 + z^2 + 2cz = c^2$ and $x^2 + y^2 + z^2 = a^2 + b^2 + c^2$.

Solution:

Let the required equation of the sphere be given as

$$x^2 + y^2 + z^2 + 2ux + 2vy + 2wz + d = 0 \qquad ...(A)$$

If this sphere cuts the first of the given spheres orthogonally, then we have

$$2ua = d - a^2 \qquad ...(i),$$

using $\quad$ "$2uu' + 2vv' + 2ww' = d + d'$".

Similarly if its cuts the second and third of the given spheres orthogonally, then we have

$$2bv = d - b^2 \qquad ...(ii)$$

and

$$2wc = d - c^2 \qquad ...(iii)$$

Also if this sphere cuts the last of the given spheres orthogonally, then we have

$$2u.0 + 2v.0 + 2w.0 = d - a^2 - b^2 - c^2$$

$\Rightarrow \quad d = a^2 + b^2 + c^2$ ...(iv)

From (i), (ii), (iii), with the help of (iv), we have

$$2u = (b^2 + c^2)/a,\ 2v = (c^2 + a^2)/b,\ 2w = (a^2 + b^2)/c.$$

Hence the required sphere from (A) is

$$x^2 + y^2 + z^2 + \left(\frac{b^2 + c^2}{a}\right)x + \left(\frac{c^2 + a^2}{b}\right)y + \left(\frac{a^2 + b^2}{c}\right)z + (a^2 + b^2 + c^2) = 0$$

Example 2:

Find the equation of the sphere which touches the plane $3x + 2y - z + 2 = 0$ at the point $(1, -2, 1)$ and cuts orthogonally the sphere $x^2 + y^2 + z^2$.

Solution:

As the plane $3x + 2y - z + 2 = 0$ is a tangent plane to the required sphere at A $(1, -2, 1)$, so the line joining the centre C of the sphere and the point A $(1, -2, 1)$ must be at right angles to this plane.

Hence, the equation of the line AC is $\frac{1}{3}(x - 1) = \frac{1}{2}(y + 2) = -(z - 1)$, ...(i)

the d.c.'s of the line AC are the coefficients of x, y, z in the given plane.

Any point on this line (i) is $(1 + 3r, -2 + 2r, 1 - r)$ and can be taken as the centre C of the sphere. Also the radius of the sphere is CA

i.e., $\sqrt{[(1 + 3r - 1)^2 + (-2 + 2r + 2)^2 + (1 - r - 1)^2]}$ *i.e.,* $r\sqrt{14}$.

Now the centre and radius of the given sphere are C' $(2, -3, 0)$ and $\sqrt{(2^2 + 3^2 - 4)} = 3$

If the two spheres cut orthogonally, then we have

$$(CP)^2 + (C'P)^2 = (CC')^2$$

i.e., $(r\sqrt{14})^2 + (3)^2 = (1 + 3r - 2)^2 + (-2 + 2r + 3)^2 + (1 - r - 0)^2$

$\Rightarrow$ $14r^2 + 9 = (3r - 1)^2 + (2r - 1)^2 + (1 - r)^2$ or $r = -3/2$.

$\therefore$ The centre C of the sphere is $(1 + 3r, -2 + 2r, 1 - r)$,

where $r = -3/2$

i.e., $[1 - (9/2), -2 -3, 1 + (3/2)]$ *i.e.,* $(-7/2, -5, 5/2)$,

and radius is $r\sqrt{(14)}$ *i.e.,* $(3/2)\sqrt{(14)}$, numerically.

$\therefore$ The equation of the sphere is

$$\{x + (7/2)\}^2 + (y + 5)^2 + \{z - (5/2)^2 = [-(3/2)\sqrt{(14)}]^2$$

$$x^2 + y^2 + z^2 + 7x + 10y - 5z + 12 = 0$$ **Ans.**

LENGTH OF THE TANGENT

Let P (x_1, y_1, z_1) be a point outside the sphere

$$S \equiv x^2 + y^2 + z^2 + 2ux + 2vy + 2wz + d = 0 \quad ...(i)$$

Its centre is C $(-u, -v, -w)$ and radius $= \sqrt{(u^2 + v^2 + w^2 - d)}$.

Now let the tangent from P (x_1, y_1, z_1) to the sphere meet at T, then the radius CT at T must be at right angles to the tangent PT. Therefore Δ PCT is a right angled triangle with $\angle T = 90°$.

$\therefore\quad PT^2 = (PC)^2 - (CT)^2$

$$= [(x_1 + u)^2 + (y1 + v)^2 + (z_1 + w)^2] - (u^2 + v^2 + w^2 - d)$$

$$= x_1^2 + y_1^2 + z_1^2 + 2ux_1 + 2wz_1 + d \equiv S_1, \text{ form (i)}$$

This gives the square of the length of the tangent PT. This is also called power of the point P with respect to the sphere S = 0.

Rule for Finding the Length of the Tangent

In the left hand expression of given equation of the sphere, (right hand side being zero) put the coordinates of the point, from which length of the tangent is to be calculated, in place of x, y, z and the result so obtained is the square of the length of the tangent.

Example:

Find the length of the tangent drawn from the point (1, 2, 3) to the sphere $5(x^2 + y^2 + z^2) - x + 10y + 20z + 8 = 0.$

Solution:

Let P (1, 2, 3) be the given point and let the tangent from P to the given sphere $x^2 + y^2 + z^2 - (1/5)\,x + 2y + 4z + (8/5) = 0$ meet at T, then square of the required length of the tangent = PT^2.

$= 1^2 + 2^2 + 3^2\ (1/5)\ (1) + 2\ (2) + 4\ (3) + (8/5)$

$= 1 + 4 + 9 - (1/5) + 4 + 12 + (8/5) = 30 + (7/5) - (157/5)$

$\Rightarrow$ Required length of the tangent = PT = $\sqrt{(157/5)}$. **Ans.**

RADICAL PLANE

Definition : *The radical plane of two spheres is defined as the locus of a point whose powers with respect to the spheres are equal i.e., the locus of a point from where the square of the lengths of the tangents to the two spheres are equal.*

Let the equations of the two spheres be

$$S_1 \equiv x^2 + y^2 + z^2 + 2u_1x + 2v_1y + 2w_1z + d_1 = 0 \quad \text{...(i)}$$

and

$$S_2 \equiv x^2 + y^2 + z^2 + 2u_2x + 2v_2y + 2w_2z + d_2 = 0 \quad \text{...(ii)}$$

Let P (x_1, y_1, z_1) be a point which is such that its powers with respect to the two spheres are equal *i.e.,* the squares of the lengths of tangents from P (x_1, y_1, z_1) to the spheres $S_1 = 0$ and $S_2 = 0$ are equal.

Then $x_1^2 + y_1^2 + z_1^2 + 2u_1x_1 + 2v_1y_1 + 2w_1z_1 + d_1$

$$= x_1^2 + y_1^2 + z_1^2 + 2u_2x_1 + 2v_2y_1 + 2w_2z_1 + d_2$$

$\Rightarrow \quad 2(u_1 - u_2)\, x_1 + 2\,(v_1 - v_2)\, y_1 + 2\,(w_1 - w_2)\, z_1 + (d_1 - d_2) = 0$

$\therefore$ The equation of the radical plane of the given spheres or the locus or $P\,(x_1, y_1, z_1)$ is

$$2(u_1 - u_2)\, x + 2\,(v_1 - v_2)\, y + 2\,(w_1 - w_2)\, z + (d_1 - d_2) = 0 \quad ...(iii)$$

$\Rightarrow \quad S_1 - S_2 = 0$

PROPERTIES OF RADICAL PLANE

Property I

The radical plane of two spheres passes through their point to intersection.

The radical plane of the spheres $S_1 = 0$ and $S_2 = 0$ is $S_1 - S_2 = 0$. The points which satisfy both $S_1 = 0$ and $S_2 = 0$, satisfy $S_1 - S_2 = 0$ hence the point of intersection of $S_1 = 0$ and $S_2 = 0$ lie on the radicle plane $S_1 - S_2 = 0$.

Also two spheres intersect in a circle which from above lies on the radical plane of the spheres.

Property II

The radical planes of two spheres is at right angles to the line joining their centres.

The direction ratios of the line joining the centres

$$(-u_1, -v_1, -w_1) \text{ and } (-u_2, -v_2, -w_2)$$

of the spheres $S_1 = 0$, $S_2 = 0$ respectively are $u_1 - u_2$, $v_1 - v_2$, $w_1 - w_2$ which are also the direction ratios of the normal to the radical plane $S_1 - S_2 = 0$ given by (iii) above. **Hence the property.**

RADICAL AXIS (OR RADICAL LINE) AND RADICAL CENTRE

Radical Axis

The radical planes of three spheres taken two by two pass through a line which is called the radical axis or the radical line of the three spheres.

Let the three spheres be given by the equations $S_1 = 0$, $S_2 = 0$ and $S_3 = 0$. The radical planes of these three spheres taken two by two are $S_1 - S_2 = 0$; $S_2 - S_3 = 0$ and $S_3 - S_1 = 0$, and these three radical planes evidently pass through the line $S_1 = S_2 = S_3$, which is defined as the radical axis of the spheres $S_1 = 0$, $S_2 = 0$ and $S_3 = 0$.

Radical Centre

The four radical axes of the four spheres taken three at a time meet in a point which is called radical centre.

If $S_1 = 0$, $S_2 = 0$, $S_3 = 0$, $S_4 = 0$ be the four spheres, then the four radical axes taking three spheres at a time are given by

$$S_1 = S_2 = S_3;\ S_1 = S_2 = S_4;\ S_1 = S_3 = S_4$$

and $$S_2 = S_3 = S_4.$$

These radical axes evidently pass through the point given by

$$S_1 = S_2 = S_3 = S_4$$

which is defined as the radical centre.

Note : This point also lies on the radical planes of the four spheres taken two by two.

COAXIAL SYSTEM OF SPHERES

Definition : *A system of spheres is said to be coaxial system of spheres if any two spheres of the system have the same radical plane.*

General Equation of Coaxial System of Spheres

If $S_1 = 0$ and $S_2 = 0$ be the equations of two spheres, then the equation $S_1 + \lambda S_2 = 0$ always represent a system of spheres for all values of λ except -1 and then it degenerates into a plane and all the spheres of this system passes, through the circle of intersection of the spheres $S_1 = 0$ and $S_2 = 0$ and hence any two spheres of this system have the same radical planes $S_1 - S_2 = 0$.

Again if $S_1 = 0$ be the equation of sphere and $P = 0$ be the equation of a plane, then the equation $S_1 + \lambda P = 0$ represents a sphere for all values of λ and passes through the intersection of $S_1 = 0$ and $P = 0$.

Let $S_1 + \lambda_1 P = 0$ and $S_1 + \lambda_2 P = 0$ be two members of this system of spheres $S_1 + \lambda P = 0$. Their radical plane is given by

$$(S_1 + \lambda_1 P) - (S_1 + \lambda_2 P) = 0$$

or $$P = 0,\ \lambda_1 \neq \lambda_2$$

Hence the system of spheres $S_1 + \lambda P = 0$ form a system of coaxial spheres as any two spheres of the system have the same radical plane $P = 0$.

THE EQUATION OF COAXIAL SPHERES IN THE SIMPLEST FORM

Let us suppose the centre of all the spheres of the system lie on x-axis, then y and z-coordinates of the centres of all the spheres are zero whereas the x-coordinates differ.

$\therefore$ The equation of such a system of sphere can be written as

$$x^2 + y^2 + z^2 + 2ux + d = 0 \quad ...(i)$$

The radical plane of two spheres

$$x^2 + y^2 + z^2 + 2u_1x + d_1 = 0 \text{ and } x^2 + y^2 + z^2 + 2u_2x + d_2 = 0$$

of the system (i) is $2(u_1 - u_2)x + (d_1 - d_2) = 0$ ...(ii)

Now if yz-plane is chosen as the radical plane then the equation (ii) should reduce to $x = 0$, the condition for which is

$$d_1 - d_2 = 0 \text{ or } d_1 = d_2$$

i.e., d is absolute constant.

Hence the equation of coaxial system of spheres can be written as

$$x^2 + y^2 + z^2 + 2\lambda x + d = 0, \quad ...(iii)$$

the common radical plane being $x = 0$ *i.e.,* yz-plane.

LIMITING POINTS

Definition : *The centre of spheres of zero radii of a coaxial system of spheres are called the Limiting points of the system.*

Let the equation of the coaxial system of spheres be

$$x^2 + y^2 + z^2 + 2\lambda x + d = 0, \quad ...(i)$$

where λ is a parameter and d an absolute constant.

This can be rewritten as $(x + \lambda)^2 + (y - 0)^2 + (z - 0)^2 = (\lambda^2 - d)$

$\therefore$ The radius of a sphere of this system is $\sqrt{(\lambda^2 - d)}$ and centre is $(-\lambda, 0, 0)$. If this radius is zero, then we get $\lambda^2 = d$ or $\lambda = \pm\sqrt{d}$. Hence the limiting points of the system as defined above are $(\pm\sqrt{d}, 0, 0)$.

Also the section of the system of the coaxial spheres [given by (i)] by the radical plane $x = 0$ is the circle

$$x^2 + y^2 + z^2 + 2\lambda x + d = 0,\ x = 0$$

or in the plane geometry by the circle $y^2 + z^2 = -d$ *i.e.,* aa circle in yz-plane of radius $\sqrt{(-d)}$.

If d is negative, then the radius of the circle is real *i.e.,* the circle is real *i.e.,* the spheres intersect.

If $d = 0$, then the radius of the circle is zero *i.e.,* the spheres intersect in a point circle *i.e.,* the spheres touch each other.

If d is positive, the radius of the circle is imaginary *i.e.,* the circle is imaginary *i.e.,* the spheres do not intersect.

Hence the two spheres of the coaxial system of spheres intersect, touch or do not intersect each other according as d is negative, zero or positive.

Rule for Finding the Limiting Points

Equate to zero the radius of the coaxial system of spheres.

Example 1:

Prove that the spheres cutting two given spheres along a great circle pass through two fixed points.

Solution:

Let x-axis be the line of centres and yz-plane be the radical plane, then the equation of the two given spheres can be taken as

$$x^2 + y^2 + z^2 + 2u_1x + d = 0 \quad ...(i)$$

and $$x^2 + y^2 + z^2 + 2u_2x + d = 0 \quad ...(ii)$$

Let the equation of another sphere be

$$x^2 + y^2 + z^2 + 2ux + 2vy + 2wz + k = 0 \quad ...(iii)$$

Now (iii) cuts (i) along a great circle *i.e.,* the radical plane of (i) and (iii) viz. $2(u - u_1)x + 2vy + 2wz + (k - d) = 0$ passes through the centre $(-u_1, 0, 0)$ of the sphere (i)

i.e., $$-2(u - u_1)u_1 + (k - d) = 0. \quad ...(iv)$$

Similarly if (iii) cuts (ii) along a great circle, we have

$$-2(u - u_2)u_2 + (k - d) = 0. \quad ...(v)$$

Subtracting (v) from (iv), we get $u = u_1 + u_2$ and so from (iv) we get

$$k = d + 2u_1u_2$$

Since (i) and (ii) spheres are given so u_1, u_2 and d are constants and therefore u and k are also constants.

Now the x-axis meets (iii) in points whose absciassae are given by $x^2 + 2ux + k = 0$, the roots of which depend upon the values of u and k and hence constants. Thus every sphere [of the form (iii)] cuts the x-axis in the same points.

Example 2:

Prove that every sphere that passes through the limiting points of a coaxial system cuts every sphere of that system orthogonally.

Solution:

Let the equation of the coaxial system of spheres be

$$x^2 + y^2 + z^2 + 2\lambda x + d = 0 \quad ...(i)$$

Its limiting points are $(\sqrt{d}, 0, 0)$ and $(-\sqrt{d}, 0, 0)$.

Let any sphere be $x^2 + y^2 + 2ux + 2vy + 2wz + k = 0$ ...(ii)

It is passes through ($\sqrt{d}$, 0, 0) and ($-\sqrt{d}$, 0, 0), then

$$d + 2u\sqrt{d} + k = 0$$

and $d - 2u\sqrt{d} + k = 0.$

Solving we get $u = 0$ and $k = -d$.

Hence (ii) becomes $x^2 + y^2 + z^2 + 2vy + 2wz - d = 0$ ...(iii)

If (i) and (iii) cut orthogonally, then $2\lambda\,(0) + 2.0\,v + 2.0\,w = d - d$ which is true for all values of v and w. **Hence proved.**

SOME SOLVED EXAMPLES

Example 1:

Find the equation of the sphere through the four points (0, 0, 0), (–a, b, c), (a, b, –c), (a, b – c) and determine its radius.

Solution:

Let the equation of the sphere be

$$x^2 + y^2 + z^2 + 2ux + 2vy + 2wz + d = 0. \qquad \text{...(i)}$$

Since it passes through (0, 0, 0), $d = 0$

Again (i) passes through the points (–a, b, c) (a, –b, c) and (a, b, –c), therefore

$$(a^2 + b^2 + c^2) - 2au + 2bv + 2cw = 0, \qquad \text{...(ii)}$$

$$(a^2 + b^2 + c^2) + 2au - 2bv + 2cw = 0, \qquad \text{...(iii)}$$

$$(a^2 + b^2 + c^2) + 2au + 2bv - 2cw = 0, \qquad \text{...(iv)}$$

Adding (iii) and (iv), we obtain

$$2\,(a^2 + b^2 + c^2) + 4au = 0 \Rightarrow u = -\frac{1}{2a}(a^2 + b^2 + c^2).$$

Similar $y = -\dfrac{1}{2b}(a^2 + b^2 + c^2)$, $w = -\dfrac{1}{2c}(a^2 + b^2 + c^2)$

Substituting in (i), the equation of the sphere is

$$x^2 + y^2 + z^2 - \left(\frac{x}{a} + \frac{y}{b} + \frac{z}{c}\right)(a^2 + b^2 + c^2) = 0$$

$$\Rightarrow \qquad \frac{x^2 + y^2 + z^2}{a^2 + b^2 + c^2} - \left(\frac{x}{a} + \frac{y}{b} + \frac{z}{c}\right) = 0.$$

Radius $\sqrt{u^2 + v^2 + w^2 - d} = \frac{1}{2}(a^2 + b^2 + c^2)\sqrt{a^{-2} + b^{-2} + c^{-2}}$.

Example 2:

Obtain the sphere having its centre on the line $5y + 2z = 0 = 2x - 3y$ and passing through the two points $(0, -2, -4)$, $(2, -1, -1)$.

Solution:

Let the equation of the sphere be

$$x^2 + y^2 + z^2 + 2ux + 2vy + 2wz + d = 0. \qquad \text{...(i)}$$

Since (i) passes through (0, –2, –4) and (2, –1, –1), so that

$$-4y - 8w + d + 20 = 0, \qquad \text{...(ii)}$$

$$4u - 2y - 2w + d + 6 = 0. \qquad \text{...(iii)}$$

Since the centre (–u, –v, –w) of (i) lies on the given line,

$$\therefore \qquad -5y - 2w = 0 \text{ and } -2u + 3y = 0$$

$$\Rightarrow \qquad w = \frac{-5}{2}v \text{ and } u = \frac{3}{2}v. \qquad \text{...(iv)}$$

Subtracting (iii) from (ii), we get

$$2u + v + 3w - 7 = 0$$

$$\Rightarrow \qquad 3v + v - \frac{15}{2}v - 7 = 0, \text{ using (iv)}$$

$$\Rightarrow \qquad -7v - 14 = 0 \Rightarrow v = -2. \text{ From (iv), } u = -3,\ w = 5.$$

Putting in (ii), we get d = 12. Hence, from (i),

$$x^2 + y^2 + z^2 - 6x - 4y + 10z + 12 = 0$$

is the required equation of the sphere.

Example 3:

Show that the equation of the sphere passing through the three points (3, 0, 2), (–1, 1, 1), (2, –5, 4) and having its centre on the plane $2x + 3y + 4z = 6$ is $x^2 + y^2 + z^2 + 4y - 6z = 1$.

Solution:

Let the equation o the sphere be

$$x^2 + y^2 + z^2 + 2ux + 2vy + 2wz + d = 0. \qquad \text{...(i)}$$

Since it passes through (3, 0, 2), (–1, 1, 1) and (2, –5, 4), therefore

$$6u + 4w + d + 13 = 0, \qquad \text{...(ii)}$$

$$-2u + 2v + 2w + d + 3 = 0, \qquad \text{...(iii)}$$

$$4u - 10v + 8w + d + 45 = 0. \qquad \text{...(iv)}$$

Since the centre (–u, –v, –w) lies on $2x + 3y + 4z - 6 = 0$, therefore

$$2u + 3v + 4w + 6 = 0. \qquad \text{...(v)}$$

Subtracting (iii) from (ii) and (iv), we respectively obtain

$$4u - v + w + 5 = 0, \qquad \text{...(vi)}$$

$$u - 2v + w + 7 = 0 \Rightarrow u = 2v - w - 7. \qquad \text{...(vii)}$$

Using (vii) in (v) and (vi), we get

$$7v + 2w - 8 = 0,$$

$$7v - 3w - 23 = 0.$$

From these equations, we get

$$5w + 15 = 0 \Rightarrow w = -3 \quad \text{and} \quad v = 2.$$

From (viii), u = 0 and from (ii), d = –1.

From (i), the required equation of the sphere is

$$x^2 + y^2 + z^2 + 4y - 6z - 1 = 0.$$

Example 4:

Find the equation of the sphere with centre at (2, 3, –4) and touching the plane $2x + 6y - 3z + 15 = 0$.

Solution:

Centre of the sphere is given as (2, 3, –4)

Radius of the sphere = length of perpendicular from the centre (2, 3, –4) on the given plane

$$= \frac{2(2) + 6(3) - 3(-4) + 15}{\sqrt{[2^2 + 6^2 + (-3)^2]}} = \frac{49}{\sqrt{(49)}} = 7$$

∴ The required equation of the sphere is

$$(x - 2)^2 + (y - 3)^2 + (z + 4)^2 = 7^2$$

$$\Rightarrow \quad x^2 + y^2 + z^2 - 4x + 6y + 8z - 20 = 0 \qquad \textbf{Ans.}$$

Example 5:

A sphere of radius k passes through the origin and meets the axes in A, B, C. Prove that the centroid of the triangle ABC lies on the sphere $9(x^2 + y^2 + z^2) = 4k^2$.

Solution:

Let the coordinates of A, B and C be (a, 0, 0), (0, b, 0) and (0, 0, c) respectively. The sphere also passes through the origin O(0, 0, 0).

We can find the equation of the sphere through O, A, B and C as

$$x^2 + y^2 + z^2 - ax - by - cz = 0 \quad \text{...(i)}$$

Its radius $= \sqrt{[(\frac{1}{2}a)^2 + (\frac{1}{2}b)^2 + (\frac{1}{2}c)^2]} = \frac{1}{2}\sqrt{(a^2 + b^2 + c^2)} = k$ (given)

$$a^2 + b^2 + c^2 = 4k^2 \quad \text{...(ii)}$$

Also if (x_1, y_1, z_1) be the coordinates of the centroid of ΔABC, then $x_1 = \frac{1}{3}a$, $y_1 = \frac{1}{3}b$, $z_1 = \frac{1}{3}c$ or $a = 3x_1$, $b = 3y_1$, $c = 3z_1$.

Substituting these values of a, b and c in (ii) we get

$$(3x_1)^2 + (3y_1)^2 + (3z_1)^2 = 4k^2$$

$\therefore$ The required locus of (x_1, y_1, z_1) is $9(x^2 + y^2 + z^2) = 4k^2$.

Example 6:

Prove that equation

$$ax^2 + ay^2 + az^2 + 2ux + 2vy + 2wz + d = 0$$

represents a sphere and find its centre and radius.

Solution:

From the given equation we find that :

(i) this equation is of second degree in x, y and z

(ii) the coefficients of x^2, y^2 and z^2 are equal, each being a and

(iii) the terms containing the product xy, yz and zx are absent

Hence the given equation represents a sphere.

Also as $a \neq 0$, so the given equation of the sphere can be rewritten as $x^2 + y^2 + z^2 + 2(u/a)x + 2(v/a)y + 2(w/a)z + (d/a) = 0$, which is of the form "$x^2 + y^2 + z^2 + 2ux + 2vy + 2wz + d = 0$"

$\therefore$ Its centre is $\left(-\frac{u}{a}, -\frac{v}{a}, -\frac{w}{a}\right)$ and radius is

$$\sqrt{\left(\frac{u^2}{a^2} + \frac{v^2}{a^2} + \frac{w^2}{a^2} - \frac{d}{a}\right)} \text{ i.e. } \frac{1}{a}\sqrt{(u^2 + v^2 + w^2 - ad)}$$

Example 7:

Find the equation of the sphere circumscribing the tetrahedron whose faces are $x = 0$, $y = 0$, $z = 0$, $x/a + y/b + z/c = 1$.

Solution:

Taking three planes at a time and solving their equations we get the coordinates of the vertices of this tetrahedron as (0, 0, 0), (a, 0, 0), (0, b, 0) and (0, 0, c).

The equation of the sphere circumscribing the tetrahedron *i.e.,* through these four points can be obtained.

Example 8:

Find the equation of a sphere which passes through the origin and intercepts lengths a, b, c on the x, y and z-axes respectively.

Or

Find the equation of the sphere which passes through (a, 0, 0) and (0, b, 0), (0, 0, c) and (0, 0, 0).

Or

Find its centre and radius.

Solution:

The sphere intercepts a length a on x-axis so it passes through the point (a, 0, 0). Similarly it passes through the points (0, b, 0) and (0, 0, c). Also it passes through the origin *i.e.,* (0, 0, 0).

Let the equation of the sphere be

$$x^2 + y^2 + z^2 + 2ux + 2vy + 2wz + d = 0 \qquad \ldots(i)$$

If it passes through (0, 0, 0) then from (i) we have $d = 0$ ...(ii)

If (i) passes through (a, 0, 0), then we get $a^2 + 2ua + d = 0$

$\Rightarrow$ from (ii), $a^2 + 2ua + 0 = 0$ or $u = -\frac{1}{2}a$, as $a \neq 0$

Similarly as (i) passes through (0, b, 0) and (0, 0, c) we get

$$y = -\tfrac{1}{2}b \text{ and } w = -\tfrac{1}{2}c$$

Example 9:

A point moves so that the sum of the squares of its distances from the six faces of a cube is constant, show that its locus is a sphere.

Solution:

Take any vertex of the cube as origin and the edges terminating at that vertex as coordinate axes. Then if a be the length of each edge of the cube, then the equations of its six faces are $x = 0$, $y = 0$, $z = 0$, $x = a$, $y = a$ and $z = a$. **(Note)**

Let $P(x_1, y_1, z_1)$ be the moving point, then according to the given problem we have the sum of squares of its distances from the faces of the cube is constant k (say).

i.e., $$\left(\frac{x_1}{1}\right)^2 + \left(\frac{y_1}{1}\right)^2 + \left(\frac{z_1}{1}\right)^2 + \left(\frac{x_1 - a}{1}\right)^2 + \left(\frac{y_1 - a}{1}\right)^2 + \left(\frac{z_1 - a}{1}\right)^2 = k.$$

$\Rightarrow\ x_1^2 + y_1^2 + z_1^2 + (x_1^2 - 2ax_1 + a^2) + (y_1^2 - 2ay_1 + a^2) + (z_1^2 - 2az_1 + a^2 = k)$

$\Rightarrow\ 2(x_1^2 + y_1^2 + z_1^2) - 2a(x_1 + y_1 + z_1) + (3a^2 - k) = 0$

$\Rightarrow\ x_1^2 + y_1^2 + z_1^2 - a(x_1 + y_1 + z_1) + d = 0$, where $d = (3a^2 - k)/2$

$\therefore$ The locus of P $(x_1 + y_1 + z_1)$ is $x^2 + y^2 + z^2 - ax - ay - az + d = 0$ which evidently represents a sphere as it is second degree equation in x, y, z; coefficients of x^2, y^2, z^2 are equal and terms containing product terms xy, yz, zx are absent. **Hence proved.**

Example 10:

OA, OB, OC are three mutually perpendicular lines through the origin having direction cosines l_1, m_1, n_1; l_2, m_2, n_2 and l_3, m_3, n_3.

If OA = a, OB = b, OC = c. Find the equation of the sphere OABC.

Solution:

Talking OA, OB and OC as coordinate axes the coordinates of A, B and C are (a, 0, 0), (0, b, 0) and (0, 0, c) respectively. Consequently the equation of the sphere OABC,

$$x^2 + y^2 + z^2 - ax - by - cz = 0 \qquad \ldots(i)$$

Now transferring to original axes by putting $x = l_1x + m_1y + n_1z$, $y - l_2x + m_2y + n_2z$, $z = l_3x + m_3y + n_3z$, we get from (i) the required equation of the sphere as $(l_1x + m_1y + n_1z)^2 + (l_2x + m_2y + n_2z)^2 + (l_3x + m_3y + n_3z)^2$

$$- a(l_1x + m_1y + n_1z) - b(l_2x + m_2y + n_2z) - c(l_3x + m_3y + n_3z) = 0$$

$$\Rightarrow\ x^2(\Sigma l_1^2) + y^2(\Sigma m_1^2) + z^2(\Sigma n_1^2) + 2xy(\Sigma l_1m_1) + 2yz(\Sigma m_1n_1) + 2zx(\Sigma l_1n_1) - (al_1 + bl_2 + cl_3)x - (am_1 + bm_2 + cm_3)y - (an_1 + bn_2 + cn_3)z = 0$$

$$\Rightarrow\ x^2 + y^2 + z^2 = (al_1 + bl_2 + cl_3)x + (am_1 + bm_2 + cm_3)y + (an_1 + bn_2 + cn_3)z.$$

since $\Sigma l_1^2 = 1$ and $\Sigma l_1m_1 = 0$ etc.

Example 11:

Find the equation of the sphere circumscribing the tetrahedron whose faces are $y/b + z/c = 0$, $z/c + x/a = 0$, $x/a + y/b = 0$, $x/a + y/b + z/c = 1$.

Solution:

Taking three planes at a time and solving their equations, we get the coordinates of the vertices of this tetrahedron as (0, 0, 0), (a, b, –c), (a, –b, c) and (–a, b, c).

$2u = -(a^2 + b^2 + c^2)/a,$

$2v = -(a^2 + b^2 + c^2)/b,\ 2w = -(a^2 + b^2 + c^2)/c$ and $d = 0$

$\therefore$ The required equation is

$$x^2 + \text{iojh}\ \frac{x^2 + y^2 + z^2}{a^2 + b^2 + c^2} - \frac{x}{a} - \frac{y}{b} - \frac{z}{c} = 0$$ **Ans.**

Example 12:

Find the equation of the sphere which passes through the points (1, –3, 4), (1, –5, 2), (1, –3, 0) and whose centre lies on the plane $x + y + z = 0$.

Solution:

Let the equation of the sphere be given as

$$x^2 + y^2 + z^2 + 2ux + 2vy + 2wz + d = 0 \quad ...(i)$$

If it passes through (1, –3, 4), then we have

$$1 + 9 + 16 + 2u - 6v + 8w + d = 0 \quad ...(ii)$$

If it passes through (1, –5, 2), then we get

$$1 + 25 + 4 + 2u - 10v + 4w + d = 0 \quad ...(iii)$$

If it passes through (1, –3, 0), then $1 + 9 + 2u - 6v + d = 0$ ...(iv)

Again the centre of the sphere is (–u, –v, –w) and this centre lies on the plane $x + y + z = 0$.

So we have $-u - v - w = 0 \Rightarrow u + v + w = 0$ ...(v)

Solving (ii) and (iv) we get $16 + 8w = 0$ or $w = -2$ ...(vi)

Solving (ii) and (iii) we get $-4 + 4v + 4w = 3$

$\Rightarrow$ $-4 + 4v - 8 = 0$, from (vi)

$\Rightarrow$ $4v = 12$ or $v = 3$ ...(vii)

$\therefore$ From (v), (vi) and (vii) we get $u = -1$ and so from (iv) $d = 10$

Hence from (i) we get the required equation as

$$x^2 + y^2 + z^2 - 2x + 6y - 4z + 10 = 0$$ **Ans.**

Example 13:

Find the centre and the radius of the sphere

$$x^2 + y^2 + z^2 - 2x + 4y - 6z = 11$$

Solution:

Here 'u' = –1, 'v' = 2, 'w' = –3 'd' = –11

∴ Centre is (–u, –v, –w) *i.e.,* (1, –2, 3)

Hence from (i), required equation is $x^2 + y^2 + z^2 - ax - by - cz = 0$

Ans.

Here '2u' = –a, '2v' = –b, '2w' = –c, 'd' = 0

and radius $= \sqrt{(u^2 + v^2 + w^2 - d)}$

$$= \sqrt{\left[\left(-\frac{1}{2}a\right)^2 + \left(-\frac{1}{2}b\right)^2 + \left(-\frac{1}{2}c\right)^2 - 0\right]}$$

$$= \frac{1}{2}\sqrt{(a^2 + b^2 + c^2)}.$$ **Ans.**

Example 14:

Find the equation of the sphere whose centre is (2, –3, 4) and radius 5.

Solution:

Let P (x_1, y_1, z_1) be any point on the sphere, then as C (2, – 3, 4) is the given centre of the sphere, so

CP = radius of the sphere = 5 *i.e.,* $CP^2 = (5)^2$

$\Rightarrow \quad (x_1 - 2)^2 + (y_1 + 3)^2 + (z_1 - 4)^2 = 5^2$

$\Rightarrow \quad x_1^2 - 4x_1 + 4 + y_1^2 + 6y_1 + 9 + z_1^2 - 8z_1 + 16 = 25$

$\Rightarrow \quad x_1^2 + y_1^2 + z_1^2 - 4x_1 + 6y_1 - 8z_1 + 4 = 0$

∴ Locus of P is $x^2 + y^2 + z^2 - 4x + 6y - 8z + 4 = 0$, which is the required equation of the sphere. **Ans.**

Example 15:

What is the equation of a sphere which passes through (0, 0, 0) and which has its centre at $(\frac{1}{2}, \frac{1}{2}, 0)$.

Solution:

Centre of the sphere is $\frac{1}{2}, \frac{1}{2}, 0)$. As it passes through (0, 0, 0), so

Its radius = distance between (0, 0, 0) and $(\frac{1}{2}, \frac{1}{2}, 0)$ **(Note)**

$$= [(\frac{1}{2} - 0)^2 + (\frac{1}{2} - 0)^2 + (0 - 0)^2] = \sqrt{\left(\frac{1}{2}\right)}$$

∴ Required equation of the sphere is

$$(x - 0)^2 + (y - 0)^2 + (z - 0)^2 = [\sqrt{(\frac{1}{2})^2}]$$

$\Rightarrow \quad x^2 + y^2 + z^2 = \frac{1}{2}$ or $2x^2 + 2y^2 + 2z^2 = 1.$ **Ans.**

Example 16:

Show that every sphere through the circle $x^2 + y^2 - 2ax + r^2 = 0$, $z = 0$ cuts orthogonally every sphere through the circle $x^2 + z^2 = r^2$, $y = 0$.

Solution:

The spheres through the given circles are

$$(x^2 + y^2 + z^2 - 2ax + r^2) + 2\lambda z = 0 \quad \text{...(i)}$$

and $$(x^2 + y^2 + z^2 - r^2) + 2\mu z = 0 \quad \text{...(ii)} \ \textbf{(Note)}$$

If (i) and (ii) cut orthogonally, then we have

$$2(-a)\,0 + 2\,(0)\,(\mu) + 2\,(\lambda)\,0 = r^2 - r^2.$$

which is true for all values of λ and μ. **Hence proved.**

Example 17:

Find the equation of a sphere which cuts four given spheres orthogonally.

Solution:

Let the sphere $x^2 + y^2 + z^2 + 2ux + 2vy + 2wz + d = 0$ ...(i)

cut the four given spheres $x^2 + y^2 + z^2 + 2u_i x + 2v_i y + 2w_i z + d_i = 0$

where i = 1, 2, 3, 4, orthogonally.

Then we have the conditions $2uu_i + 2vv_i + 2ww_i = d + d_i$, $i = 1, 2, 3, 4$ which can be rewritten as

$$-d_1 + 2uu_1 + 2vv_1 + 2ww_1 - d = 0, \quad \text{...(ii)}$$

$$-d_2 + 2uu_2 + 2vv_2 + 2ww_2 - d = 0, \quad \text{...(iii)}$$

$$-d_3 + 2uu_3 + 2vv_3 + 2ww_3 - d = 0, \quad \text{...(iv)}$$

and $$-d_4 + 2uu_4 + 2vv_4 + 2ww_4 - d = 0. \quad \text{...(v)}$$

Eliminating u, v, w and d from (i), (ii), (iii), (iv) and (v), we get the required equation as

$$\begin{vmatrix} x^2 + y^2 + z^2 & x & y & z & -1 \\ -d_1 & u_1 & v_1 & w_1 & -1 \\ -d_2 & u_2 & v_2 & w_2 & -1 \\ -d_3 & u_3 & v_3 & w_3 & -1 \\ -d_4 & u_4 & v_4 & w_4 & -1 \end{vmatrix} = 0$$

Ans.

Example 18:

Find the limiting points of the coaxial system defined by the spheres

$$x^2 + y^2 + z^2 + 3x - 3y + 6 = 0,$$

$$x^2 + y^2 + z^2 - 6y - 6z + 6 = 0.$$

Solution:

The equation of the coaxial system determined by the given spheres is

$$x^2 + y^2 + z^2 + 3x - 3y + 6 + \lambda\,(x^2 + y^2 + z^2 - 6y - 6z + 6) = 0$$

$$\Rightarrow\ x^2 + y^2 + z^2 + \frac{3x}{1+\lambda} - \frac{3(1+2\lambda)y}{1+\lambda} - \frac{6\lambda z}{1+\lambda} + 6 = 0, \qquad \text{...(i)}$$

λ being a parameter.

The centre of (i) is

$$\left[-\frac{3}{2(1+\lambda)}, \frac{3(1+2\lambda)}{2(1+\lambda)}, \frac{3\lambda}{1+\lambda}\right], \qquad \text{...(ii)}$$

$$\text{and radius} = \left[\left\{-\frac{3}{2(1+\lambda)}\right\}^2 + \left\{\frac{3(1+2\lambda)}{2(1+\lambda)}\right\}^2 + \left\{\frac{3\lambda}{1+\lambda}\right\}^2 - 6\right]^{1/2},$$

Equating the radius to zero, we obtain

$$9 + 9(1 + 2\lambda)^2 + 36\lambda^2 - 24\,(1 + \lambda)^2 = 0$$

$$\Rightarrow\qquad 0 = 8\lambda^2 - 2\lambda - 1 = (2\lambda - 1)\,(4\lambda + 1). \ \therefore \lambda = \frac{1}{2}, -\frac{1}{4}.$$

Substituting these values of λ in (ii), the limiting points are $(-1, 2, 1)$ and $(-2, 1, -1)$.

Example 19:

Three sphers of radii r_1, r_2, r_3 have their centres A, B, C at the points (a, 0, 0), (0, b, 0), (0, 0, c) and $r_1^2 + r_2^2 + r_3^2 ++ a^2 + b^2 + c^2$. A fourth sphere passes through the origin and the points A, B, C. Show that the radical centre of four spheres lies on the plane $ax + by + cz = 0$.

Solution:

The equations of the spheres with their centres at A(a, 0, 0), B (0, b, 0), C (0, 0, c) respectively, are

$$S_1 : (x - a)^2 + y^2 + z^2 = r_1^2,$$

$$S_2 : x^2 + (y - b)^2 + z^2 = r_2^2,$$

$$S_3 : x^2 + y^2\,(z - c)^2 = r_3^2.$$

Now the equation of the sphere OABC is

$$S_4 : x^2 + y^2 + z^2 - ax - by - cz = 0.$$

We know that the radical centre of four squares is given by

$$S_1 = S_2 = S_3 = S_4$$

$$\therefore \begin{cases}(x-a)^2 + y^2 + z^2 - r_1^2 = x^2 + (y-b)^2 + z^2 - r_2^2 \\ x^2 + y^2 + (z-c)^2 - r_3^2 = x^2 + y^2 + z^2 - ax - by - cz.\end{cases}$$

Cancelling out $x^2 + y^2 + z^2$ throughout, we obtain

$$\begin{cases} -2ax + a^2 - r_1^2 = -2by + b^2 - r_2^2 \\ = -2cz + c^2 - r_3^2 = -ax - by - cz \end{cases}$$

$$\Rightarrow \quad (-2ax - 2by - 2cz) + a^2 + b^2 + c^2 - (r_1^2 + r_2^2 + r_3^2)$$
$$= 3(-ax - by - cz).$$

Hence $ax + by + cz = 0$, since $r_1^2 + r_2^2 + r_3^2 = a^2 + b^2 + c^2$.

Example 20:

Prove that a sphere $S = 0$ which cuts the two spheres $S_1 = 0$ and $S_2 = 0$ at right angles will also cut the sphere $\lambda_1 S_1 + \lambda_2 S_2 = 0$ at right angles.

Solution:

Let $\quad S \equiv x^2 + y^2 + z^2 + 2ux + 2vy + 2wz + d = 0$

$\quad S_1 \equiv x^2 + y^2 + z^2 + 2u_1x + 2v_1y + 2w_1z + d_1 = 0$

and $\quad S_2 \equiv x^2 + y^2 + z^2 + 2u_2x + 2v_2y + 2w_2z + d_2 = 0$

Then if $S = 0$ cuts $S_1 = 0$ and $S_2 = 0$ orthogonally (*i.e.*, at right angles), we get $\quad 2uu_1 + 2vv_1 + 2ww_1 = d + d_1 \quad$ ...(i)

and $\quad 2uu_2 + 2vv_2 + 2ww_2 = d + d_2 \quad$ ...(ii)

Now the equation of the sphere $\lambda_1 S_1 + \lambda_2 S_2 = 0$ reduces to

$$x^2 + y^2 + z^2 + 2\left(\frac{\lambda_1 u_1 + \lambda_2 u_2}{\lambda_1 + \lambda_2}\right)x + 2\left(\frac{\lambda_1 v_1 + \lambda_2 v_2}{\lambda_1 + \lambda_2}\right)y$$
$$+ 2\left(\frac{\lambda_1 w_1 + \lambda_2 w_2}{\lambda_1 + \lambda_2}\right)z + 2\left(\frac{\lambda_1 d_1 + \lambda_2 d_2}{\lambda_1 + \lambda_2}\right) = 0$$

If $S = 0$ cuts this sphere $l_1 S_1 + l_2 S_2 = 0$ orthogonally, then we have

$$2u\left(\frac{\lambda_1 u_1 + \lambda_2 u_2}{\lambda_1 + \lambda_2}\right) + 2v\left(\frac{\lambda_1 v_1 + \lambda_2 v_2}{\lambda_1 + \lambda_2}\right)$$
$$+ 2w\left(\frac{\lambda_1 w_1 + \lambda_2 w_2}{\lambda_1 + \lambda_2}\right) = d + \left(\frac{\lambda_1 d_1 + \lambda_2 d_2}{\lambda_1 + \lambda_2}\right)$$

$$\Rightarrow \lambda_1 (2uu_1 + 2vv_1 + 2ww_1) + \lambda_2 (2uu_2 + 2vv_2 + 2ww_2)$$
$$= \lambda_1 (d + d_1) + \lambda_2 (d + d_2)$$

$\Rightarrow$ λ_1 (d + d_1) + λ_2 (d + d_2) = λ_1 (d + d_1) + λ_2 (d + d_2), from (i) and (ii) and is evidently true for all values of λ_1 and λ_2. **Hence proved.**

Example 21:

If the variable sphere $x^2 + y^2 + z^2 + 2ux + 2vy + 2wz + 2 = 0$ always cuts the sphere $3(x^2 + y^2 + z^2) - 6x + 10y + z = 8$ at right angles, then show that the point (u, v, w) moves on a fixed plane.

Solution:

The given sphers are $x^2 + y^2 + z^2 + 2ux + 2vy + 2wz + 2 = 0$ ...(i)

and $x^2 + y^2 + z^2 - 2x + (10/3)y + (1/3)z - (8/3) = 0$ ...(ii)

If the spheres (i) and (ii) cut each other at right angles, then

$$2u(-1) + 2v(5/3) + 2w(1/6) = 2 + (-8/3),$$

using $2uu' + 2vv' + 2ww' = d + d'$

$\Rightarrow$ $u - 10v - w = 2$, on simplifying.

$\therefore$ The locus of the point (u, v, w) is $6x - 10y - z = 2$, which represents a fixed plane. **Hence proved.**

Example 22:

The plane $x + 2y + 3z = 12$ cuts the axes of coordinates in A, B, C. Find the equation of the circle circumscribing the triangle ABC.

Solution:

The given plane $x + 2y + 3z = 12$ ...(i)

meets the x-axis *i.e.*, $y = 0$, $z = 0$ in the point A whose coordinates are (12, 0, 0).

Similarly the coordinates of B and C where the given plane meets y-axis are (0, 6, 0) and (0, 0, 4) respectively.

Thus the point A, B, C are (12, 0, 0), (0, 6, 0) and (0, 0, 4) respectively.

Let the equation of the circle circumscribing the triangle ABC be

$$x^2 + y^2 + z^2 + 2ux + 2vy + 2wz + d = 0 \quad ...(ii)$$

If it passes through A, B, C, then we have

$$(12)^2 + 2u(12) + d = 0 \Rightarrow 24u + d + 144 = 0 \quad ...(iii)$$

$$(6)^2 + 2v(6) + d = 0 \Rightarrow 12v + d + 36 = 0 \quad ...(iv)$$

and $$(4)^2 + 2w(4) + d = 0 \Rightarrow 8w + d + 16 = 0 \quad ...(v)$$

From (iii), (iv) and (v) we get

$$2u = -[12 - (d/12)],\ 2v = -[6 - (d/6)] \text{ and } 2w = -[4 - (d/4)].$$

Substituting these values of 2u, 2v, 2w in (ii), we get the required equation as

$$x^2 + y^2 + z^2 - [12 - (d/12)]\,x - [6 - (d/6)]y - [4 - (d/4)]z + d = 0.$$

where d can take any value. **Ans.**

Example 23:

Find the general equation of all sphere through the given point A(a, 0, 0), B (0, b, 0) C (0, 0, c).

Also find the condition that this sphere may cut orthogonally the ***sphere*** $x^2 + y^2 + z^2 - 2ax - 2by - 2cz = 0.$

Solution:

Let the fourth point on the sphere be taken as origin. Then the equation of the sphere through O, A, B and C is given as

$$S \equiv x^2 + y^2 + z^2 - ax + by - cz = 0 \quad \ldots(i)$$

Also the equation of the plane ABC is

$$P \equiv (x/a) + (y/b) + (z/c) - 1 = 0 \quad \ldots(ii)$$

Now the equation of any sphere through the intersection of (i) and (ii) is given by

$$S + \lambda P = 0$$

i.e., $(x^2 + y^2 + z^2 - ax - by - cz) + \lambda(x/a + y/b + z/c - 1) = 0 \quad \ldots(iii)$

which is the required general equation.

Also (iii) can be rewritten as

$$x^2 + y^2 + z^2 - (a - \lambda/a)\,x - (b - \lambda/b)\,y - (c - \lambda/c)\,z - \lambda = 0 \ \ldots(iv)$$

If the given sphere and sphere (iv) cut orthogonally, then we have

$$-(a - \lambda/a)\,(-a) - (b - \lambda/b)\,(-b) - (c - \lambda/c)\,(-c) = -\lambda + 0$$

$$1 = \frac{1}{4}\,(a^2 + b^2 + c^2), \text{ which is the required condition.}$$

Example 24:

Find the limiting points of coaxial systems defined by the spheres $x^2 + y^2 + z^2 + 2x + 2y + 4z + 2 = 0$ *and* $x^2 + y^2 + z^2 + x + y + 2z + 2 = 0.$

Solution:

The equation of the coaxial system of spheres is

$$(x^2 + y^2 + z^2 + 2x + 2y + 4z + 2) + \lambda\,(x^2 + y^2 + z^2 + x + y + 3z + 2) = 0$$

$$\Rightarrow\ x^2 + y^2 + z^2 + \left(\frac{2+\lambda}{1+\lambda}\right)x + \left(\frac{2+\lambda}{1+\lambda}\right)y + \left(\frac{4+2\lambda}{1+\lambda}\right)z + 2 = 0$$

Its centre is $\left(-\frac{2+\lambda}{1+\lambda} - \frac{2+\lambda}{1+\lambda} - \frac{4+2\lambda}{1+\lambda}\right)$ ...(i)

and equating its radius to zero, we get

$$\frac{(2+\lambda)^2}{(1+\lambda)^2} + \frac{(2+\lambda)^2}{(1+\lambda)^2} + \frac{(4+2\lambda)^2}{(1+\lambda)^2} - 2 = 0$$

$$\Rightarrow\quad 6(2+\lambda)^2 = 2(1+\lambda)^2 \Rightarrow 3\,(\lambda^2 + 4\lambda + 4) = \lambda^2 + 2\lambda + 1$$

$$\Rightarrow\quad 2\lambda^2 + 10\lambda + 11 = 0 \Rightarrow \lambda = \frac{-10 \pm \sqrt{(100-88)}}{4} = \frac{-5 \pm \sqrt{3}}{2}$$

$$\Rightarrow\quad \lambda = \frac{-5 \pm \sqrt{3}}{2}, \frac{-5 - \sqrt{3}}{2}$$

Substituting these values of l in the coordinates of the above centre, the required limiting points are

$(1/\sqrt{3}, 1/\sqrt{3}, 2/\sqrt{3})$ and $(-1/\sqrt{3}, -1/\sqrt{3}, -2/\sqrt{3})$ **Ans.**

Example 25:

A sphere of radius r passes through the origin. Show that the ends of the diameter which is parallel to x-axis lie on each of the spheres

$$x^2 + y^2 + z^2 \pm 2rx = 0.$$

Solution:

The equation of any sphere of radius r through the origin is

$$x^2 + y^2 + z^2 + 2ux + 2vy + 2wz = 0 \quad \text{...(i)}$$

where $r^2 = u^2 + v^2 + w^2$ ...(ii)

Its centre is $(-u, -v, -w)$. The equations of its diameter parallel to x-axis, whose d.c.'s are $(1, 0, 0)$ are

$$\frac{x-(-u)}{1} = \frac{y-(-v)}{0} = \frac{z-(-w)}{0} = \pm r.$$

since its extremities are at distance $+r$ or $-r$ from the centre.

Co-ordinates of its extremities are $(r - u, -v, -w)$ and $(-r - u, -v, -w)$.

Let one extremity be (x_1, y_1, z_1) then $x_1 = r - u$, $y_1 = -v$ and $z_1 = -w$

$\Rightarrow \qquad u = r - x_1,\ v = -y_1$ and $w = -z_1$.

Substituting these values of u, v and w in (ii), we get

$$r^2 = (r - x_1)^2 + (-y_1)^2 + (-z_1)^2 \Rightarrow x_1^2 + y_1^2 + z_1^2 - 2rx_1 = 0.$$

$\therefore$ The locus of this extremity is $x^2 + y^2 + z^2 = 2rx$. Similarly the locus of the other extremity is $\qquad x^2 + y^2 + z^2 = -2rx$. **Hence proved.**

Example 26:

Obtain the radical axis for the spheres.

$$(x - 2)^2 + y^2 + z^2 = 1;\ x^2 + (y - 3)^2 + z^2 = 6;$$
$$(x + 2)^2 + (y + 1)^2 + (z - 2)^2 = 6.$$

Solution:

The given spheres are

$$(x - 2)^2 + y^2 + z^2 = 1 \qquad \text{...(i)}$$
$$x^2 + (y - 3)^2 + z^2 = 6 \qquad \text{...(ii)}$$

and

$$(x + 2)^2 + (y + 1)^2 + (z - 2)^2 = 6 \qquad \text{...(iii)}$$

The radical plane of spheres (i) and (ii) is

$$(x - 2)^2 + y^2 + z^2 - 1 = x^2 + (y - 3)^2 + z^2 - 6$$
$$\Rightarrow \quad 4x - 6y = 0 \Rightarrow 2x - 3y + 0.z = 0 \qquad \text{...(iv)}$$

The radical plane of spheres (i) and (iii) is

$$(x - 2)^2 + y^2 + z^2 - 1 = (x + 2)^2 + (y + 1)^2 + (z - 2)^2 - 6$$
$$\Rightarrow \quad 8x + 2y - 4z = 0 \Rightarrow 4x + y - 2z = 0 \qquad \text{...(v)}$$

From (iv) and (v) it is evident (as none contains a constant term) that both the radical planes pass through (0, 0, 0) *i.e.*, the required radical axis passes through (0, 0, 0).

If l, m, n be the d.c.'s of this radical axis, then from (iv) and (v) we get

$$2l - 3m + 0.n = 0,\ 4l + m - 2n = 0$$

Solving these we have $\dfrac{l}{6} = \dfrac{m}{4} = \dfrac{n}{14} \Rightarrow \dfrac{l}{3} = \dfrac{m}{2} = \dfrac{n}{7}$

Hence the required axis is given by

$$\frac{1}{3}(x - 0) = \frac{1}{2}(y - 0) = \frac{1}{7}(z - 0) \Rightarrow \frac{x}{3} = \frac{y}{2} = \frac{z}{7}.$$ **Ans.**

Example 27:

Find the equation of the radical axis in the symmetric form of the spheres

$$S_1 \equiv x^2 + y^2 + z^2 + 2x + 2y + 2z + 2 = 0$$

$$S_2 \equiv x^2 + y^2 + z^2 + 4x + 4z + 4 = 0$$

and $$S_3 \equiv x^2 + y^2 + z^2 + x + 6y - 4z - 2 = 0.$$

Solution:

The radical plane of $S_1 = 0$ and $S_2 = 0$ is given as

$S_1 - S_2 = 0$ *i.e.,* $2x - 2y + 2z + 2 = 0$ *i.e.,* $x - y + z + 1 = 0$ ...(i)

And the radical plane of $S_2 = 0$ and $S_3 = 0$ is

$$S_2 - S_3 = 0 \text{ i.e., } 3x - 6y + 8z + 6 = 0 \quad \text{...(ii)}$$

Now putting $z = 0$ in (i) and (ii), we get

$$x - y + 1 = 0,\ 3x - 6y + 6 = 0.$$

Solving these we get $x = 0$, $y = 1$. Therefore the planes (i) and (ii) pass through the point (0, 1, 0) *i.e.,* (0, 1, 0) is a point on the radical axis given by (i) and (ii) of the given spheres. Also from (i) and (ii) we find that

$l - m + n = 0$ and $3l - 5m + 8n = 0$, where l, m, n are the d.c.'s of the radical axis.

Solving these we get $l/2 = m/5 = n/3$.

∴ The required equation is $\frac{1}{2}(x - 0) = \frac{1}{5}(y - 1) = \frac{1}{3}(z - 0)$ **Ans.**

Example 28:

Prove that the tangent planes to the spheres

$$x^2 + y^2 + z^2 + 2ux + 2vy + 2wz + d = 0$$

and $$x^2 + y^2 + z^2 + 2u_1x + 2v_1y + 2w_1z + d_1 = 0$$

at any common point are at right angles if

$$2uu_1 + 2vv_1 + 2ww_1 = d + d_1$$

Solution:

The given spheres are $x^2 + y^2 + z^2 + 2ux + 2vy + 2wz + d = 0$ ...(i)

and $x^2 + y^2 + z^2 + 2u_1x + 2v_1y + 2w_1z + d_1 = 0$ ...(ii)

The centre and radius of (i) are $(-u, -v, -w)$ and $\sqrt{(u^2 + v^2 + w^2 - d)}$ and those of (ii) are $(-u_1, -v_1, -w_1)$ and $\sqrt{(u_1^2 + v_1^2 + w_1^2 - d_1)}$.

Now if the tangent plane to the spheres (i) and (ii) at any common point are at right angles then the tangent plane of each sphere at such a point must pass through the centre of the other sphere (students may draw the figure themselves) and as such the sum of the squares of the radii of the two spheres must be equal to the square of the distance between their centres.

i.e., $(u^2 + v^2 + w^2 - d) + (u_1^2 + v_1^2 + w_1^2 - d_1)$

$$= (-u + u_1)^2 + (-v + v_1)^2 + (-w + w_1)^2$$

$$2uu_1 + 2vv_1 + 2ww_1 = d + d_1.$$ **Hence proved.**

Example 29:

Find the equation of the circumcircle of the triangle ABC, whose vertices are A (a, 0, 0), B (0, b, 0) and C (0, 0, c). Find also the coordinates of its centre.

Or

The equation of the plane ABC is x/a + y/b + z/c = 1. Obtain the equations of the circle passing through A, B and C, where A, B, C are the points of intersection of the plane with coordinates axes.

Solution:

Equation of the plane ABC is $(x/a) + (y/b) + (z/c) = 1$. ...(i)

Let the fourth point be O(0, 0, 0). Then the equation of the sphere through O, A, B, C is

$$x^2 + y^2 + z^2 - ax - by - cz = 0. \qquad ...(ii)$$

The equations (i) and (ii) together give the required circumcircle of Δ ABC.

The equations of a line through the centre $(\frac{1}{2}a, \frac{1}{2}b, \frac{1}{2}c)$ of the sphere and perpendicular to (i) is

$$\frac{x - \frac{1}{2}a}{1/a} = \frac{y - \frac{1}{2}b}{1/b} = \frac{z - \frac{1}{2}c}{1/c} = r \text{ (say)} \qquad \textbf{(Note)}$$

Any point on this line is $(\frac{1}{2}a + r/a, \frac{1}{2}b + r/b, \frac{1}{2}c + r/c)$. If this is the centre of the circle, then it must lie on (i) and so we get

$$\left(\frac{1}{2} + \frac{r}{a^2}\right) + \left(\frac{1}{2} + \frac{r}{b^2}\right) + \left(\frac{1}{2} + \frac{r}{c^2}\right) = 1 \Rightarrow r = \frac{-1}{2(a^{-2} + b^{-2} + c^{-2})}$$

$$\therefore \quad \frac{1}{2}a + \frac{r}{a} = \frac{1}{2}a - \frac{1}{2a(a^{-2} + b^{-2} + c^{-2})}$$

$$= \frac{a^2(a^{-2} + b^{-2} + c^{-2}) - 1}{2a(a^{-2} + b^{-2} + c^{-2})} = \frac{a(b^{-2} + c^{-2})}{2(a^{-2} + b^{-2} + c^{-2})}$$

Similarly $\frac{1}{2}b + \frac{r}{b} = \frac{b(c^{-2} + a^{-2})}{2(a^{-2} + b^{-2} + c^{-2})}; \frac{1}{2}c + \frac{r}{2} = \frac{a(a^{-2} + b^{-2})}{2(a^{-2} + b^{-2} + c^{-2})}$

$\therefore$ The required centre of the circle is

$$\left[\frac{a(b^{-2} + c^{-2})}{2(a^{-2} + b^{-2} + c^{-2})}, \frac{b(c^{-2} + a^{-2})}{2(a^{-2} + b^{-2} + c^{-2})}, \frac{c(a^{-2} + b^{-2})}{2(a^{-2} + b^{-2} + c^{-2})}\right]$$ **Ans.**

Example 30:

Show that the locus of the centre of the circle of radius of which always intersects the co-ordinates axes (rectangular), is

$$x\sqrt{(a^2 - y^2 - z^2)} + y\sqrt{(a^2 - z^2 - x^2)} + z\sqrt{(a^2 - z^2 - y^2)} = a^2.$$

Solution:

Let C (x_1, y_1, z_1) be the centre of the circle which cuts the co-ordinates axes at A (p, 0, 0), B (0, q, 0) and C (0, 0, r).

Then the centre (x_1, y_1, z_1) lies on the plane ABC, whose equation is

$$x/p + y/q + z/r = 1$$

$$\therefore \quad (x_1/p) + (y_1/q) + (z_1/r) = 1. \quad \ldots(i)$$

Also a = radius CA = $\sqrt{[(x_1 - p)^2 + y_1^2 + z_1^2]}$

$$\Rightarrow \quad (x_1 - p)^2 = a^2 - y_1^2 - z_1^2 \Rightarrow p = x_1 - \sqrt{(a^2 - y_1^2 - z_1^2)}$$

Similarly $q = y_1 - \sqrt{(a^2 - z_1^2 - x_1^2)}$, $r = z_1 - \sqrt{(a^2 - x_1^2 - y_1^2)}$

$$\therefore \quad \frac{1}{p} = \frac{1}{x_1 - \sqrt{(a^2 - y_1^2 - z_1^2)}} = \frac{x_1 + \sqrt{(a^2 - y_1^2 - z_1^2)}}{x_1^2 - a^2 - y_1^2 - z_1^2}$$

Similarly $$\frac{1}{q} = \frac{y_1 + \sqrt{(a^2 - z_1^2 - x_1^2)}}{x_1 + y_1^2 + z_1^2 + a^2}; \frac{z_1 + \sqrt{(a^2 - x_1^2 - y_1^2)}}{x_1^2 + y_1^2 + z_1^2 - a^2}$$

Substituting these values of 1/p, 1/q, 1/r in (i) and simplifying we can get the required locus of (x_1, y_1, z_1).

Example 31:

A sphere of constant radius r passes through the origin o and cuts the axis in A, B, C. Find the locus of the foot of the perpendicular from O to the plane ABC.

Solution:

Let A, B and C be (a, 0, 0), (0, b, 0) and (0, 0, c).

Then the equation of the sphere through O, A, B, C is

$$x^2 + y^2 + z^2 - ax - by - cz = 0, \quad \ldots(i)$$

where $\quad 4r^2 = a^2 + b^2 + c^2 \quad \ldots(ii)$

Also the equation of the plane ABC is $x/a + y/b + z/c = 1 \quad \ldots(iii)$

The equation of the line through O perpendicular to the plane (iii) is

$$\frac{x}{(1/a)} = \frac{y}{(1/b)} = \frac{z}{(1/c)} \quad \ldots(iv)$$

Any point on it (k/a, k/b, k/c) and if it is the foot (x_1, y_1, z_1) of the perpendicular then $x_1 = k/a$, $y_1 = k/b$, $z_1 = k/c$

$\Rightarrow \quad a = k/x_1,\ b = k/y_1,\ c = k/z_1$

Therefore from (ii) we get $4r^2 = k^2 (x_1^{-2} + y_1^{-2} + z_1^{-2}) \quad \ldots(v)$

Also from (iv) we get $\dfrac{x}{(1/a)} = \dfrac{y}{(1/b)} = \dfrac{z}{(1/c)} = k$

$$\Rightarrow \quad k = \frac{x^2}{x/a} = \frac{y^2}{y/b} = \frac{z^2}{z/c} \quad \textbf{(Note)}$$

$$= \frac{x^2 + y^2 + z^2}{(x/a + y/b + z/c)} = \frac{x^2 + y^2 + z^2}{1}, \text{ from (iii)}$$

$$\Rightarrow \quad k = x^2 + y^2 + z^2. \quad \ldots(vi)$$

Now from (v) the locus of (x_1, y_1, z_1) is $4r^2 = k^2 (x^{-2} + y^{-2} + z^{-2})$

$$\Rightarrow \quad 4r^2 = (x^2 + y^2 + z^2)^2 (x^{-2} + y^{-2} + z^{-2}), \text{ from (vi)} \quad \textbf{Ans.}$$

Example 32:

Obtain the equations of the spheres which pass through the circle $y^2 + z^2 = 4$, $x = 0$ and are cut by the plane $2x + 2y + z = 0$ in a circle of radius 3.

Solution:

The equation of any sphere through the given circle is given as

$$(x^2 + y^2 + z^2 - 4) + \lambda r = 0 \quad \textbf{(Note } \mathbf{x^2}\textbf{)}$$

$$\Rightarrow \quad x^2 + y^2 + z^2 + \lambda r - 4 = 0 \quad \ldots(i)$$

Its centre C is $(-\frac{1}{2}\lambda, 0, 0)$ and radius $\sqrt{(\frac{1}{4}\lambda^2 + 4)}$.

Let A be any point on the circumference of the circle in which the sphere (i) is cut by the plane $2x + 2y + z = 0$ and B be the centre of this circle.

Then CA = radius of sphere (i) = $\sqrt{(\frac{1}{4}\lambda^2 + 4)}. \quad \ldots(ii)$

and CB = length of perp. from C to the plane $2x + 2y + z = 0$

i.e., $$CB = \frac{2(-\frac{1}{2}\lambda) + 2.0 + 1.0}{\sqrt{(2^2 + 2^2 + 1)}} = \frac{-\lambda}{3}. \qquad \text{...(iii)}$$

Also given that the radius of the circle is 3 *i.e.,* BA = 3.

Since ABC is a right angled triangle, with $\angle B = \frac{1}{2}\pi$,

so $CA^2 = CB^2 + BA^2$ or $(\frac{1}{2}\lambda^2 + 4) = \frac{1}{9}\lambda^2 + 9$ from (ii), (iii)

$$\Rightarrow \quad \left[\frac{1}{4} - \frac{1}{9}\right]\lambda^2 = 9 - 4 \quad \text{or} \quad \lambda^2 = 36 \quad \text{or} \quad \lambda = \pm 6$$

$\therefore$ From (i), the required equations of the spheres are

$$x^2 + y^2 + z^2 \pm 6x - 4 = 0$$ **Ans.**

Example 33:

If O be the centre of a sphere of radius unity and A and B be two points in a line with O such that OA . OB = 1 and if P be any variable point on the sphere, show that PA : PB = constant.

Solution:

The equation of the sphere is $x^2 + y^2 + z^2 = 1$ ...(i)

Let A be the point (x_1, y_1, z_1) then the direction ratios of OA are x_1, y_1, z_1 so that the equation of line OA is $x/x_1 = y/y_1 = z/z_1$.

It B be the point on this line at a distance r from O, then the coordinates of B are (rx_1, ry_1, rz_1).

Given that the OA . OB = 1, which reduces to

$$\sqrt{\left(x_1^2 + y_1^2 + z_1^2\right)} \cdot \sqrt{\left(r^2x_1^2 + r^2y_1^2 + r^2z_1^2\right)} = 1$$

$$\Rightarrow \quad r(x_1^2 + y_1^2 + z_1^2) = 1. \qquad \text{...(ii)}$$

If the coordinates of any variable point P on the sphere be (x_1, y_2, z_2), then from (i) we get $x_2^2 + y_2^2 + z_2^2 = 1$. ...(iii)

$$\text{Now } \frac{PA^2}{PB^2} = \frac{(x_1 - x_2)^2 + (y_1 - y_2)^2 + (z_1 - z_2)^2}{(rx_1 - x_2)^2 + (ry_1 - y_2)^2 + (rz_1 - z_2)^2}$$

$$= \frac{(x_1^2 + y_1^2 + z_1^2) - 2(x_1x_2 + y_1y_2 + z_1z_2) + (x_2^2 + y_2^2 + z_2^2)}{r^2(x_1^2 + y_1^2 + z_1^2) - 2r(x_1x_2 + y_1y_2 + z_1z_2) + (x_2^2 + y_2^2 + z_2^2)}$$

$$= \frac{(1/r) - 2(x_1x_2 + y_1y_2 + z_1z_2) + 1}{r^2(1/r) - 2r(x_1x_2 + y_1y_2 + z_1z_2) + 1}, \text{ from (ii) and (iii)}$$

$$= \frac{[1 - 2r(x_1x_2 + y_1y_2 + z_1z_2) + r]/r}{r - 2r(x_1x_2 + y_1y_2 + z_1z_2) + 1} = \frac{1}{r} = \text{constant..}$$

Hence PA : PB is constant.

Example 34:

Find the equation of the sphere which touches the sphere $x^2 + y^2 + z^2 - x + 3y + 2z - 3 = 0$ at the point (1, 1, –1) and passes through the origin.

Solution:

Equation of the tangent plane to the given sphere

$$x^2 + y^2 + z^2 - x + 3y + 2z - 3 = 0 \qquad \text{...(i)}$$

at (1, 1, –1) is $x.1 + y.1 + z(-1) - \frac{1}{2}(x + 1) + \frac{3}{2}(y + 1) + (z - 1) - 3 = 0$

$$\Rightarrow \qquad x + 5y - 6 = 0 \qquad \text{...(ii)}$$

The required sphere is the sphere through the point circle (or a circle of zero radius) of intersection of the sphere (i) and the plane (ii) and also passing through the point (0, 0, 0). **(Note)**

The equation of any sphere through the circle of intersection (point circle) is $(x^2 + y^2 + z^2 - x + 3y + 2z - 3) + \lambda(x + 5y - 6) = 0.$ ...(iii)

If it passes through (0, 0, 0), then $-3 - 6\lambda = 0$ or $\lambda = -1/2$

Substituting this value of λ in (iii), the required equation is

$$(x^2 + y^2 + z^2 - x + 3y + 2z - 3) - (1/2)(x + 5y - 6) = 0$$

$$\Rightarrow \qquad 2(x^2 + y^2 + z^2 - x + 3y + 2z - 3) - (x + 5y - 6) = 0$$

$$\Rightarrow \qquad 2(x^2 + y^2 + z^2) - 3x + y - 4z = 0. \qquad \textbf{Ans.}$$

Example 35:

Prove that the plane $x + 2y - z = 0$ cuts the sphere $x^2 + y^2 + z^2 - x + z - 2 = 0$ in a circle of radius unity and find the equation of the sphere which has this circle for one of its great circles.

Solution:

If O be the centre of the sphere, then O is $\left(\frac{1}{2}, 0, -\frac{1}{2}\right)$ and radius OA of the sphere $= \sqrt{(\frac{1}{2})^2 + (0)^2 + (-\frac{1}{2})^2 - (-2)]} = \sqrt{\left(\frac{5}{2}\right)}$.

Also OC = length of perpendicular from O on the plane

$$x + 2y - z = 4 \quad i.e., \quad -x - 2y + z + 4 = 0 \qquad \textbf{(Note)}$$

i.e.,
$$OC = \frac{-\frac{1}{2} - 2(0) + (-\frac{1}{2}) + 4}{\sqrt{[(-1)^2 + (-2)^2 + (1)^2]}} = \frac{3}{\sqrt{6}} = \frac{\sqrt{3}}{\sqrt{2}}$$

∴ The radius of the circle $= CA = \sqrt{(OA^2 - OC^2)}$

$= \sqrt{[(5/2) - (3/2)]} = 1.$ **Hence proved.**

Now the equation of any sphere through the given circle is

$$(x^2 + y^2 + z^2 - x + z - 2) - \lambda (x + 2y - z - 4) = 0 \qquad ...(i)$$

$$\Rightarrow \quad x^2 + y^2 + z^2 + (\lambda - 1) x + 2\lambda y + (1 - \lambda)z - (2 + 4\lambda) = 0 \qquad ...(ii)$$

Its centre is $[-\frac{1}{2} (\lambda - 1), -\lambda, -\frac{1}{2} (1 - \lambda)]$

Now if the given circle is a great circle of the sphere (i), then the centre of (i) must lie on the plane $x + 2y - z = 4$ of the circle.

$$\therefore \quad -\frac{1}{2} (\lambda - 1) + 2(-\lambda) + \frac{1}{2} (1 - \lambda) = 4$$

$$\Rightarrow -3\lambda = 3 \Rightarrow \lambda = -1$$

∴ From (i) the required sphere is

$$(x^2 + y^2 + z^2 - x + z - 2) - (x + 2y - z - 4) = 0$$

$$\Rightarrow \quad x^2 + y^2 + z^2 - 2x - 2y + 2z + 2 = 0 \qquad \textbf{Ans.}$$

Example 36:

Find the plane and centre and radius of the circle common to the two spheres $x^2 + y^2 + z^2 - 4z + 1 = 0$ and $x^2 + y^2 + z^2 - 4x - 2y - 1 = 0$.

Solution:

The equation of the plane of the circle in which the given spheres intersect is $(x^2 + y^2 + z^2 - 4z + 1) - (x^2 + y^2 + z^2 - 4x - 2y + 1) = 0$

$$\Rightarrow \quad 4x + 2y - 4z + 2 = 0 \quad \text{or} \quad 2x + y - 2z + 1 = 0 \quad ...(i)$$

Also the centre of the sphere $x^2 + y^2 + z^2 - 4z + 1 = 0$ ...(ii)

is C (0, 0, 2) and its radius $= \sqrt{(2^2 - 1)} = \sqrt{3}$.

Let A and B be respectively any point on the circumference and the centre of the circle in which the plane (i) intersect the sphere (ii), then CA = $\sqrt{3}$ and CB = length of perpendicular from C to plane (i)

$$= \frac{2(0) + 0 - 2\,(2) + 1}{\sqrt{[2^2 + 1^2 + (-2)^2]}} = -\frac{3}{3} = 1, \text{ numerically}$$

$\therefore$ The required radius of the circle = BA = $\sqrt{(CA^2 - CB^2)}$

$= \sqrt{(3 - 1)} = \sqrt{2}$. **Ans.**

Also CB is a line through C (0, 0, 2) perpendicular to the plane given by (i) and so its equation is $\frac{x - 0}{2} = \frac{y - 0}{1} = \frac{z - 2}{-2}$ **(Note)**

Any point of this line is (2r, r, 2 – 2r)

If this point is B, then it must lie on (i) and so we have

$$2\,(2r) + (r) - 2\,(2 - 2r) + 1 = 0$$

$\Rightarrow$ $4r + r - 4 + 4r + 1 = 0$ or $9r = 3$

$\Rightarrow$ $r = \frac{1}{3}$

The centre B is $\left(\frac{2}{3}, \frac{1}{3}, 2 - \frac{2}{3}\right)$ or $\left(\frac{2}{3}, \frac{1}{3}, \frac{4}{3}\right)$. **Ans.**

Example 37:

A point moves so that the ratio of its two distances from two fixed points is constant. Show that its locus is a sphere.

Solution:

Let the moving point be P(x, y, z) and the fixed points A and B be (a, 0, 0) and (–a, 0, 0) respectively.

Given PA/PB = k or $PA^2 = k^2 \,.\, PB^2$

$\Rightarrow$ $(x - a)^2 + (y - 0)^2 + (z - 0)^2 = k^2\,[(x + a)^2 + (y - 0)^2 + (z - 0)^2]$

$\Rightarrow$ $x^2(1 - k^2) + y^2(1 - k^2) + z^2(1 - k^2) - 2ax(1 + k^2) + a^2(1 - k^2) = 0$

$\Rightarrow$ $x^2 + y^2 + z^2 - \frac{2a(1 + k^2)}{(1 - k^2)}\, x + a^2 = 0$, which evidently represents a sphere.

Example 38:

Find the equation of the sphere which passes through the points (1, 0, 0), (0, 0, 1) and (0, 0, 1) and has its radius as small as possible.

Solution:

Let the equation of the sphere be

$$x^2 + y^2 + z^2 + 2ux + 2vy + 2wz + c = 0 \qquad \text{...(i)}$$

If it passes through (1, 0, 0), (0, 1, 0) and (0, 0, 1) then

$$1 + 2u + c = 0,\ 1 + 2v + c = 0,\ 1 + 2w\ \ c = 0$$

$$\Rightarrow \qquad u = v - w = -\tfrac{1}{2}(1 + c) \qquad \text{...(ii)}$$

∴ If r be the radius of the sphere (i), then

$$r^2 = u^2 + v^2 + w^2 - c = R \text{ (say)}$$

$$\Rightarrow \qquad R = \tfrac{3}{4}(1 + c)^2 = c, \text{ from (ii)}$$

If r is least then R least.

Now $\dfrac{dR}{dc} = \dfrac{3}{2}(1 + c) - 1$ and $\dfrac{d^2R}{dc^2} = \dfrac{3}{2}$ = positive

Equating $\dfrac{dR}{dc}$ to zero we get $\dfrac{3}{2}c + \dfrac{1}{2} = 0$

$\Rightarrow c = -\dfrac{1}{3}$ and $\dfrac{d^2R}{dc^2}$ being positive R is least when $c = -\dfrac{1}{3}$.

∴ From (ii) when R *i.e.,* r^2 is least we have

$$u = v = w = -\frac{1}{2}\left(1 - \frac{1}{3}\right) = -\frac{1}{3}$$

∴ From (i), the required equation is

$$x^2 + Y^2 + z^2 - \frac{2}{3}(x + y + z) - \frac{1}{3} = 0$$

$$\Rightarrow \qquad 3(x^2 + Y^2 + z^2) - 2(x + y + z) - 1 = 0 \qquad \textbf{Ans.}$$

Example 39:

A plane passes through a fixed point (p, q, r) and cuts the axes in A, B, C. Show that the locus of the centre of the sphere OABC is

$$\frac{p}{x} + \frac{q}{y} + \frac{r}{z} = 2.$$

Solution:

Let the coordinates of A, B and C be (a, 0, 0), (0, b, 0) and (0, 0, c) respectively.

The equation of the plane ABC is $\frac{x}{a} + \frac{y}{b} + \frac{z}{c} = 1$ ...(i)

If it passes through (p, q, r), then $\frac{p}{a} + \frac{q}{b} + \frac{r}{c} = 1$ ...(ii)

Also the equation of the sphere OABC can be found

$$x^2 + y^2 + z^2 - ax - by - cz = 0.$$

If its centre be (x_1, y_1, z_1) then $x_1 = ½a$, $y_1 = ½b$ and $z_1 = ½c$.

$\Rightarrow$ $a = 2x_1$, $b = 2y_1$, and $c = 2z_1$

Substituting these values of a, b and c in (ii), we get

$$(p/2x_1) + (q/2y_1) + (r/2z_1) = 1$$

$\therefore$ Locus of the centre (x_1, y_1, z_1) of the sphere OABC is

$$(p/x) + (q/y) + r/z) = 2.$$ **Hence proved.**

Example 40:

A sphere of constant radius 2k passes through origin and meets the axes in A, B and C. Find the locus of the centroid of the tetrahedron OABC.

Solution:

The sphere OABC is

$$x^2 + y^2 + z^2 - ax - by - cz = 0$$

Its radius $= \sqrt{[(½a)^2 + (½b)^2 + (½c)^2]} = ½\sqrt{(a^2 + b^2 + c^2)} = 2k$ (given)

$\Rightarrow$ $a^2 + b^2 + c^2 = 16k^2$ ...(i)

If (x_1, y_1, z_1) be the centroid of the tetrahedron OABC, then

$$x_1 = \frac{1}{4}\ [0 + a + 0 + 0\} = \frac{1}{4}\ a \text{ or } a = 4x_1$$

Similarly $b = 4y_1$ and $c = 4z_1$

Substituting these values of a, b, c in (i) we get

$$(4x_1)^2 + (4y_1)^2 + (4z_1)^2 = 16k^2 \Rightarrow r\ x_1^2 + y_1^2 + z_1^2 = k^2$$

$\therefore$ The required locus of (x_1, y_1, z_1) is $x^2 + y^2 + z^2 = k^2$. **Ans.**

Example 41:

A sphere of radius k passes through the origin and the axes in A, B, C. Prove that the centroid of the triangle lies on the sphere

$$9(x^2 + y^2 + z^2) = 4k^2.$$

Solution:

Let the equation of the sphere be

$$x^2 + y^2 + z^2 + 2ux + 2vy + 2wz + c = 0 \quad \ldots(i)$$

Since (i) passes through (0, 0, 0), so d = 0,

Now (i) meets the x-axis, so $x^2 + 2ux = 0$ Þ $x = -2u$.

Thus (i) meets the x-axis at A(–2u, 0, 0).

Similarly (i) meets the y-axis and z-axis at B(0, –2v, 0) and C(0, 0, –2w) respectively. If (x, y, z) is the centroid of the triangle ABC,

then $x = \dfrac{-2u + 0 + 0}{3}$ or $u = -\dfrac{3}{2}x$.

Similarly $v = -\dfrac{3y}{2}$ and $w = -\dfrac{3z}{2}$.

It is given that k is the radius of (i). So

$$k = \sqrt{u^2 + v^2 + w^2 - d} \Rightarrow k^2 = u^2 + v^2 + w^2 \ (\because d = 0) \quad \ldots(ii)$$

Substituting the values of u, v, w in (ii), we obtain

$$9(x^2 + y^2 + z^2) = 4k^2.$$

MISCELLANEOUS EXAMPLES

Example 1:

A sphere of constant radius 2k passes through the origin and meets the axes in A, B, C. Show that the locus of the centroid of the tetrahedron OABC is the sphere $x^2 + y^2 + z^2 = k^2$.

Solution:

As done in Example 21, we have d = 0 and the points are O(0, 0, 0), A(–2u, 0, 0), B(0, –2v, 0), C(0, 0, –2w).

The centroid (x_1, y_1, z_1) of the tetrahedron, OABC is given by

$$x_1 = \frac{0 - 2u + 0 + 0}{4} \text{ or } u = -2x_1.$$

Similarly $v = -2y_1$ and $w = -2z_1$.

Since the radius of the given sphere is 2k, so

$$2k = \sqrt{u^2 + v^2 + w^2} \text{ or } 4k^2 = 4x_1^2 + 4y_1^2 + 4z_1^2.$$

Hence the locus of (x_1, y_1, z_1) is the sphere $x^2 + y^2 + z^2 = k^2$.

Example 2:

A sphere of constant radius r passes through the origin O and cuts the axes in A, B, C. Prove that the locus of the foot of the perpendicular from O to the plane ABC is

$$(x^2 + y^2 + z^2)^2 (x^{-2} + y^{-2} + z^{-2}) = 4r^2.$$

Solution:

We have d = 0 and the points are A(–2u, 0, 0), B(0, –2v, 0), C(0, 0, –2w).

Since the radius of the given sphere is r, so

$$r = \sqrt{u^2 + v^2 + w^2} \text{ or } u^2 + v^2 + w^2 = r^2. \quad \text{...(i)}$$

Clearly the equation of the plane ABC is

$$\frac{x}{-2u} + \frac{y}{-2v} + \frac{z}{-2w} = 1. \quad \text{...(ii)}$$

Let $P(x_1, y_1, z_1)$ be the foot of the perpendicular from O on the plane (ii). Let OP = k.

Since OP is perpendicular to (ii), the equations of the line OP are

$$\frac{x-0}{-\frac{1}{2u}} = \frac{y-0}{-\frac{1}{2v}} = \frac{z-0}{-\frac{1}{2w}} (= k)$$

Thus P has coordinates $(-k/2u, -k/2v, -k/2w) = (x_1, y_1, z_1)$.

$$\Rightarrow \quad -\frac{k}{2u} = x_1, \frac{-k}{2v} = y_1 \text{ and } \frac{-k}{2w} = z_1$$

$$\Rightarrow \quad u = -k/2x_1, y = -k/2y_1 \text{ and } w = -k/2z_1. \quad \text{...(iii)}$$

Since $P(x_1, y_1, z_1)$ lies on (ii), therefore

$$\frac{x_1}{-2u} + \frac{y_1}{-2v} + \frac{z_1}{-2w} = 1$$

$$\Rightarrow \quad k = x_1^2 + y_1^2 + z_1^2, \text{ using (iii)} \quad \text{...(iv)}$$

Substituting (iii) in (i), we get

$$r^2 = \frac{k^2}{4}(x_1^{-2} + y_1^{-2} + z_1^{-2})$$

$$\Rightarrow \quad 4r^2 = (x_1^2 + y_1^2 + z_1^2)^2 (x_1^{-2} + y_1^{-2} + z_1^{-2}), \text{ using (iv)}$$

Hence the locus of $P(x_1, y_1, z_1)$ is

$$4r^2 = (x^2 + y^2 + z^2)^2 (x^{-2} + y^{-2} + z^{-2}).$$

Example 3:

A variable plane passes through a fixed point (a, b, c) and cuts the coordinate axes in the points A, B, C. Show that the locus of the centre of the sphere OABC is $a/x + b/y + c/z = 2$.

Solution:

Let the equation of the sphere OABC be

$$x^2 + y^2 + z^2 + 2ux + 2vy + 2wz = 0 \qquad \ldots(i)$$

It cuts the coordinates axes at A(–2u, 0, 0),B(0, –2v, 0) and C(0, 0, –2w). The equation of the plane ABC is

$$\frac{x}{-2u} + \frac{y}{-2v} + \frac{z}{-2w} = 1.$$

It passes through (a, b c). So

$$\frac{a}{-2u} + \frac{b}{-2v} + \frac{c}{-2w} = 1 \text{ or } \frac{a}{2x_1} + \frac{b}{2y_1} + \frac{c}{2z_1} = 1,$$

where $(x_1, y_1, z_1) = (-u, -v, -w)$ is the centre of (i).

Hence the locus of the centre (x_1, y_1, z_1) is

$$\frac{a}{2x} + \frac{b}{2y} + \frac{c}{2z} = 1 \Rightarrow \frac{a}{x} + \frac{b}{y} + \frac{c}{z} = 2.$$

Example 4:

Find the equation of the sphere that passes through the point (4, 1, 0), (2, –3, 4), (1, 0, 0) and touches the plane $2x + 2y - z = 11$.

Solution:

Let the equation of the sphere be given as

$$x^2 + y^2 + z^2 + 2ux + 2vy + 2wz + d = 0 \qquad \ldots(i)$$

Its centre is (–u, –v, –w) and radius = $\sqrt{(u^2 + v^2 + w^2 - d)}$

Since the sphere (i) passes through the point (4, 1, 0), so

$$4^2 + 1^2 + 0^2 + 2u(4) + 2v(0) + 2w(0) + d = 0$$

$$\Rightarrow \qquad 8u + 2v + d + 17 = 0 \qquad \ldots(ii)$$

Similarly if the sphere (i) passes through (2, –3, 4) and (1, 0, 0), then we have

$$4u - 6v + 8w + d + 29 = 0 \qquad \ldots(iii)$$

and

$$2u + d + 1 = 0 \qquad \ldots(iv)$$

Also as the sphere (i) touches the given plane, so the length of perpendicular from its centre (–u, –v, –w) to the given plane 2x + 2y – z – 11 = 0 must be equal to the radius $\sqrt{(u^2 + v^2 + w^2 - d)}$ of the sphere (i).

i.e., $$\frac{2(-u) + 2(-v) - (-w) - 11}{\sqrt{[2^2 + 2^2 + (-1)^2]}} = \sqrt{(u^2 + v^2 + w^2 - d)}$$

$\Rightarrow$ $(-2u - 2v + w - 11)^2 = 9(u^2 + v^2 + w^2 - d)$

$\Rightarrow$ $5u^2+5v^2+5w^2-8vw+4uw-44u-44v+22w-9d-121=0$ ...(v)

From (iv), $u = \frac{1}{2}(-d - 1) = -\frac{1}{2}(d + 1)$ ...(vi)

From (ii), $2v = -8u - d - 17 = 4d + 4 - d - 17 = 3d - 13$

$\Rightarrow$ $v - \frac{1}{2}(3d - 13)$ ...(vii)

From (iii), $8w = -4u + 6v - d - 29$

$= (2d + 2) + (9d - 39) - d - 29$, from (vi), (vii)

$\Rightarrow$ $8w = 10d - 66$ or $w \frac{1}{4}(5d - 33)$ ...(viii)

Substituting values of u, v, w from (vi), (vii), (viii) in (v) and simplifying we get $72d^2 - 747d + 1935 = 0$ which gives d = 5.

$\therefore$ From (vi), (vii) and (viii) we get u = –3, v = 1, w = –2.

$\therefore$ From (i) the required equation is

$x^2 + y^2 + z^2 - 6x + 2y - 4z + 5 = 0$. **Ans.**

Example 5:

Find the equation of the sphere on the join of (2, –3, 4) also (–5, 6, –7) as diameter.

Solution:

The required equation of the sphere is

$(x - 2)(x + 5) + (y + 3)(y - 6) + (z - 4)(z + 7) = 0$

$\Rightarrow$ $x^2 + y^2 + z^2 + 3x - 3y + 3z - 49 = 0$.

Example 6:

Find the equation of the sphere through the circle $x^2 + y^2 + z^2 = 9$, $x + y - 2z + 4 = 0$ and the origin.

Solution:

The equation of any sphere through the given circle is given by

$$(x^2 + y^2 + z^2 - 9) + \lambda (x + y - 2z + 4) = 0 \quad \text{...(i)}$$

If this sphere passes through the origin (0, 0, 0), then we have

$$(0 + 0 + 0 - 9) + \lambda (0 + 0 + 0 + 4) = 0 \text{ and } \lambda = 9/4$$

$\therefore$ From (i), the required equation of the sphere is given by

$$(x^2 + y^2 + z^2 - 9) + (9/4)(x + y - 2z + 4) = 0$$

$\Rightarrow$ $4x^2 + 4y^2 + 4z^2 + 9x + 9y - 18z = 0.$ **Ans.**

Example 7:

Prove that the centres of all sections of the sphere $x^2 + y^2 + z^2 = r^2$ by planes through point (x', y', z') lie on the sphere

$$\Sigma x (x - x') = 0.$$

Solution:

Let C (x_1, y_1, z_1) be the centre of the section and the centre of the sphere $x^2 + y^2 + z^2 = a^2$ is O(0, 0, 0).

$\therefore$ The d.c.'s of the line OC are $x_1 - 0, y_1 - 0, z_1 - 0$ *i.e.*, x_1, y_1, z_1 and so the equation of the plane cutting the given sphere in a circle with centre (x_1, y_1, z_1) is the equation of the plane through (x_1, y_1, z_1) and at right angles to OC and its equation therefore is $x_1 (x - x_1) + y_1 (y - y_1) + z_1 (z - z_1) = 0$

If it passes through (x', y', z'), then we get

$$x_1 (x' - x_1) + y_1 (y' - y_1) + z_1 (z' - z_1) = 0.$$

$\therefore$ The locus of the centre (x_1, y_1, z_1) is

$$x (x' - x) + yl (y' - y_1) + z (z' - z_1) = 0$$

$$\Sigma x (x - x') = 0.$$ **Hence proved.**

Example 8:

Find the radius and centre of the circle, determined by the equations $x^2 + y^2 + z^2 = ax + by + cz$, and $(x/a) + (y/b) + (z/c) = 1$.

Solution:

Centre of the sphere $x^2 + y^2 + z^2 - ax - by - cz = 0$ is O (a/2, b/2, c/2) and radius OA $= \sqrt{[(a/2)^2 + (b/2)^2 + (c/2)^2]} = \sqrt{(a^2 + b^2 + c^2)}/2$

$\therefore$ Length of perpendicular from the centre O (a/2, b/2, c/2) to the given plane $(x/a) + (y/b) + (z/c) - 1 = 0$

$$= OC = \frac{(1/a)(a/2) + (1/b)(b/2) + (1/c)(c/2) - 1}{\sqrt{[(1/a)^2 + (1/b)^2 + (1/c)^2]}} \quad \textbf{(Note)}$$

$= abc/[2 \sqrt{(b^2c^2 + c^2a^2 + a^2b^2)}]$

$\therefore$ **Radius of the circle** $= CA = \sqrt{(OA^2 - OC^2)}$

$$= \sqrt{\left[\frac{a^2 + b^2 + c^2}{4} - \frac{a^2b^2c^2}{4(b^2c^2 + c^2a^2 + a^2b^2)}\right]}$$

$$= \frac{1}{2}\sqrt{\left[\frac{(a^2 + b^2 + c^2)(b^2c^2 + c^2a^2 + a^2b^2) - a^2b^2c^2}{b^2c^2 + c^2a^2 + a^2b^2}\right]}$$

$$= \frac{1}{2}\sqrt{\left[\frac{(b^2 + c^2)(c^2 + a^2) - (a^2 + b^2)}{b^2c^2 + c^2a^2 + a^2b^2}\right]}, \text{ on simplifying}$$

Coordinate of the Centre C : As OC is perpendicular to the plane $(x/a) + (y/b) + (z/c) = 1$, so the direction ratios of the line OC are the same as those of the normal to this plane *i.e.,* 1/a, 1/b, 1/c *i.e.,* the coefficients of x, y and z in the equation of this plane.

$\therefore$ Equations of line OC are

$$\frac{x - (a/2)}{1/a} = \frac{y - (b/2)}{1/b} = \frac{z - (c/2)}{1/c} = r \text{ (say)}$$

Let OC = r, then C is $\left(\frac{a}{2} + \frac{r}{a}, \frac{b}{2} + \frac{r}{b}, \frac{c}{2} + \frac{r}{c}\right)$

But C lies on the plane $(x/a) + (y/b) + (z/c) = 1$

$$\therefore \quad \frac{1}{a}\left(\frac{a}{2} + \frac{r}{a}\right) + \frac{1}{b}\left(\frac{b}{2} + \frac{r}{b}\right) + \frac{1}{c}\left(\frac{c}{2} + \frac{r}{c}\right) = 1$$

$$\Rightarrow \quad r\left(\frac{1}{a^2} + \frac{1}{b^2} + \frac{1}{c^2}\right) = -\frac{1}{2} \Rightarrow r = -\frac{a^2b^2c^2}{2(b^2c^2 + c^2a^2 + a^2b^2)} \quad \text{...(i)}$$

$\therefore$ C is $\left(\frac{a}{2} + \frac{r}{a}, \frac{b}{2} + \frac{r}{b}, \frac{c}{2} + \frac{r}{c}\right)$, where r is given by (i).

Example 9:

Find the equations of the spheres which pass through circle $x^2 + y^2 + z^2 = 5$, $x + 2y + 3z = 3$ *and touch the plane* $4x + 3y = 15$.

Solution:

The equation of any sphere through the given circle is

$$(x^2 + y^2 + z^2 - 5) + \lambda (x + 2y + 3z - 3) = 0$$

$$\Rightarrow \quad x^2 + y^2 + z^2 + \lambda x + 2\lambda y + 3\lambda z - (5 + 3\lambda) = 0 \qquad \text{...(i)}$$

Its centre is $(-\frac{1}{2}\lambda, -\lambda, -\frac{3}{2}\lambda)$ and its radius

$$= \sqrt{[(-\tfrac{1}{2}\lambda)^2 + (-\lambda)^2 + (-\tfrac{3}{2}\lambda)^2 + (5 + 3\lambda)]} = \sqrt{(\tfrac{7}{2}\lambda^2 + 3\lambda + 5)}$$

If the sphere (i) touches the given plane $4x + 3y - 15 = 0$ ...(ii)

then the length of the perpendicular ffrom the centre of (i) to (ii) must be equal to the radius of (i).

$$\therefore \quad \frac{4(-\frac{1}{2}\lambda) + 3(-\lambda) - 15}{\sqrt{(4^2 + 3^2)}} = \sqrt{(\tfrac{7}{2}\lambda^2 + 3\lambda + 5)}$$

$$\Rightarrow \quad (-5\lambda - 15)^2 - 25\ (\tfrac{7}{2}\lambda^2 + 3\lambda + 5)$$

$$\Rightarrow \quad \lambda^2 + 6\lambda + 9\ \tfrac{7}{2}\lambda^2 + 3\lambda + 5 \quad \Rightarrow \quad 5\lambda^2 - 6\lambda - 8 = 0$$

$$\Rightarrow \quad (5\lambda + 4)(l - 2) = 0 \qquad \Rightarrow \quad \lambda = 2, -\tfrac{4}{5}$$

Substituting these values of l in (i), we get required equation as

$$x^2 + y^2 + z^2 + 2x + 4y + 6z - 11 = 0$$

and $$x^2 + y^2 + z^2 - \frac{4}{5}x - \frac{8}{5}y - \frac{12}{5}z - \frac{13}{5} = 0.$$ **Ans.**

Example 10:

Find the radius and centre of the circle

$$x^2 + y^2 + z^2 - 8x + 4y + 8z - 45 = 0,\ x - 2y + 2z = 3.$$

Solution:

Centre of this sphere $x^2 + y^2 + z^2 - 8x + 4y + 8z - 45 = 0$ is O $(4, -2, -4)$ and radius OA $= \sqrt{[4^2 + (-2)^2 + (-4)^2 + 45]} = 9$.

$\therefore$ Length of perpendicular from O $(4, -2, -4)$ to the given plane $x - 2y + 2z = 3$

$$= OC = \frac{4 - 2(-2) + 2(-4) - 3}{\sqrt{[1^2 + (-2)^2 + 2^2]}} = 1, \text{ numerically}$$

$\therefore$ *Radius of the circle* $= CA = \sqrt{(OA^2 - OC^2)}$

$$= \sqrt{[(9)^2 - (1)^2]} = 4\sqrt{5}.$$ **Ans.**

Coordinates of the Centre : As OC is perpendicular to the plane $x - 2y + 2z = 3$, so the direction ratios of the line OC are the same as those of the normal to the plane *i.e.*, 1, –2, 2 *i.e.*, the coefficients of x, y, z in the equation of this plane.

∴ The equations of this line OC are

$$\frac{x-4}{1} = \frac{y+2}{-2} = \frac{z+4}{2} = r \text{ (say)}$$

Let OC = r, then the coordinates of C are (4 + r, –2 –2r, –4 + 2r). But C lies on the plane $x - 2y + 2z = 3$.

∴ $(4 + r) - 2(-2 - 2r) + 2(-4 + 2r) = 3 \Rightarrow 9r = 3$ or $r = \frac{1}{3}$

∴ Coordinates of centre C of the circle C of the circle are

(13/3, –8/3, –10/3) **Ans.**

Example 11:

If r is the radius of the circle

$$x^2 + y^2 + z^2 + 2ux + 2vy + 2wz + d = 0,\ lx + my + nz = 0$$

prove that $(r^2 + d)(l^2 + m^2 + n^2) = (mw - nv)^2 + (nu - lw)^2 + (lv - mu)^2$.

Solution:

The radius of the sphere $x^2 + y^2 + z^2 + 2ux + 2vy + 2wz + d = 0$ is $\sqrt{(u^2 + v^2 + w^2 - d)} = R$ (say) and its centre is (–u, –v, –w).

The length of perpendicular from (–u, –v, –w) to the given plane

$lx + my + nz = 0$ is $\dfrac{l(-u) + m(-v) + n(-w)}{\sqrt{(l^2 + m^2 + n^2)}} = p$ (say)

Then the radius of the circle is given as

$$r^2 = R^2 - p^2 = (u^2 + v^2 + w^2 - d) - \frac{(lu + mv + nw)^2}{(l^2 + m^2 + n^2)}$$

$$\Rightarrow r^2 (l^2 + m^2 + n^2) = (u^2 + v^2 + w^2 - d)(l^2 + m^2 + n^2) - (lu + mv + nw)^2$$

$$\Rightarrow (r^2 + d)(l^2 + m^2 + n^2) = (u^2 + v^2 + w^2)\ )\ l^2 + m^2 + n^2) - (lu + mv + nw)^2$$

$$= \Sigma (mw - nv)^2, \text{ by Lagrange's Identity.}$$

Example 12:

Prove that the plane $x + 2y - z = 4$ *cuts the sphere* $x^2 + y^2 + z^2 - x$

$+ z + 2 = 0$ in a circle of radius unity and find the equations of the sphere which has this circle for one of its great circles.

Solution:

The centre of the given sphere is $\left(\frac{1}{2}, 0, -\frac{1}{2}\right)$ and its radius is given as

$$= \sqrt{[(\tfrac{1}{2})^2 + (0)^2 + (-\tfrac{1}{2})^2 - (-2)]} = \sqrt{(\tfrac{5}{2})} = R \text{ (say)}$$

Also length of perpendicular from $(\frac{1}{2}, 0, -\frac{1}{2})$ to $x + 2y - z - 4 = 0$ is

$$\frac{\frac{1}{2} + 2(0) - (-\frac{1}{2}) - 4}{\sqrt{[2^2 + 2^2 + (-1)^2]}} = \frac{3}{\sqrt{6}} = \frac{1}{2}\sqrt{6} = p \text{ (say)}$$

Then radius of the circle $= \sqrt{(R^2 - P^2)}$

$$= \sqrt{\left(\frac{5}{2} - \frac{6}{4}\right)} = 1.$$ **Hence proved.**

The equations of the circle are $x^2 + y^2 + z^2 - x + z - 2 = 0$,

$$x + 2y - z - 4 = 0$$

$\therefore$ The equation of a sphere through this circle is

$$(x^2 + y^2 + z^2 - x + z - 2) + 1(x + 2y - z - 4) = 0$$

$$\Rightarrow \quad x^2 + y^2 + z^2 + (\lambda - 1)x + 2\lambda y + (1 - \lambda)z - (2 + 4\lambda) = 0 \quad \text{...(i)}$$

Its centre is $[-\frac{1}{2}(\lambda - 1), -\lambda, -\frac{1}{2}(1 - \lambda)]$. If this circle is a great circle of the sphere (i), then the centre of (i) should lie on the plane of the circle *i.e.*, the plane $x + 2y - z - 4 = 0$.

$$\therefore \quad -\tfrac{1}{2}(\lambda - 1) + 2(-\lambda) + \tfrac{1}{2}(1 - \lambda) - 4 = 0$$

$$\Rightarrow \quad -3\lambda - 3 = 0 \Rightarrow \lambda = -1.$$

$\therefore$ From (i), the equation of the required sphere is

$$x^2 + y^2 + z^2 - 2x - 2y + 2 = 0.$$ **Ans.**

Example 13:

Find the equation of the sphere whose great circle is $x^2 + y^2 + z^2 + 10y - 4z = 8$, $x + y + z = 3$.

Solution:

Equation of any sphere through the given circle is given by

$$(x^2 + y^2 + z^2 + 10y - 4z - 8) + \lambda(x + y + z - 3) = 0 \quad \text{...(i)}$$

$\Rightarrow\ x^2 + y^2 + z^2 + \lambda x + (10 + \lambda)\, y - (4 - \lambda)\, z - (8 + 3\lambda) = 0$

Its centre is $[-\frac{1}{2}\lambda, -5 - \frac{1}{2}, 2 - \frac{1}{2}\lambda]$ ...(ii)

Now if the given circle is a great circle of the sphere (i), then the centre of sphere (i) must lie on the plane of the circle *i.e.*, on the plane $x + y + z = 3$.

$\therefore$ From (ii), we get $(-\frac{1}{2}\lambda) + (-5 - \frac{1}{2}\lambda) = 3$

$\Rightarrow\ -(3/2)\,\lambda - 6 = 0 \quad \Rightarrow \lambda = -4$

Substituting this value of l in (i), we get the required equation as

$$(x^2 + y^2 + z^2 + 10y - 4z - 8) - 4\,(x + y + z - 3) = 0$$

$\Rightarrow\ x^2 + y^2 + z^2 - 4x + 6y - 8z + 4 = 0$ **Ans.**

Example 14:

Find the equation of the sphere for which the circle $x^2 + y^2 + z^2 - 2z + 2 = 0$ $2x + 3y + 4z = 8$ is a great circle.

Solution:

The equation of any sphere through the given circle is given by

$$(x^2 + y^2 + z^2 + 7y - 2z + 2) + \lambda\,(2x + 3y + 4z - 8) = 0 \quad ...(i)$$

$\Rightarrow\ x^2 + y^2 + z^2 + 2\lambda x + (7 + 3\lambda)\, y + (4\lambda - 2)\, z + (2 - 8\lambda) = 0$

Its centre is $[-\lambda, -\frac{1}{2}(7 + 3\lambda), 1 - 2\lambda]$ ...(ii)

Now if the given circle (which is the section of the sphere $x^2 + y^2 + z^2 - 7y - 2z + 2 = 0$ by the plane $2x + 3y + 4 = 8$) is a great circle of the sphere (i), then the centre of the sphere (i) must lie on the plane of the circle *i.e.*, on the plane $2x + 3y + 4z = 8$.

$\therefore$ From (ii) we get $2(-\lambda) + 3(-\frac{1}{2}\ (7 + 3\lambda)] + 4(1 - 2\lambda) = 8$ **(Note)**

$\Rightarrow\ -2\lambda - \frac{21}{2} - \frac{9}{2}\lambda + 4 - 8\lambda = 8 \Rightarrow \lambda = -1$

Substituting the value of l in (i), we get the required equation as

$$(x^2 + y^2 + z^2 + 7y - 2z + 2) - (2x + 3y + 4z - 8) = 0$$

$\Rightarrow\ x^2 + y^2 + z^2 - 2x + 4y - 6z + 10 = 0.$ **Ans.**

Example 15:

Find the centre and the radius of the circles:

(i) $x + 2y + 2z = 15, x^2 + y^2 + z^2 - 2y - 4z = 11.$

(ii) $x^2 + y^2 + z^2 - 4x - y + 3z + 2 = 0, 2x + 3y - 6z = 10.$

Solution:

(i) The centre of the given sphere is O(0, 1, 2) and radius is

$$OA = \sqrt{0 + 1 + 4 + 11} = 4$$

Now ON = ⊥ distance from O on $x + 2y + 2z - 15 = 0$

$$\Rightarrow \quad ON = \frac{|0 + 2.1 + 2.2 - 15|}{\sqrt{1 + 4 + 4}} = 3.$$

Radius of the circle is $NA = \sqrt{(OA)^2 - (ON)^2} = \sqrt{4^2 - 3^2} = \sqrt{7}$.

Since ON is perpendicular to the piane $x + 2y + 2z = 15$, the equations of ON are

$$\frac{x - 0}{1} = \frac{y - 1}{2} = \frac{z - 2}{2} = r.$$

The point N(r, 2r + 1, 2r + 2) lies on the plane $x + 2y + 2z = 15$.

$\therefore \quad r + 2(2r + 1) + 2(2r + 2) = 15 \Rightarrow r = 1.$

Hence the centre is N(1, 2.1 + 1, 2.1 + 2) = N (1, 3, 4).

(ii) The centre of the given sphere is O(2, 1/2, –3/2) and radius is

$$OA = \sqrt{4 + \frac{1}{4} + \frac{9}{4} - 2} = \sqrt{\frac{9}{2}}.$$

Now ON = ⊥ distance from O on $2x + 3y - 6z - 10 = 0$.

$$\Rightarrow \quad ON = \frac{2.2 + 3.\frac{1}{2} + 6.\frac{3}{2} - 10}{\sqrt{4 + 9 + 36}} = \frac{9}{14}.$$

Radius of the circle is $NA = \sqrt{(OA)^2 - (ON)^2}$

$$i.e., \; NA = \sqrt{\frac{9}{2} - \left(\frac{9}{14}\right)^2} = \frac{1}{14}\sqrt{801}.$$

Since ON is perpendicular to $2x + 3y - 6x = 10$, the equations of ON are

$$\frac{x - 2}{2} = \frac{y - \frac{1}{2}}{3} = \frac{z + \frac{8}{3}}{-6} = r.$$

The point $N(2r + 2, 3r + \frac{1}{2}, - 6r - \frac{8}{3})$ lies on the plane

$2x + 3y - 6x = 10.$

$\therefore \quad 2(2r + 2) + 3(3r + \frac{1}{2}) + 6(6r + \frac{8}{3}) - 10 = 0$ *i.e.*, $r = -9/98$.

Hence the centre is N(178/98, 22/98, –93/98).

Example 16:

Find the equations of the circle circumscribing the triangle formed by the three points (a, 0, 0), (0, b, 0), (0, 0, c). Obtain also the coordinates of the centre of this circle.

Solution:

The equation of the plane through the given points is

$$x/a + y/b + z/c = 1. \qquad \ldots(i)$$

To obtain the equation of the sphere, we need a fourth point which for the sake of convenience is taken as the origin. The equation of the sphere through the points (0, 0, 0), (a, 0, 0), (0, b, 0) and (0, 0, c) is

$$x^2 + y^2 + z^2 - ax - by - cz = 0. \qquad \ldots(ii)$$

Hence (i) and (ii) taken together represent the equations of the required circle. The centre of the sphere (ii) is C(a/2, b/2, c/2). Let N be the centre of the circle.

Since CN is perpendicular to the plane (i), the equations of the CN are

$$\frac{x - a/2}{1/a} = \frac{y - b/2}{1/b} = \frac{z - c/2}{1/c} = r.$$

The point N $\left(\frac{r}{a} + \frac{a}{2}, \frac{r}{b} + \frac{b}{2}, \frac{r}{c} + \frac{c}{2}\right)$ lies on (i),

$$\therefore \quad r\left(\frac{1}{a^2} + \frac{1}{b^2} + \frac{1}{c^2}\right) + \frac{3}{2} = 1 \Rightarrow r = -\frac{1}{2(a^{-2} + b^{-2} + c^{-2})}$$

Now $\frac{r}{a} + \frac{a}{2} = -\frac{1}{2(a^{-2} + b^{-2} + c^{-2})a} + \frac{a}{2}$

$$= \frac{-1 + a^2(a^{-2} + b^{-2} + c^{-2})}{2(a^{-2} + b^{-2} + c^{-2})a}$$

$$= \frac{a^2(b^{-2} + c^{-2})}{2a(a^{-2} + b^{-2} + c^{-2})} = \frac{a(b^{-2} + c^{-2})}{2\Sigma a^{-2}} \text{ etc.}$$

Hence the centre of the circle is

$$\left(\frac{a(b^{-2} + c^{-2})}{2\Sigma a^{-2}}, \frac{b(c^{-2} + a^{-2})}{2\Sigma a^{-2}}, \frac{c(a^{-2} + b^{-2})}{2\Sigma a^{-2}}\right).$$

Example 17:

Find the equations of the section of the sphere $x^2 + y^2 + z^2 = a^2$ of which a given internal point (x_1, y_1, z_1) is the centre.

Solution:

Obviously he section of the sphere is a circle. The centres of the sphere and circle are given as O)0, 0, 0) and $N(x_1, y_1, z_1)$ respectively. So the direction ratios of ON are $x_1 - 0, y_1 - 0, z_1 - 0$ *i.e.*, x_1, y_1, z_1. Now the equation of the plane through $N(x_1, y_1, z_1)$ whose normal has direction ratios x_1, y_1, z_1 is

$$x_1 (x - x_1) + y (y - y_1) + z_1 (z - z_1) = 0$$

$$\Rightarrow xx_1 + yy_1 + zz_1 = x_1^2 + y_1^2 + z_1^2.$$

Hence the equations of the circle are

$$x^2 + y^2 + z^2 = a^2, xx_1 + yy_1 + zz_1 = x_1^2 + y_1^2 + z_1^2.$$

Example 18:

Find the equations of the circle lying on the sphere $x^2 + y^2 + z^2 - 2x + 4y + 3 = 0$ and having centre at (2, 3, –4).

Solution:

The centre C of the given sphere is (1, –2, 3) and that of circle is given to bc Λ (2, 3, –4).

The equation of any plane through A (2, 3, –4) is given as

$$\alpha (x - 2) + \beta (y - 3) + \gamma (z + 4) = 0 \quad \text{...(i)}$$

If this plane is perpendicular to the line Ac, whose d.r.'s are 1 – 2, –2 –3, 3 + 4 *i.e.*, –1, –5, 7, then from (i) we have

$$\frac{\alpha}{-1} = \frac{\beta}{-5} = \frac{\gamma}{7}$$

∴ From (i), the equation of the plane through A (2, 3, –4) at right angles to AC (*i.e.*, the line joining the centres of the circle and sphere) is

$$-(x - 2) - 5 (y - 3) + 7 (z + 4) = 0$$

i.e., $$x + 5y - 7z - 45 = 0 \quad \text{...(ii)}$$

Also the given sphere is $x^2 + y^2 + z^2 - 2x + 4y - 6z + 3 = 0$ ...(iii)

∴ Equation (ii) and (iii) together represent the required circle. **Ans.**

Example 19:

Find the equation of the circle whose centre is (x_1, y_1, z_1) and which lies on the sphere $x^2 + y^2 + z^2 = a^2$.

Solution:

The centre of the given plane is O(0, 0, 0) and that of the circle is given to be $C(x_1, y_1, z_1)$.

Now the equation of a plane through $C(x_1, y_1, z_1)$ is

$$\alpha(x - x_1) + \beta(y - y_1) + \gamma(z - z_1) = 0 \quad \text{...(i)}$$

If this plane is perpendicular to the line OC whose d.r.'s are $x_1 = 0$, $y_1 = 0$, $z_1 = 0$ *i.e.*, x_1, y_1, z_1 then we have $\frac{\alpha}{x_1} = \frac{\beta}{y_1} = \frac{\gamma}{z_1}$ and hence from (i) the equation of plane through $C(x_1, y_1, z_1)$ at right angles to OC is

$$x_1(x - x_1) + y_1(y - y_1) + z_1(z - z_1) = 0$$

i.e.,
$$xx_1 + yy_1 + zz_1 = x_1^2 + y_1^2 + z_1^2 \quad \text{...(ii)}$$

Also the equation of the given sphere is $x^2 + y^2 + z^2 = a^2$ ...(iii)

The equation (ii) and (iii) together represent the circle whose centre is (x_1, y_1, z_1) and which lies on the sphere (iii). **Ans.**

Example 20:

Find the equations of the circle which has its centre at (–2, 2, 1) and lies on the sphere $x^2 + y^2 + z^2 + 5x - 7y + 2z - 8 = 0$.

Solution:

Do your self.

$$x^2 + y^2 + z^2 + 5x - 7y + 22 + 8x - 3y + 4z + 4 = 0 \quad \textbf{Ans.}$$

Example 21:

P is the variable point on the given line and A, B, C are its projections on the axes. Show that the sphere O, ABC passes through a fixed circle.

Solution:

Let the equations of the given line be $\frac{x - \alpha}{l} = \frac{y - \beta}{m} = \frac{z - \gamma}{n}$.

Any variable point on this line may be taken as $P(\alpha + lr, \beta + mr, \gamma + nr)$.

It is given that the points A, B, C are the projections of P on the axes, so the coordinates of A, B and C are $(\alpha + lr, 0, 0)$, $(0, \beta + mr, 0)$ and $(0, 0, \gamma + nr)$ respectively.

Now the equation of the sphere through (0, 0, 0), (a, 0, 0), (0, b, 0) and (0, 0, c) is $x^2 + y^2 + z^2 - ax - by - cz = 0$

∴ In this problem the equation of the sphere O, ABC is

$$x^2 + y^2 + z^2 - (\alpha + lr)\,x - (\beta + mr)\,y - (\gamma + nr)\,z = 0$$

$$\Rightarrow (x^2 + y^2 + z^2 - \alpha x - \beta y - \gamma z) - r\,(lx + my + nz) = 0,$$

which is of the form $S + \lambda p = 0$ where $\lambda = -r$.

$\therefore$ The sphere for all values of r passes through the fixed circle

$$x^2 + y^2 + z^2 - \alpha x + \beta y - \gamma z = 0,\ lx + my + nz = 0.$$

Example 22:

Find the equation of the sphere which passes through the point (α, β, γ) and the circle $x^2 + y^2 = a^2$, $z = 0$.

Solution:

The equations of the circle are given as $x^2 + y^2 = a^2$, $z = 0$.

These can be written as $x^2 + y^2 + z^2 - a^2 = 0$, $z = 0$

$\therefore$ The equation of the sphere through given circle is

$(x^2 + y^2 + z^2 - a^2) + \lambda z = 0$, where λ is constant. ...(i) **(Note)**

If it passes through (α, β, γ) then we have

$$\alpha^2 + \beta^2 + \gamma^2 - a^2 + \lambda g = 0 \text{ of } \lambda = (a^2 - \alpha^2 - \beta^2 + \gamma^2)/\gamma.$$

Substituting this value of λ in (i) the required equation is

$(x^2 + y^2 + z^2 - a^2)\,\lambda + (a^2 - \alpha^2 - \beta^2 - \gamma^2)\,z = 0.$ **Ans.**

Example 23:

POP' is a variable diameter of the ellipse $z = 0$, $x^2/a^2 + y^2/b^2 = 1$ and a circle is described in the plane PP' zz' on PP' as diameter, prove that as PP' varies, the circle generates the surface.

$$(x^2 + y^2 + z^2)\,[(x^2/a^2) + (y^2/b^2)] = x^2 + y^2.$$

Solution:

We known that the parametric coordinates of extremities of a diameter of the ellipse $(x^2/a^2) + (y^2/b^2) = 1$ can be taken as $(a \cos \theta, b \sin \theta)$ and $(-a \cos \theta, -b \sin \theta)$.

$\therefore$ Here in this problem (of 3 dimensions), since POP' is a diameter of the ellipse on the plane $z = 0$, so the coordinates of P and P' can be taken as

$(a \cos \theta, b \sin \theta, 0)$ and $(-a \cos \theta, -b \sin \theta, 0)$ **(Note)**

$\therefore$ The equation of the sphere drawn on PP' as diameter is

$$(x - a \cos \theta)(x + a \cos \theta) + (y - b \sin \theta)(y + b \sin \theta) + (z - 0)(z - 0) = 0$$

$\Rightarrow\ x^2 + y^2 + z^2 = a^2 \cos^2 \theta + b^2 \sin^2 \theta$ ...(i)

The circle referred in the problem is the intersection of the sphere (i) and the plane PP' zz'.

Now the equation of any plane through zz' *i.e.,* z-axis *i.e.,* $x = 0; y = 0$ is

$x + \lambda y = 0$ ...(ii)

If this plane passes through P $(a \cos \theta, b \sin \theta, 0)$ then

$$a \cos \theta + \lambda\, b \sin \theta = 0$$

$$\Rightarrow \quad \lambda = -(a \cos \theta)/(b \sin \theta)$$

$\therefore$ From (ii), the equation of the plane PP' zz' is

$$x - \left(\frac{a \cos \theta}{b \sin \theta}\right) y = 0 \quad \text{or} \quad \frac{x}{a \cos \theta} = \frac{y}{b \sin \theta} \qquad \text{...(iii)}$$

Evidently the coordinates of P' $(-a \cos \theta, -b \sin \theta, 0)$ satisfy it, so (iii) represents the equation of the plane PP' zz'.

$\therefore$ The equations of circle described in the plane PP' zz' on PP' as diameter are given by the sphere (i) and the plane (iii). Now in order to find the locus of this circle, we are to eliminate θ from (i) and (ii). From (iii) we get

$$\frac{(x/a)}{\cos \theta} = \frac{(y/b)}{\sin \theta} = \frac{\sqrt{[(x/a)^2 + (y/b)^2]}}{\sqrt{(\cos^2 \theta + \sin^2 \theta)}} = \sqrt{\left[\left(\frac{x}{a}\right)^2 + \left(\frac{y}{b}\right)^2\right]}$$

The give $a \cos \theta = \dfrac{x}{\sqrt{[(x/a)^2 + (y/b)^2]}}.\ b \sin \theta = \dfrac{y}{\sqrt{[(x/a)^2 + (y/b)^2]}}$

Substituting these values in (i) we get the required locus as

$$(x^2 + y^2 + z^2) = \frac{x^2 + y^2}{(x/a)^2 + (y/b)^2}$$

$$\Rightarrow \quad (x^2 + y^2 + z^2)\left(\frac{x^2}{a^2} + \frac{y^2}{b^2}\right) = x^2 + y^2$$ **Hence proved.**

Example 24:

A sphere S has points (0, 1, 0), (3, –5, 2) at opposite ends of a diameters. Find the equation of the sphere having the intersection of the sphere S with the plane $5x - 2y + 4z + 7 = 0$ as a great circle.

Solution:

The equation of the sphere S is

$$(x - 0)(x - 3) + (y - 1)(y + 5) + (z - 0)(z - 2) = 0$$

$$\Rightarrow x^2 + y^2 + z^2 - 3x - 4y - 2z - 5 = 0 \quad ...(1)$$

Now the equation of the sphere through the circle having equations (1) and $5x - 2y + 4z + 7 = 0$ is

$$x^2 + y^2 + z^2 - (3 - 5k)x - 2(k - 2)y - 2(1 - 2k)z + (7k - 5) = 0 \quad ...(2)$$

Its centre is C $((3 - 5k)/2, k - 2, 1 - 2k)$.

By definition of a great circle, the centres of the sphere and circle are same. Thus C lies on the plane

$$5x - 2y + 4z + 7 = 0$$

$$\therefore \quad \frac{5}{2}(3 - 5k) - 2(k - 2) + 4(1 - 2k) + 7 = 0 \text{ or } k = 1.$$

Putting $k = 1$ in (2), we obtain

$$x^2 + y^2 + z^2 + 2x + 2y + 2z + 2 = 0,$$

which is the required sphere.

Example 25:

Obtain the equation of the sphere which passes through the circle $x^2 + y^2 = 4$, $z = 0$ and is cut by the plane

$x + 2y + 2z = 0$ in a circle of radius 3.

Solution:

The equation of the given circle may be written as

$$x^2 + y^2 = 4, z = 0$$

The equation of the sphere through this circle is

$$x^2 + y^2 - 4 + kz = 0. \quad ...(1)$$

Its centre is $O(0, 0, -k/2)$ and radius $OA = \sqrt{(k^2,4) + 4}.$

Refer to, let N be the centre of the circle. Then

$$ON = \perp \text{ distance from O on } x + 2y + 2z = 0.$$

$$\Rightarrow \quad ON = \frac{|0 + 0 - k|}{\sqrt{1 + 4 + 4}} = \frac{k}{3}.$$

The radius of the circle is

$$NA = \sqrt{OA^2 - ON^2} = \sqrt{(k^2/4) + 4 - (k^2/9)}$$

It is given that $NA = 3$ *i.e.*, $NA^2 = 9$.

$\therefore \quad \dfrac{k^2}{4} + 4 - \dfrac{k^2}{9} \qquad \Rightarrow \qquad 5k^2 + 144 = 324$

$= k^2 = 36,\ k = \pm\ 5$

Putting $k = \pm\ 6$ in (1) we have

$x^2 + y^2 + z^2 + 6z - 4 = 0$ **Ans.**

Example 26:

Find the equation of the sphere having its centre on the plane $4x - 5y - z = 3$ and passing through the circle

$$x^2 + y^2 + z^2 - 2x - 3y + 4z + 8 = 0,$$

$$x^2 + y^2 + z^2 + 4x + 5y - 6z + 2 = 0.$$

Solution:

The equation of the sphere through the given circle is $x^2 + y^2 + z^2 - 2x - 3y + 4z + 8 + k\ (x^2 + y^2 + z^2 + 4x + 5y - 6z + 2) = 0$

$$\Rightarrow \ x^2 + y^2 + z^2 - 2\left(\frac{1-2k}{1+k}\right)x - \left(\frac{3-5k}{1+k}\right)y - 2\left(\frac{3k-2}{1+k}\right)z + \frac{8+2k}{1+k} = 0. \quad ...(1)$$

Its centre $\left(\dfrac{1-2k}{1+k}, \dfrac{3-5k}{2\,(1+k)}, \dfrac{3k-2}{1+k}\right)$ lie on the plane

$4x - 5y - z = 3.$

$\therefore \quad 8(1 - 2k) - 5\ (3 - 5k) - 2\ (3k - 2) = 6\ (1 + k) \Rightarrow k = -3.$

Putting $k = -3$ in (1), we obtain

$x^2 + y^2 + z^2 + 7x + 9y - 11z - 1 = 0$, as the required sphere.

Example 27:

If a plane cuts the sphere $x^2 + y^2 + z^2 - 8x + 4y + 8z - 45 = 0$ in a circle whose centre is (13/3, –8/3, –10/3), then obtain the equations of the circle and hence its radius.

Solution:

The centre O is the given sphere is (4, –2, –4) and radius

$= \sqrt{(4^2 + 2^2 + 4 + 45)} = \sqrt{(81)} = 9 = OA.$

The centre C of the circle is given as (13/3, –8/3, –10/3)

$\therefore \quad OC^2 = \left(\frac{13}{3} - 4\right)^2 + \left(-\frac{8}{3} + 2\right)^2 + \left(-\frac{10}{3} + 4\right)^2$

$\Rightarrow \quad OC2 = (1/9) + (4/9) + (4/9) = 1 \Rightarrow OC = 1.$

Also d. ratios of Oc are (13/3) –4, –4, (–8/3) + 2, (–10/3) + 4

i.e., 1/3, –2/3, 2/3 *i.e.,* 1, –2, 2

$\therefore$ d. cosines of OC are $\frac{1, -2, 2}{\sqrt{[1^2 + (-2)^2 + 2^2]}}$ *i.e.,* $\frac{1}{3}, -\frac{2}{3}, \frac{2}{3}$

$\therefore$ Equations of the plane through C (13/3, –8/3, –10/3) and perpendicular to OC is

$$(1/3)\,[x - (13/3)] - (2/3)\,[y + (8/3)] + (2/3)\,[z + (10/3)] = 0$$

$\Rightarrow \quad (3x - 13) - 2\,(3y + 8) + 2\,(3z + 10) = 0$

$\Rightarrow \quad 3x - 6y + 6x + 4y + 8z - 45 = 0, \Rightarrow x - 2y + 2z = 3$

$\therefore$ Required equations of the circle is

$$x^2 + y^2 + z^2 - 8x + 4y + 8z - 45 = 0 \quad x - 2y + 2z = 3$$

Also its radius = CA = $\sqrt{(OA^2 - OC^2)} = \sqrt{[9^2 = 1^2]} = \sqrt{(80)} = 4\sqrt{5}$. **Ans.**

Example 28:

Find the equation of the sphere through the circle $x^2 + y^2 + z^2 = 9$, $2x + 3y + 4z = 5$ and the point (1, 2, 3).

Solution:

The equation of the sphere through the given circle is given as

$$(x^2 + y^2 + z^2 - 9) + \lambda\,(2x + 3y + 4z - 5) = 0, \quad \text{...(1)}$$

where λ is a constant.

If it passes through the given point (1, 2, 3) then we have

$$(1^2 + 2^2 + 3^2 - 9) + \lambda\,(2 + 6 + 12 - 5) = 0$$

$\Rightarrow \quad 5 + 15\lambda = 0 \quad \Rightarrow \quad \lambda = -1/3$

$\therefore$ From (i), the required equation of the sphere is given by

$$(x^2 + y^2 + z^2 - 9) - \frac{1}{3}\,(2x + 3y + 4z - 5) = 0$$

$$3\,(x^2 + y^2 + z^2) - (2x + 3y + 4z) - 22 = 0.$$ **Ans.**

Example 29:

A variable plane is parallel to the given plane x/a + y/b + z/c = ***0*** ***and*** *meets the axes in A, B, C respectively. Prove that the circle ABC lies on the cone*

$$yz\left(\frac{b}{c}+\frac{c}{b}\right)+ax\left(\frac{c}{a}+\frac{a}{c}\right)+xy\left(\frac{a}{b}+\frac{b}{a}\right)=0$$

Solution:

The equation of any plane parallel to the given plane is given as

$$(x/a) +)y/b) + (z/c) = k \quad ...(1)$$

The plane (1) meets the axes in A, B and C, so the coordinates of A, B and C are (ak, 0, 0); (0, bk, 0) and (0, 0, ck) respectively.

$\therefore$ The equation of the sphere O, ABC is

$$x^2 + y^2 + z^2 - akx - bky - ckz = 0$$

$$\Rightarrow (x^2 + y^2 + z^2) - k(ax - by - cz) = 0 \quad ...(2)$$

The equation of the circle ABC are given by (1) and (2) hence eliminating k between (1) and (2), we get the required locus of circle ABC as

$$(x^2 + y^2 + z^2) - \left(\frac{x}{a}+\frac{y}{b}+\frac{z}{c}\right)(ax + by + cz) = 0$$

$$\Rightarrow \frac{x}{a}(by + cz) + \frac{y}{b}(ax + cz) + \frac{z}{c}(ax + by) = 0$$

$$\Rightarrow yz\left(\frac{c}{b}+\frac{b}{c}\right) zx\left(\frac{c}{a}+\frac{a}{c}\right)+xy\left(\frac{b}{a}+\frac{a}{b}\right)=0$$

Ans.

Example 30:

Find the plane, the centre and the radius of the circle common to the two spheres

$$x^2 + y^2 + z^2 + 1 = 0 \text{ and } x^2 + y^2 + z^2 - 4x - 2y - 1 = 0$$

Solution:

The required equation of the plane is given as

$$(x^2 + y^2 + z^2 - 4z + 1) - (x^2 + y^2 + z^2 - 4x - 2y - 1) = 0 \textbf{ (Note)}$$

$$\Rightarrow 4x + 2y = 4z + 2 = 0 \Rightarrow 2x + y - 2z + 1 = 0 \quad ...(1)$$

The centre of the sphere $x^2 + y^2 + z^2 - 4z + 1 = 0$ ...(2)

is C (0, 0, 2) and radius = $\sqrt{[(0)]^2 + (0)^2 + (2)^2 - (1)]} = \sqrt{3} \doteq R$ (say).

$\therefore$ Length of perpendicular from C (0, 0, 2) to the plane (1)

$$= \frac{2.0 + 0 - 2.2 + 1}{\sqrt{[2^2 + 0^2 + (-2)^2]}} = \frac{3}{2\sqrt{2}} \text{ numerically} = p \text{ (say)}$$

$\therefore$ Required radius of the circle $= \sqrt{(R^2 - p^2)} = \sqrt{[3 - (9/8)]}$.
$= \sqrt{(15/8)}$. **Ans.**

Let C_1 be the centre of the circle, then CC_1 is perpendicular to the plane (1); so the d.c.'s of the line CC_1 are the same as those of the normal to the plane (1) *i.e.*, 2, 1, –2.

$\therefore$ The equation of line CC_1 is $\frac{x - 0}{2} = \frac{y = 0}{1} = \frac{z - 2}{-2} = r$ (say).

Let $CC_1 = r$, then the coordinates of C_1 are (2r, r, 2 – 2r) and C_1 lies on (1) so we get $2(2r) + (r) - 2(2 - 2r) + 1 = 0 \Rightarrow 9r - 3 = 0 \Rightarrow r = 1\ 3$.

$\therefore$ The coordinates of C_1 are (2/3, 1/3, 4/3). **Ans.**

Example 31:

Prove that the circle

$$x^2 + y^2 + z^2 - 2x + 3y + 4z - 5 = 0,\ 5y + 6z + 1 = 0;$$

$$x^2 + y^2 + z^2 - 3x - 4y + 5z - 6 = 0,\ x + 2y - 7z = 0;$$

lie on the same sphere and find its equation

Solution:

The equation of two spheres through the given circles are

$$x^2 + y^2 + z^2 - 2x + 3y + 4z - 5 + k(5y + 6z + 1) = 0 \quad ...(1)$$

and $$x^2 + y^2 + z^2 - 3x - 4y + 5z - 6 + k'(x + 2y - 7z) = 0 \quad ...(2)$$

The equations (1) and (2) will represent the same sphere if the coefficients of x, y, z and the constants in (1) and (2) are same.

$\therefore$ $-2 = k' - 3,\ 5k + 3 = 2k' = 4,$ ...(3)

$6k + 4 = 5 = 7k',\ k - 5 = -6.$ ...(4)

From (3), we get $k' = 1$ and $k - 1$.

These values of k and k' also satisfy the equations (4).

Thus the two circles lie on the same sphere whose equation is obtained by putting either $k = -1$ in (1) or $k' = 1$ in (2).

Putting $k = -1$ in (1), the required sphere is

$$x^2 + y^2 + z^2 - 2x - 2y - 2z - 6 = 0.$$

Example 32:

Show that the two circles

$$2(x^2 + y^2 + z^2) + 8x - 13y + 17z - 17 = 0,\ 2x + y - 3z + 1 = 0;$$

$$x^2 + y^2 + z^2 + 3x - 4y + 3z = 0,\ x - y + 2z - 4 = 0;$$

lie on the same sphere and find its equation.

Solution:

The equation of the sphere thought the first circle is $x^2 + y^2 + z^2 + 4x - \frac{13}{2}y + \frac{17}{2}z - \frac{17}{2} + k(2x + y - 3z + 1) = 0$,

$$\Rightarrow \quad x^2 + y^2 + z^2 + (4 + 2k)x + \left(-\frac{13}{2} + k\right)y + \left(\frac{17}{2}\ 3k\right)z + \left(-\frac{17}{2} + k\right) = 0. \quad ...(1)$$

Similarly the equation of the sphere through the second circle is

$$x^2 + y^2 + z^2 + 3x - 4y + 3z + k'(x - y + 2z - 4) = 0$$

$$\Rightarrow \quad x^2 + y^2 + z^2 + (3 + k')x + (-4 - k')y + (3 + 2k')z - 4k' = 0. \quad ...(2)$$

The equations (1) and (2) will represent the same sphere if the coefficients of x, y, z and the constants in (1) and (2) are same.

$$\therefore \quad 4 + 2k = 3 + k' - (13/2) + k = -4 - k', \quad ...(3)$$

$$(17/2) - 3k = 3 + 2k', -(17/2) + k = -4k'. \quad ...(4)$$

From (3), we obtain $2k = k' + 1 = 0$ and $2k + 2k' - 5 = 0$.

These equations give $k' = 2$ and $k = 1/2$. These values of k and k' also satisfy the equation (4). Thus the two circles lie on the same sphere whose equation is obtained by putting either $k = 1/2$ in (1) or $k' = 2$ in (2). Putting $k' = 2$ in (2), we get

$$x^2 + y^2 + z^2 + 5x - 6y + 7z - 8 = 0,$$ as the required sphere.

Example 33:

P is a variable point on a given line and points A, B, C are its projections on the axes. Show that the sphere OABC passes through a fixed circle

Solution:

Let the given line be

$$\frac{x - \alpha}{l} = \frac{y - \beta}{m} = \frac{z - \gamma}{n} \ (= r).$$

Any variable point on this line is $P(lr + \alpha, mr + \beta, nr + \gamma)$.

The projections of the point P on the axes are

$$A(lr + \alpha, 0, 0), B(0, mr + \beta, 0), C(0, 0, nr + \gamma).$$

Let $a = lr + \alpha$, $b = mr + \beta$, $c = nr + \gamma$.

The equation of the sphere through O, A, B, C is

$$x^2 + y^2 + z^2 - ax - by + cz = 0,$$

$$\Rightarrow x^2 + y^2 + z^2 - (lr + \alpha) x - (mr + \beta) - (nr + \gamma) z = 0$$

$$\Rightarrow (x^2 + y^2 + z^2 - \alpha x - \beta y - \gamma z) - r (lx + my + nz) = 0$$

The sphere clearly passes through the fixed circle

$$x^2 + y^2 + z^2 - \alpha x - \beta y + \gamma z = 0, lx + my + zn = 0.$$

Example 34:

A sphere of radius R passes through the origin, show that the extremities of the diameter parallel to the x-axis lie one on each of the spheres

$$x^2 + y^2 + z^2 \pm 2Rx = 0.$$

Solution:

Let AB be any diameter of the given sphere. Since AB is parallel to x-axis, we may take $A(x_1, 0, 0)$, $B(x_2, 0, 0)$. The equation of the sphere with AB as diameter is

$$(x - x_1)(x - x_2) + (y - 0) + (y - 0) + (z - 0) = 0$$

$$\Rightarrow x^2 + y^2 + z^2 - (x_1 + x_2)x + x_1x_2 = 0. \quad ...(1)$$

Since (1) passes through the origin, $x_1x_2 = 0$

Thus (1) reduces to $x^2 + y^2 + z^2 - (x_1 + x_2) x = 0$. ...(2)

The centre of (2) is $(\frac{1}{2}(x_1 + x_2), 0, 0)$. Since the radius of (2) is given to be R,

$$\therefore \quad R = \sqrt{\left(\frac{x_1 + x_2}{2}\right)^2 + 0 + 0} \Rightarrow R^2 = \left(\frac{x_1 + x_2}{2}\right)^2$$

$\Rightarrow R = \pm 1/2 (x_1 + x_2)$. Thus $x_1 + x_2 = \pm 2R$.

Substituting in (2), we obtain

$$x^2 + y^2 + z^2 \pm 2Rx = 0.$$

Example 35:

A sphere passes through the circle $z = 0$, $x^2 + y^2 = a^2$. Prove that the locus of the extremities of its diameter parallel to x-axis is the rectangular hyperbola $y = 0$, $x^2 = z^2 = a^2$.

Solution:

Any sphere through the given circle $x^2 + y^2 + z^2 = a^2$, $z = 0$ is $x^2 + y^2 + z^2 - a^2 + lz = 0$. ...(1)

The centre of this sphere is $(0, 0, -\lambda/2)$. So the equations of its diameter parallel to the x-axis are

$$y = 0,\ z = -\lambda/2 \text{ (or } \lambda = -2z). \quad ...(2)$$

Eliminating λ between (1) and (2), we obtain

$$y = 0,\ x^2 + z^2 - a^2 - 2z^2 = 0.$$

Hence $\quad y = 0,\ x^2 - z^2 = a^2.$

Example 36:

A variable plane is drawn parallel to a given plane x/a + y/b + z/c = 0 and meets the coordinate axes in A, B, C. Prove that the circle ABC lies on the cone

$$yz\left(\frac{b}{c}+\frac{c}{b}\right) + zx\left(\frac{c}{a}+\frac{a}{c}\right) + xy\left(\frac{a}{b}+\frac{b}{a}\right) = 0$$

Solution:

Any plane parallel to the plane $x/a + y/b + z/c = 0$ is $\frac{x}{a}+\frac{y}{b}+\frac{z}{c} = k$, k being a constant. ...(1)

The plane (1) meets the coordinate axes at A(ka, 0, 0), B(0, kb, 0), C(0, 0, kc).

Now the equation of the sphere through O, A, B, C is

$$x^2 + y^2 + z^2 - k(ax + by + cz) = 0. \quad ...(2)$$

The equations (1) and (2) represent the circle ABC.

Eliminating k between (1) and (2), we obtain

$$x^2 + y^2 + z^2 - \left(\frac{x}{a}+\frac{y}{b}+\frac{z}{c}\right)(ax + by + cz) = 0$$

$$\Rightarrow \quad -\frac{x}{a}(by + cz) - \frac{y}{b}(ax + cz) - \frac{z}{c}(ax + by) = 0$$

Hence yz $yz\left(\frac{b}{c}+\frac{c}{b}\right) + zx\left(\frac{c}{a}+\frac{a}{c}\right) + xy\left(\frac{a}{b}+\frac{b}{a}\right) = 0.$

Example 37:

A circle, centre (2, 3, 0) and radius 1, is drawn in the plane z = 0. Find the equation of the sphere which passes through this circle and the point (1, 1, 1)

Solution:

The equation of the given circle are

$$(x - 2)^2 + (y - 3)^2 = 1^2, z = 0. \qquad \textbf{(Note)}$$

which can be rewritten as $(x - 2)^2 + (y - 3)^2 + z^2 - 1 = 0, z = 0$

(Note the introduction of term z^2)

Equation of any sphere through this circle is

$$[(x - 2)^2 + (y - 3)^2 + (z^2 - 1)] + \lambda z = 0. \qquad ...(1)$$

where λ is constant.

If it passes through (1, 1, 1), then we have

$$[(1 - 2)^2 + (1 - 3)^2 + (1^2 - 1)] + \lambda (1) = 0 \Rightarrow \lambda = -5$$

Substituting this value of λ in (1), the required equation is

$$(x - 2)^2 + (y - 3)^2 + z^2 - 1 - 5z = 0$$

$$\Rightarrow x^2 + y^2 + z^2 - 4x - 6y - 5z + 12 = 0. \qquad \textbf{Ans.}$$

Example 38:

Spheres are described to contain the circle $z = 0.\ x^2 + y^2 = a^2$. Prove that the locus of the extremities of their diameters which are parallel to the x-axis is the rectangular hyperbola $x^2 - z^2 = a^2, y = 0$

Solutlon:

The equation of the sphere through the given circle

$$x^2 + y^2 = a^2, z = 0 \text{ is given as } (x^2 + y^2 = a^2) + \lambda z = 0 \qquad ...(1)$$

Its centre is $(0, 0, -\lambda/2)$ and radius $= \sqrt{[(-l/2)^2 - (-a^2)]}$

$$= [\sqrt{(\lambda^2 + 4a^2)}]/2$$

Now the equation of the diameter of the sphere (1) and parallel to x-axis *i.e.,* the line through the enter $(0, 0 -\lambda/2)$ and parallel to the line with d.c.'s 1, 0, 0 are $\dfrac{x - 0}{1} = \dfrac{y - 0}{0} = \dfrac{z + (\lambda/2)}{0}$

The coordinates of any point on it at a distance r from the centre $(0, 0 -\lambda/2)$ of the sphere (1) are $(r, 0 -\lambda/2)$.

If we take $r = \pm \dfrac{1}{2} \sqrt{(\lambda^2 + 4a^2)} = \pm$ radius of the sphere then we find that the coordinates of the extremities of the diameter parallel to x-axis are given

by $x = \pm \dfrac{1}{2} \sqrt{(\lambda^2 + 4a^2)}, y = 0, z = -\lambda/2$ ...(2)

Required locus is obtained by eliminating λ from (2).

From (2) we have $4x^2 = \lambda^2 + 4a^2$, $y = 0$, $2z = -\lambda$

$\Rightarrow$ $4x^2 = (-2z)^2 + 4a^2$, $y = 0$ on eliminating λ

$\Rightarrow$ $x^2 - z^2 = a^2$, $y = 0$ which is the required locus and is a rectangular hyperbola on the plane $y = 0$. **Hence proved.**

Example 39:

A plane passes through a fixed point (a, b, c), show that the locus of the foot of the perpendicular to it from the origin is the sphere OABC i.e., $x^2 + y^2 + z^2 - ax - by - cz = 0$.

Solution:

The equation of any plane through (a, b, c) is

$$l(x - a) + m(y - b) + n(z - c) = 0 \qquad \text{...(i)}$$

The direction ratios of the normal to this plane are l, m, n

$\therefore$ The equations of the line through (0, 0, 0) and perpendicular to the plane (i) are $\quad x/l = y/m = z/n$ ...(ii)

The point of intersection of (i) and (ii) is the foot of the perpendicular whose locus is required and to get the same we should eliminate l, m, n between (i) and (ii).

Hence the required equation is $x(x - a) + y(y - b) + z(z - c) = 0$

$$\Rightarrow \quad x^2 + y^2 + z^2 - ax - by - cz = 0$$ **Hence proved.**

Example 40:

Show that the circle in which the spheres $x^2 + y^2 + z^2 - 2x - 4y - 11 = 0$ *and* $x^2 + y^2 + z^2 + 2x - y + 12z + 5 = 0$ *intersect each other lie in the plane* $4x + 3y + 12z + 16 = 0$.

Show also that this plane is perpendicular to the line joining the centres of these spheres.

Solution:

The equation of the plane common to given spheres is

$(x^2 + y^2 + z^2 - 2x - 4y + 11) - (x^2 + y^2 + z^2 + 2x - y + 12z + 5) = 0$

$\Rightarrow$ $-4x - 3y - 12z - 16 = 0 \Rightarrow 4x + 3y + 12z + 16 = 0$ ...(i)

The d. ratios of the normal to this plane are 4, 3, 12. ...(ii)

Also the centres of the given spheres are (1, 2, 0), $\left(-1, \frac{1}{2}, -6\right)$.

$\therefore$ The direction ratios of the line joining these centres are

$$1 + 1,\ 2 - \frac{1}{2},\ 0 + 6 \quad \Rightarrow \quad 2,\ 3/2,\ 6 \quad \Rightarrow \quad 4,\ 3,\ 12$$

which are the same as given by (ii).

Hence the normal to the plane (i) is parallel to the line joining the centres of the given spheres *i.e.*, the plane (i) is perpendicular to the line joining the centres of the given spheres. **Hence proved.**

Example 41:

Find the diameter of the circumcircle of ΔABC.

Solution:

We can prove the equation of the plane O, A, B, C is $\frac{1}{2}\ \sqrt{(a^2 + b^2 + c^2)}$ and the centre of this sphere is $(\frac{1}{2}a,\ \frac{1}{2}b,\ \frac{1}{2}c)$.

$\therefore$ The length of perpendicular from the centre of this sphere to the plane

$$ABC = \frac{[(\frac{1}{2}a/a) + (\frac{1}{2}b/b) + (\frac{1}{2}c/c) - 1}{\sqrt{[(1/a)^2 + (1/b)^2 + (1/c)^2]}} = \frac{1}{2\sqrt{(a^{-2} + b^{-2} + c^{-2})}} \text{ (Note)}$$

Now if R be the radius of the circumcircle of ΔABC we have

R^2 = (radius of the sphere)^2 – (length of the perpendicular from centre of the sphere on the plane of the circle)^2

$$i.e.,\ R^2 = [\frac{1}{2}\ \sqrt{a^2 + b^2 + c^2})] - [1/2\ \sqrt{(a^{-2} + b^{-2} + c^{-2})}]^2$$

$$= \frac{a^2 + b^2 + c^2}{4} - \frac{1}{4(a^{-2} + b^{-2} + c^{-2})}$$

$$\Rightarrow \quad R^2 = \frac{a^2 + b^2 + c^2}{4} - \frac{a^2b^2c^2}{4(b^2c^2 + c^2a^2 + a^2b^2)}$$

$$\Rightarrow \quad 4R^2 = \frac{(a^2 + b^2 + c^2)(b^2c^2 + c^2a^2 + a^2b^2) - a^2b^2c^2}{(b^2c^2 + c^2a^2 + a^2b^2)}$$

$$\Rightarrow \quad (2R)^2 = \frac{(b^2 + c^2)(c^2 + a^2) + (a^2 + b^2)}{(b^2c^2 + c^2a^2 + a^2b^2)}, \text{ on simplifying}$$

$\Rightarrow$ the required diameter $= 2R = \sqrt{\dfrac{(b^2 + c^2)(c^2 + a^2) + (a^2 + b^2)}{(b^2c^2 + c^2a^2 + a^2b^2)}}$.

Example 42:

If d be the distance between the centres of two spheres of radii r_1 and r_2, prove that the angle between them is

$$\cos^{-1}[(r_1^2 + r_2^2 - d^2)/2r_1r_2].$$

Hence find the angle of intersection of the sphere $x^2 + y^2 + z^2 - 2x - 4y - 6z + 10 = 0$ with the sphere, the extremities of whose diameter are (1, 2, –3) and (5, 0, 1).

Solution:

Let C_1, C_2 be the centres of the spheres and P be their point of intersection. Then the angle between the spheres is the angle between their radii C_1P and C_2P.

$\therefore$ In $\Delta\, C_1PC_2$, $C_1P = r_1$, $C_2P = r_2$ and $C_1C_2 = d$.

$\therefore$ If θ be the required angle, then

$$\cos\theta = \cos\angle C_1PC_2 \; \frac{C_1P^2 + C_2P^3 - C_1C_2^2}{2C_1P \,.\, C_2P} = \frac{r_1^2 + r_2^2 - d^2}{2r_1r_2}.$$

Now for the second part, the given spheres are

$$x^2 + y^2 + z^2 - 2x - 4y - 6z + 10 = 0 \qquad \text{...(i)}$$

and $\quad (x - 1)(x - 5) + (y - 2)(y - 0) + (z + 3)(z - 1) = 0$

$\Rightarrow \quad x^2 + y^2 + z^2 - 6x - 2y + 2z + 2 = 0 \qquad \text{...(ii)}$

Centre and radius of (i) are (1, 2, 3) and 2.

Centre and radius of (ii) are (3, 1, –1) and 3.

$\therefore$ Here $r_1 = 2$, $r_2 = 3$, $d^2 = [(3 - 1)^2 + (1 - 2)^2 + (-1 - 3)^2] = 21$.

The required angle $= \cos^{-1}\left[\dfrac{r_1^2 + r_2^2 - d^2}{2r_1r_2}\right] = \cos^{-1}\left[\dfrac{4 + 9 - 21}{2.2.3}\right]$

$= \cos^{-1}(-2/3)$ **Ans.**

Example 43:

Prove that the sum of the squares of the intercepts by a given sphere on any three mutually perpendicular lines through a fixed point is constant.

Solution:

Let the three mutually perpendicular lines be the co-ordinate axes which pass through a fixed point (0, 0, 0). **(Note)**

Now the equations of the x-axis are $y = 0, z = 0$. ...(ii)

$\therefore$ From (i) and (ii), we find that x-axis meets the sphere (i) at points whose abscissae are given by $x^2 + 2ux + d = 0$ ...(iii)

If these points be $(x_1, 0, 0)$ and $(x_2, 0, 0)$, then from (iii), we have

$$x_1 + x_2 = -2u \text{ and } x_1x_2 = d.$$

$$\therefore \quad (x_1 - x_2)^2 = (x_1 + x_2)^2 - 4x_1x_2$$

$$= (-2u)^2 - 4d = 4(u^2 - d)$$

i.e., the square of the intercept by the sphere (i) on the x-axis

$$= 4(u^2 - d).$$

Similarly the squares of the intercepts by the sphere (i) on y and z axes are $4(v^2 - d)$ and $4(w^2 - d)$ respectively.

$\therefore$ Required sum of squares of the intercepts

$$= 4(u^2 - d) + 4(v^2 - d) + (w^2 - d)$$

$4(u^2 + v^2 + w^2 - 3d) =$ constant. **Hence proved.**

Example 44:

Show that the two spheres each of which pass through (0, 0, 0), (2a, 0, 0) and (0, 2b, 0) touch the line $\frac{x-a}{l} = \frac{y-b}{m} = \frac{z-c}{n}$.

Also find the distance between the centres of these spheres.

Solution:

Let the given points be O(0, 0, 0), A (2a, 0, 0) and B (0, 2b, 0).

Let us suppose that the spheres cut z-axis at C $(0, 0, 2\lambda)$.

Then the equation of the sphere through O, A, B, and C is

$$x^2 + y^2 + z^2 - 2ax - 2by - 2\lambda z = 0. \quad ...(i)$$

Any point on the given line is $(a + lr, b + mr, c + nr)$. If this lies on (i), then we have

$$(a + lr)^2 + (b + mr)^2 + (c + nr)^2 - 2a(a + lr)$$
$$- 2b(b + mr) - 2\lambda(c + nr) = 0$$

$$\Rightarrow \quad r^2 + 2rn(c - \lambda) - (a^2 + b^2 + 2\lambda c - c^2) = 0, \quad \because \Sigma l^2 = 1.$$

If the given line touches the sphere (i), then the roots of this equation must be coincident and the condition for the same is

$$\text{'}b^2 = 4ac\text{'} \; i.e., \; [2n\,(c - \lambda)]^2 = -4\,(a^2 + b^2 + 2\lambda c - c^2)$$

$$\Rightarrow \quad n^2\lambda^2 - 2c\,(n^2 - 1)\lambda + [a^2 + b^2 + c^2\,(n^2 - 1)] = 0, \quad \ldots\text{(ii)}$$

which being a quadratic equation in λ gives the values λ_1 and λ_2 of λ. Corresponding to these two values of λ we have two spheres from (i) whose centres are (a, b, λ_1) and (a, b, λ_2).

$\therefore$ The required distance between these centres

$$= (\lambda_1 - \lambda_2) = [(\lambda_1 + \lambda_2)^2 - 4\lambda_1\lambda_2]$$

$$= \sqrt{\left[\left\{\frac{2c(n^2 - 1)}{n^2}\right\}^2 - 4\left\{\frac{a^2 + b^2 + c^2\,(n^2 - 1)}{n^2}\right\}\right]}, \text{ from (ii)}$$

$$= (2/n^2)\,\sqrt{[c^2\,(n^2 - 1)^2 - n^2\,[a^2 + b^2 + c^2\,(n^2 - 1)]}$$

$$= (2/n^2)\,\sqrt{[c^2 - n^2\,(a^2 + b^2 + c^2)]}, \text{ on simplifying.}$$ **Ans.**

Example 45:

Prove that the plane $(x - \alpha)\,(u + \alpha) + (y - \beta)\,(v + \beta) + (z - \gamma)\,(w + \gamma) = 0$ *cuts the sphere* $x^2 + y^2 + z^2 + 2ux + 2vy + 2wz + d = 0$ *in circle whose center is* (α, β, γ) *and the equation of the sphere which has this circle as a great circle is*

$$x^2 + y^2 + z^2 - 2\alpha\,(x - u) - 2\beta\,(y - v) - 2\gamma\,(z - w) + 2\alpha^2 + 2\beta^2 + 2\gamma^2 - d = 0$$

Solution:

The co-ordinates of the centre O of the given sphere are $(-u, -v, -w)$.

The line OC being perpendicular to the given plane, its d.r.'s will be the same as those of the normal to the given plane.

i.e., d.r.'s of OC are $(u + \alpha, v + \beta, w + \gamma)$. **(Note)**

$\therefore$ The equations of OC (a line through O and perpendicular to the given plane) are

$$\frac{x + u}{(u + \alpha)/[\Sigma(u + \alpha)^2]} = \frac{x + v}{(v + \beta)/\sqrt{[\Sigma(u + \alpha)^2]}} = \frac{z + w}{(w + \gamma)/\sqrt{[\Sigma(u + \alpha)^2]}}$$

$\therefore$ Any point on this line is

$$\left[-u + \frac{(u+\alpha)r}{\sqrt{[\Sigma(u+\alpha)^2]}}, -v + \frac{(v+\beta)r}{\sqrt{[\Sigma(u+\alpha)^2]}}, -w + \frac{(w+\gamma)r}{\sqrt{[\Sigma(u+\alpha)^2]}},\right]$$

If it lies on the given plane, then

$$\Sigma\left[(u+\alpha)\left\{-u+\frac{(u+\alpha)r}{\sqrt{[\Sigma(u+\alpha)^2]}}-\alpha\right\}\right]=0$$

$$\Rightarrow \quad \Sigma\left[(u+\alpha)\left\{\frac{r}{\sqrt{[\Sigma(u+\alpha)^2]}}-1\right\}\right]=0$$

$$\Rightarrow \quad r\left[\frac{(u+\alpha)^2+(v+\beta)^2+(w+\gamma)^2}{\sqrt{[\Sigma(u+\alpha)^2]}}\right]$$

$$-\ [(u+\alpha)^2+(v+\beta)^2+(w+\gamma)^2]=0$$

$$\Rightarrow \quad r=\sqrt{[\Sigma(u+\alpha)^2]}.$$

$\therefore$ From (i) the co-ordinates of C, the centre of the circle are $\{-u + (u+\alpha), -v+(v+\beta), -w+(w+\gamma)$ *i.e.,* (α, β, γ).

The equation of the sphere through the given circle is

$$[x^2+y^2+z^2+2ux+2vy+2wz+d]+\lambda\,[(x-\alpha)(u+\alpha)+(y-\beta)$$
$$(v+\beta)+(z-\gamma)(w+\gamma)]=0 \quad \text{...(ii)}$$

$$\Rightarrow \quad x^2+y^2+z^2+x\,\{2u+\lambda(u+\alpha)+y\,\{\ldots\}+z\,\{\ldots\}+\{\ldots\}=0.$$

$\therefore$ Its centre is

$$[-\frac{1}{2}\ \{2u+\lambda(u+\alpha)\}, -\frac{1}{2}\ \{2v+\lambda(v+\beta)\}, -\frac{1}{2}(2w+\lambda(w+\gamma)\}]$$

$$\Rightarrow \quad [-\frac{1}{2}\ (2u+\lambda u+\alpha\lambda), -\frac{1}{2}\ (2v+\lambda v+\beta\lambda), -\frac{1}{2}(2w+\lambda w+\gamma\lambda)]$$

If the given circle is a great circle of (ii), then the centre of (ii) must lie on the given plane of the circle.

$$\therefore \quad \Sigma[\{-\frac{1}{2}\ (2u+\lambda u+\alpha\lambda)-\alpha\}\ (u+\alpha)]=0$$

$$\Rightarrow \quad \Sigma[\{-u-\frac{1}{2}\lambda u-\frac{1}{2}\alpha\lambda-\alpha\}\ (u+\alpha)]=0$$

$$\Rightarrow \quad \Sigma[\{(u+\alpha)+\frac{1}{2}\lambda(u+\alpha)\}\ (u+\alpha)] \Rightarrow \Sigma[(u+\alpha)^2\ (1+\frac{1}{2}\lambda)]=0$$

$$\Rightarrow \quad 1+(1/2)\lambda=0 \quad \Rightarrow \lambda=-2$$

$\therefore$ From (ii), required sphere is

$$[x^2+y^2+z^2+2ux+2vy+2wz+d]-2\ [\Sigma\ (x-\alpha)\ (u+\alpha)]=0$$

$$\Rightarrow \quad x^2+y^2+z^2+2\ \Sigma\ [x\ \{u-(u+\alpha)\}]+2\ \Sigma\ [\alpha\ (u+\alpha)]+d=0$$

$\Rightarrow\ x^2 + y^2 + z^2 + 2\Sigma\,(-x\alpha) + \Sigma\,(\alpha u + \alpha^2) + d = 0$

$\Rightarrow\ x^2 + y^2 + z^2 - 2\,(\alpha x + \beta y + \gamma z)$

$\qquad + 2\,(\alpha u + \beta v + \gamma w + \alpha^2 + \beta^2 + \gamma^2) + d = 0$

$\Rightarrow\ x^2 + y^2 + z^2 - 2\alpha\,(x - u) - 2\beta\,(y - v) - 2\gamma\,(z - w)$

$\qquad + 2\alpha^2 + 2\beta^2 + 2\gamma^2 + d = 0$

Example 46:

A is point on OX and B on OY, so that the angle OAB is constant and equal to α, On AB as diameter a circle is drawn whose plane is parallel to OZ. Prove that as AB varies the circle generates the cone

$$2xy - z^2 \sin 2\alpha = 0.$$

Solution:

Let the coordinates of A and B be (a, 0, 0) and (0, b, 0) respectively. Also ∠AOB = α (given), so in right angled triangle AOB with ∠AOB = π/2. then we have $\tan\alpha = OB/OA = b/a$ ...(i)

Now the equation of the plane containing z-axis *i.e.,* x = 0, y = 0 is of the form $x + \lambda y = 0$, which does not contain z.

The equation of the plane through A and B and parallel to OZ (*i.e.,* z-axis) is given as $(x/a) + (y/b) = 1$ ...(ii)

which also does not contain z.

Now the equation of the sphere drawn on AB as diameter is

$$(x - a)\,(x - 0) + (y - 0)\,(y - b) + (z - 0)\,(z - 0) = 0$$

$$\Rightarrow\quad x^2 + y^2 + z^2 = ax + by$$

∴ The equations of the circle drawn on AB as diameter and plane parallel to z-axis are $x^2 + y^2 + z^2 = ax + by,\ (x/a) + (y/b) = 1$ ...(iii)

The required locus of the circle as AB varies (*i.e.,* as a and b vary), is obtained by eliminating a and b between (i) and (iii).

From (iii) we have $x^2 + y^2 + z^2 = (ax + by)\,[(x/a) + (y/b)]$ **(Note)**

$$\Rightarrow\quad x^2 + y^2 + z^2 = x^2 + y^2 + xy\,[(a/b) + (b/a)]$$

$$\Rightarrow\quad z^2 = xy\,[\cot\alpha + \tan\alpha],\ \text{from (i)}$$

$$\Rightarrow\quad z^2 = xy\left[\frac{\cos\alpha}{\sin\alpha} + \frac{\sin\alpha}{\cos\alpha}\right] = xy\left[\frac{\cos^2\alpha + \sin^2\alpha}{\sin\alpha\cos\alpha}\right]$$

$\Rightarrow\quad 2z^2 \sin\alpha\cos\alpha = 2xy \Rightarrow z^2 \sin 2\alpha = 2xy.$ **Hence proved.**

Example 47:

Show that the plane $2x - 2y + z + 16 = 0$ touches the sphere $x^2 + y^2 + z^2 + 2x - 4y + 2z - 3 = 0$ and find the coordinates of the point of contact

Solution:

Do your self **Ans.** (–3, 4, –2)

Example 48:

Obtain the equation of the sphere inscribed on the line joining the points A (3, 4, 1) and B (–1, 0, 5) as diameter. Find also the equation of the tangent plane at B.

Solution:

The equation of the sphere described on the line joining the given points A (3, 4, 1) and B (–1, 0, 5) is given as

$$(x - 3)\,[x - (-1)] + (y - 4)\,(y - 0) + (z - 1)\,(z - 5) = 0$$

$\Rightarrow$ $x^2 + y^2 + z^2 - 2x - 4y - 6z + 2 = 0$, on simplifying. **Ans.**

The tangent plane to this sphere at any point (α, β, γ) is

$$x\alpha + y\beta + z\gamma - (x + \alpha) - 2\,(y + \beta) - 3\,(z + \gamma) + 2 = 0 \text{ (Note)}$$

$\therefore$ The required tangent plane to this sphere at B (–1, 0, 5) is given by

$$x\,(-1) + y.0 + z\,(5) - [x + (-1)] - 2\,[y + 0] - 3\,[z + 5] + 2 = 0$$

$\Rightarrow$ $-x + 5z - x + 1 - 2y - 3z - 15 + 2 = 0$

$\Rightarrow$ $x + y - z + 6 = 0$ **Ans.**

Example 49:

Find the equation of the tangent plane to the sphere $x^2 + y^2 + z^2 - 3x + 4y + z - 6 = 0$ at the point (1, –2, 3).

Solution:

The equation of the tangent plane at (1, –2, 3) is

$$x.1 - y.2 + z.3 - \frac{3}{2}\,(x + 1) + 2(y - 2) + \frac{1}{2}\,(z + 3) - 6 = 0.$$

$\Rightarrow$ $x - 7z + 20 = 0$ (after simplification).

Example 50:

Find the equation of the sphere described on the line joining (2, –1, 4), (–2, 2, –2) as diameter and determine the equation of the tangent plane at (2, –1, 4).

Solution:

The equation of the required sphere is

$$(x - 2)\,(x + 2) + (y + 1)\,(y - 2) + (z - 4)\,(z + 2) = 0$$

$$\Rightarrow \qquad x^2 + y^2 + z^2 - y - 2z - 14 = 0.$$

The equation of the tangent plane to the above sphere at (2, –1, 4) is

$$x.2 + y(-1) + z.4 - \frac{1}{2}(y - 1) - (z + 4) - 14 = 0.$$

$$\Rightarrow \qquad 4x - 3y + 6z - 35 = 0.$$

Example 51:

Find the equation of the sphere which touches the sphere $4(x^2 + y^2 + z^2) + 10x - 25y - 2z = 0$ at the point (1, 2, –2) and passes through the point (–1, 0, 0).

Solution:

The equation of the tangent plane at (1, 2, –2) to the given sphere $x^2 + y^2 + z^2 + \frac{10}{4}x - \frac{25}{4}y - \frac{2}{4}z = 0$ is

$$x.1 + y.2 - z.2 + \frac{10}{8}(x + 1) - \frac{25}{8}(y + 2) - \frac{2}{8}(z - 2) = 0$$

$$\Rightarrow \; 2x - y - 2z - 4 = 0$$

The equation of the required sphere is

$$x^2 + y^2 + z^2 + \frac{10}{4}x - \frac{25}{4}y - \frac{2}{4}z + k(2x - y - 2z - 4) = 0 \quad \text{...(i)}$$

Since (i) passes through (–1, 0, 0),

$$1 - \frac{10}{4} + k(-2, -4) = 0 \;\Rightarrow\; k = -\frac{1}{4}.$$

Putting this values of k in (i), the required sphere is

$$x^2 + y^2 + z^2 + 2x - 6y + 1 = 0.$$

Example 52:

Tangent plane at any point of the sphere $x^2 + y^2 + z^2 = r^2$ meets the coordinate axes at A, B, C. Show that the locus of the point of intersection of planes drawn parallel to the coordinate planes through A, B, C is the surface $x^{-2} + y^{-2} + z^{-2} = r^{-2}$.

Solution:

The equation of the tangent plane to $x^2 + y^2 + z^2 = r^2$ at any point (x_1, y_1, z_1) is $xx_1 + yy_1 + zz_1 = r^2$. ...(i)

The plane (i) meets the coordinate axes at $A(r^2/x_1, 0, 0)$, $B(0, r^2/y_1, 0)$, $C(0, 0, r^2/z_1)$. The equations of the planes parallel to the coordinate planes through A, B, C are

$$x = \frac{r^2}{x_1},\ y = \frac{r^2}{y_1},\ z = \frac{r^2}{z_1} \text{ respectively.}$$

$$\Rightarrow \qquad x^{-2} + y^{-2} + z^{-2} = \frac{1}{r^2}(x_1^2 + y_1^2 + z_1^2) = \frac{r^2}{r^4},$$

[since (x_1, y_1, z_1) lies on $x^2 + y^2 + z^2 = r^2$]

Hence $x^{-2} + y^{-2} + z^{-2} = r^{-2}$.

Example 53:

Find the equations of the spheres through the circle $x^2 + y^2 + z^2 = 5$, $x + 2v + 3z = 3$ and touching the plane $4x + 3y = 15$.

Solution:

The equation of any sphere through the given circle is

$$x^2 + y^2 + z^2 - 5 + k(x + 2y + 3z - 3) = 0$$

$$\Rightarrow \qquad x^2 + y^2 + z^2 + kx + 2ky + 3kz - (5 + 3k) = 0. \qquad \text{...(i)}$$

Its centre is $(-k/2, -k, -3k/2)$ and

$$\text{radius} = \sqrt{\frac{k^2}{4} + k^2 + \frac{9}{4}k^2 + 5 + 3k}.$$

The sphere (i) touches the plane $4x + 3y = 15$ if

$$\sqrt{\frac{k^2}{4} + k^2 + \frac{9}{4}k^2 + 5 + 3k} = \frac{4(-k/2) + 3(-k) - 15}{\sqrt{16+9}} = -(k+3)$$

$$4(k + 3)^2 = 14k^2 + 12k + 20 \Rightarrow 5k^2 - 6k - 8 = 0$$

$$\Rightarrow \qquad (5k + 4)(k - 2) = 0 \quad \Rightarrow k = 2,\ -4/5.$$

Putting these values of k in (i), the two sphers are

$$x^2 + y^2 + z^2 + 2x + 4y + 6z - 11 = 0,$$

$$5(x^2 + y^2 + z^2) - 4x - 8y - 12z - 13 = 0.$$

Example 54:

A sphere whose centre lies in the positive octant passes through the origin and cuts the planes $x = 0$, $y = 0$, $z = 0$ in circles of radii $a\sqrt{2}$, $b\sqrt{2}$, $c\sqrt{2}$ respectively. Find the equation of this sphere.

Solution:

The sphere passes through origin, so its equation can be taken as

$$x^2 + y^2 + z^2 + 2ux + 2vy + 2wz = 0 \qquad \text{...(1)}$$

As this sphere meets the plane $z = 0$; so putting $z = 0$ in (1) we get $x^2 + y^2 + z^2 + 2ux + 2vy = 0$; which is evidently a circle on xy-plane and its radius $= \sqrt{(u^2 + y^2)}$.

But we are given that the sphere meets the plane $z = 0$ in a circle of radius $c\sqrt{2}$, so we have $\sqrt{(u^2 + v^2)} = c\sqrt{2}$.

$$\Rightarrow \quad u^2 + v^2 = 2c^2 \qquad \text{...(2)}$$

Similarly as the sphere (1) meets the planes $x = 0$, $y = 0$, in circle of radii $a\sqrt{2}$, $c\sqrt{2}$, so we have $v^2 + w^2 = 2a^2$ ... (3) and $w^2 + u^2 = 2b^2$ ...(4)

Adding (2), (3) and (4) we get $2(u^2 + v^2 + w^2) = 2a^2 + 2b^2 + 2c^2$

$$\Rightarrow \quad u^2 + v^2 + w^2 = a^2 + b^2 + c^2. \qquad \text{...(5)}$$

Subtracting (3), (4) and (2) from (5) by turn we get

$$u^2 = b^2 + c^2 - a^2,\ v^2 = c^2 + a^2 - b^2 \text{ and } w^2 = a^2 + b^2 - c^2$$

$$\Rightarrow \quad u^2 = \pm \sqrt{(b^2 + c^2 - a^2)},\ v = \sqrt{(c^2 + a^2 - b^2)},\ w = \pm \sqrt{(a^2 + b^2 - c^2)}$$

Also it is given that the centre of the sphere (1) viz. $(-u, -v, -w)$ lies in the positive octant, so the values of u, v, w must be negative. **(Note)**

$$\therefore \quad u = -\sqrt{(b^2 + c^2 - a^2)},\ v = \sqrt{(c^2 + a^2 - b^2)},\ w = -\sqrt{(a^2 + b^2 - c^2)}$$

Substituting these values of u, v, w in (1), the required equation is

$$x^2 + y^2 + z^2 - 2x\sqrt{(b^2 + c^2 - a^2)} - 2y\sqrt{(c^2 + a^2 - b^2)}$$
$$- 2z\sqrt{(a^2 + b^2 - c^2)} = 0. \qquad \textbf{Ans.}$$

Example 66:

Show that the plane $2x - 2y + z = 12 = 0$ touches the sphere $x^2 + y^2 + z^2 - 2x - 4y + 2z - 3 = 0$ and find the point of contact.

Solution:

If the plane $\qquad 2x - 2y - z + 12 = 0 \qquad$ ...(1)

touches the sphere $\qquad x^2 + y^2 + z^2 - 2x - 4y + 2z - 3 = 0 \qquad$ ...(2)

then the length of the perpendicular from the centre $(1, 2, -1)$ of the sphere (2) to the plane (1) must be equal to the radius

$$\sqrt{[(-1)^2 + 2^2 + (-1)^2 - (-3)]} = \sqrt{(9)} = 3 \text{ of the sphere (2)}$$

$$i.e., \quad \frac{2(1) - 2(2) + 1(-1) + 12}{\sqrt{[2^2 + (-2)^2 + (1)^2]}} = 3 \quad \text{or} \quad 9/3 = 3,$$

which being true the plane (1) touches the sphere (2).

Also if C be the centre of the sphere and P the required point of contact then the d.r.'s of the line CP are same as those of the normal to the plane

(1) *i.e.,* 2, –2, 1. Also C is (1, 2, –1).

Hence the equation of the line CP is $\frac{x-1}{2} = \frac{y-2}{-2} = \frac{z+1}{1} = r$ (say)

If CP = r, the coordinates of P are (2r + 1, –2r + 2, r – 1) and P lies on (1), so we have

$2(2r + 1) - 2(-2r + 2) + (r - 1) + 12 = 0$ or $9r + 9 = 0$ or $r = -1$.

∴ The coordinates of P are

$[2(-1) + 1, -2(-1) + 2, -1 -1]$ or $(-1, 4, -2)$ **Ans.**

Example 56:

Find the points of intersection of the line $\frac{1}{2}(x-1) = \frac{1}{3}(y-2) = \frac{1}{4}(z-3)$ *with the sphere* $x^2 + y^2 + z^2 - 4y - 7 = 0$

Solution:

Any point on the given line is (1 + 2r, 2 + 3r, 3 + 4r)

If this point lies on the given sphere, then we have

$$(1 + 2r)^2 + (2 + 3r)^2 + (3 + 4r)^2 - 4(2 + 3r) - 7 = 0$$

$$\Rightarrow \quad 29r^2 + 28r - 1 = 0 \text{ or } (29r - 1) - (r + 1) = 0$$

$$\Rightarrow \quad r = -1, 1/29$$

∴ From (i) the required points are

$$(1 - 2, 2 - 3, 3 - 4) \text{ and } \left(1 + \frac{2}{29}, 2 + \frac{2}{29}, 3 + \frac{2}{29}\right)$$

$$\Rightarrow \quad (-1, -1, -1) \text{ and } \left(\frac{31}{29}, \frac{61}{29}, \frac{91}{29}\right).$$ **Ans.**

Example 57:

Find the equations of the tangent line to the circle $3x^2 + 3y^2 + 3z^2 - 2x - 3y - 4z - 22 = 0$, $3x + 4y + 5z - 26 = 0$ *at the point (1, 2, 3).*

Solution:

The equation of the tangent plane to the sphere

$$x^2 + y^2 + z^2 - \frac{2}{3}x - y - \frac{4}{3}z - \frac{22}{3} = 0 \text{ at } (1, 2, 3) \text{ is}$$

$$x(1) + y(2) + z(3) - \frac{1}{3}(x + 1) - \frac{1}{2}(y + 2) - \frac{2}{3}(z + 3) - \frac{22}{3} = 0$$

$$\Rightarrow \qquad 4x + 9y + 14z - 64 = 0.$$

The required tangent line is the line of intersection of the planes

$$4x + 9y + 14z - 64 = 0 \text{ and } 3x + 4y + 5z - 26 = 0$$

∴ If l, m, n be the direction ratios of this line [which evidently passes through (1, 2, 3)], then we have $4l + 9m + 14n = 0$, $3l + 4m + 5n = 0$

$$\therefore \qquad \frac{l}{45 - 46} = \frac{m}{42 - 20} = \frac{n}{16 - 27} \text{ or } \frac{l}{-1} = \frac{m}{2} = \frac{n}{-1}$$

The required tangent line at (1, 2, 3) is

$$x - 1 = \frac{1}{2}(y - 2) = z - 3. \qquad \textbf{Ans.}$$

Example 58:

Find the equation of the sphere which touches the sphere $4(x^2 + y^2 + z^2) + 10x - 25y - 2z = 0$ at $(1, 2, -2)$ and passes through the point $(-1, 0, 0)$.

Solution:

The tangent plane at (1, 2, –2) to the given sphere

$$x^2 + y^2 + z^2 + \frac{5}{2}x - \frac{25}{4}y - \frac{1}{2}z = 0 \qquad \textbf{(Note)}$$

$$\Rightarrow \quad x.1 + y.2 + z.(-2) + \frac{5}{4}(x + 1) - \frac{25}{8}(y + 2) - \frac{1}{4}(z - 2) = 0$$

$$\Rightarrow \quad 2x - y - 2z - 4 = 0, \text{ on simplifying.}$$

∴ The equation of the sphere touching the given sphere at (1, 2, –2) is

$$(x^2 + y^2 + z^2 + \frac{5}{2}x - \frac{25}{4}y - \frac{1}{2}z) = \lambda\,(2x - y - 2z - 4) = 0 \quad \text{...(i)}$$

If it passes through (–1, 0, 0), then we have

$$(1 - \frac{5}{2}) + l\,(-2 -4) = 0 \text{ or } \lambda = -\frac{1}{4}.$$

∴ From (i), the required equations is

$$(x^2 + y^2 + z^2 + \frac{5}{2}x - \frac{25}{4}y - \frac{1}{2}z)$$

$$= \frac{1}{4}(2x - y - 2z - 4) = 0$$

$$(x^2 + y^2 + z^2) + 2x - 6y + 1 = 0. \qquad \textbf{Ans.}$$

Example 59:

Find the condition for the plane $lx + my + nz = p$ to touch the sphere $x^2 + y^2 + z^2 = a^2$.

Solution:

The centre of the given sphere is (0, 0, 0) and its radius is a.

Now if the given plane $lx + my + nz - p = 0$ touches the given sphere, then the length of the perpendicular from the centre (0, 0, 0) of the sphere to this plane must be equal to the radius a of the sphere.

i.e., $$\frac{l.0 + m.0 + n.0 - p}{\sqrt{(l^2 + m^2 + n^2)}} = a \quad \text{or} \quad -p = a\sqrt{(l^2 + m^2 + n^2)}$$

$\Rightarrow$ $p^2 = a^2 (l^2 + m^2 + n^2)$, which is the required condition. **Ans.**

Example 60:

Find the condition for the plane $ax + by + cz + k = 0$ to be a tangent plane to the sphere $x^2 + y^2 + z^2 + 2ux + 2vy + 2wz + d = 0$.

Solution:

The centre of the given sphere is $(-u, -v, -w)$ and its radius

$$= \sqrt{(u^2 + v^2 + w^2 - d)}$$

If the given plane touches the given sphere, then the length of perpendicular from the centre $(-u, -v, -w)$ of the sphere to the given plane must be equal to the radius of the sphere.

i.e., $$\frac{a(-u) + b(-v) + c(-w) + k}{\sqrt{(a^2 + b^2 + c^2)}} = \sqrt{(u^2 + v^2 + w^2 - d)}$$

$$\Rightarrow \quad (au + bv + cw - k)^2 = (a^2 + b^2 + c^2)(u^2 + v^2 + w^2 - d),$$

which is the required condition. **Ans.**

Example 61:

Find the locus of the centres of spheres which pass through a given point an touch a given plane.

Solution:

Take $z = 0$ as the given plane and (0, 0, a) as the given point. **(Note)**

Let the equation of the sphere be

$$x^2 + y^2 + z^2 + 2ux + 2vy + 2wz = 0 \quad \ldots(i)$$

As it passes through (0, 0, a), so we have $a^2 + 2wa + d = 0$ ...(ii)

Also if the sphere (i) touches the plane $x = 0$ then the length of perpendicular from its centre $(-u, -v, -w)$ to the plane $z = 0$ must be equal to its radius $\sqrt{(u^2 + v^2 + w^2 - d)}$

i.e., $$\frac{(-w)}{\sqrt{(1^2)}} = \sqrt{(u^2 + v^2 + w^2 - d)} \text{ i.e., } u^2 + v^2 = d \quad \text{...(iii)}$$

Eliminating d from (ii) and (iii) we get $a^2 + 2wa + u^2 + v^2 = 0$ ...(iv)

∴ Locus of the centre $(-u, -v, -w)$ of the sphere (i) from (iv) is $a^2 + 2(-z)\,a + (-x)^2 + (-y)^2 = 0 \;\Rightarrow\; x^2 + y^2 - 2az + a^2 = 0.$

Example 62:

(a) Show that the plane $2x + y - z = 12$ touches the sphere $x^2 + y^2 + z^2 = 24$ and find the point of constant.

(b) Show that the plane $2x - 2y + z + 12 = 0$ touches the sphere $x^2 + y^2 + z^2 - 2x - 4y + 2x - 3 = 0$. Also find the point of contact.

Solution:

(a) Let the required point of contact be (x_1, y_1, z_1). Then the equation of the tangent plane to the given sphere at (x_1, y_1, z_1) is given by

$$xx_1 + yy_1 + zz_1 = 24 \quad \text{...(i)}$$

This should be the same as the given plane $2x + y - z = 12$ ...(ii)

Comparing (i) and (ii) we get $\frac{x_1}{2} = \frac{y_1}{1} = \frac{z_1}{-1} = \frac{24}{12}$

These give $x_1 = 4,\ y_1 = 2,\ z_1 = -2$ ...(iii)

Also (x_1, y_1, z_1) is a point on the given sphere so $x_1^2 + y_1^2 + z_1^2 = 24$

and the same is satisfied by the values of x_1, y_1, z_1 given by (iii).

Hence the plane (ii) touches the given sphere at the point

(x_1, y_1, z_1) *i.e.,* $(4, 2, -2)$ **Ans.**

Example 63:

Find the locus of the centres of spheres of constant radius which pass through a given point and touch a given line.

Solution:

Take x-axis as the given line and (0, 0, a) as the given point. **(Note)**

Let the equation of the sphere by given as

$$x^2 + y^2 + z^2 + 2ux + 2vy + 2wz = 0 \qquad \ldots(i)$$

As it passes through the given point (0, 0, a) so $a^2 + 2wa + d = 0$ ...(ii)

Also the radius of the sphere (i) is given as constant k (say).

Then we have $u^2 + v^2 + w^2 - d = k^2$ ...(iii)

The sphere (i) touches the given line which we have chosen as x-axis *i.e.,* $y = 0 = z$ at the points given by $x^2 + 2ux + d = 0$ ...(iv)

Since the sphere (i) touches the line $y = 0 = z$, so the roots of (iv) must be equal and therefore using '$b^2 = 4ac$' we have

$$(2u)^2 = 4 \,.\, 1 \,.\, d \quad \text{or} \quad u^2 = d \qquad \ldots(v)$$

Eliminating d from (ii), (iii) and (v) we get

$$a^2 + 2wa + u^2 = 0, \; v^2 + w^2 = k^2$$

∴ The required locus of the centre $(-u, -v, -w)$ of the sphere (i) is given by the equations $a^2 + 2(-z)\,a + (-x)^2 = 0 \; (-y)^2 + (-z)^2 = k^2$

$$\Rightarrow \qquad x^2 - 2az + a^2 = 0, \; y^2 + z^2 = k^2$$

which is the curve of intersection of two quadratic surfaces

$$x^2 - 2az + a^2 = 0 \quad \text{and} \quad y^2 + z^2 = k^2$$ **Ans.**

Example 64:

Prove that the sphere

$S \equiv x^2 + y^2 + z^2 + 2ux + 2vy + 2wz + d = 0$ cuts the sphere

$S' \equiv x^2 + y^2 + z^2 + 2u'x + 2v'y + 2w'z + d' = 0$ in the great circle if

$$2(u'^2 + v'^2 + w'^2) - d' = 2(uu' + vv' + ww') - d$$

or If $2(uu' + vv' + ww') = 2r'^2 + d + d'$, where r' is the radius of the second sphere.

Solution:

The equation of the plane through the circle of intersection of the given sphere is $S - S' = 0$

i.e., $2(u - u')\,x + 2(v - v')\,y + 2(w - w')\,z + (d - d') = 0$ ...(i)

If the sphere $S = 0$ cuts the sphere $S' = 0$ in a great circle, then the centre $(-u', -v', -w')$ of the sphere $S' = 0$ should lie on the plane (i) **(Note)**

$$\therefore \quad -2(u - u')\,u' - 2(v - v')\,v' - 2(w - w')\,w' + (d - d') = 0$$

$$\Rightarrow \quad 2(u'^2 + v'^2 + w'^2) - d' = 2(uu' + vv' + ww') - d \qquad \ldots(ii)$$

Hence proved.

$\Rightarrow$ $2(r'^2 + d') - d' = 2(uu' + vv' + ww') - d$, where r' is the radius of the second sphere and so $r'^2 + u'^2 + v'^2 + w'^2 - d'$

$\Rightarrow$ $2r'^2 + d' = 2(uu' + vv' + ww')$ **Hence proved.**

Example 65:

Show that the two spheres $x^2 + y^2 + z^2 + 6y + 2z + 8 = 0$ and $x^2 + y^2 + z^2 + 6x + 8y + 4z + 20 = 0$ are orthogonal. Find their plane of intersection.

Solution:

Here we have u = 0, v = 3, w = 1, d = 8 and u' = 3, v' = –4. w' = 2. d' = 20

$\therefore$ $2uu' + 2vv' + 2ww' = 2(0)(3) + 2(3)(4) + 2(1)(2) = 24 + 4$
$= 28 = 8 + 20 = d + d'$

Hence the spheres cut orthogonally.

Also the plane of intersection of two spheres $S_1 = 0$ and $S_2 = 0$ is given by $S_1 - S_2 = 0$, the coefficients of x^2, y^2, z^2 in each of S_1 and S_2 must be unity.

$\therefore$ Required equation of plane of intersection is

$$(x^2 + y^2 + z^2 + 6x + 8y + 4z + 20) - (x^2 + y^2 + z^2 + 6y + 2z + 8) = 0$$

$\Rightarrow$ $6x + 2y + 2z + 12 = 0$

$\Rightarrow$ $3x + y + z + 6 = 0$ **Ans.**

Example 66:

Find the equation of a sphere touching the three co-ordinate planes. How many such spheres can be drawn?

Solution:

Let the equation of the sphere be given as

$$x^2 + y^2 + z^2 + 2ux + 2vy + 2wz + d = 0. \quad \text{...(i)}$$

If the sphere touches the yz-plane *i.e.,*, x = 0, then the length of the perpendicular from its centre (–u, –v, –w) to the plane x = 0 must be equal to its radius = $(u^2 + v^2 + w^2 - d)$.

i.e., $\dfrac{-u}{2} = \sqrt{(u^2 + v^2 + w^2 - d)}$

$\Rightarrow$ $u^2 = u^2 + v^2 + w^2 - d$

$\Rightarrow$ $v^2 + w^2 = d. \quad \text{...(ii)}$

Similarly if the sphere (i) touches zx and xy-planes then we shall have

$$w^2 + u^2 = d \quad \text{...(iii)}$$

and $$u^2 + v^2 = d \quad \text{...(iv)}$$

Adding (ii), (iii) and (iv) we get $2(u^2 + v^2 + w^2) = 3d$

$\Rightarrow \quad u^2 + v^2 + w^2 = \frac{3}{2}d$

$\Rightarrow \quad u^2 = \frac{1}{2}d$, from (ii)

Similarly from (iii), (iv) and (v),

we get $v^2 = \frac{1}{2}d = w^2$

$\therefore \quad u^2 = \frac{1}{2}d = v^2 = w^2 = \lambda^2$ (sy)

$\Rightarrow u = \pm\lambda = v = w.$

Hence from (i) the required equation is

$$x^2 + y^2 + z^2 \pm 2\lambda(x + y + z) + 2\lambda^2 = 0. \quad \textbf{Ans.}$$

Since λ can take an infinite number of values, so an infinite number of such spheres can be drawn but if the radius of the sphere is given then λ can be expressed in terms of the given radius and then only eight such spheres can be possible as the sets of values of u, v and w can be taken in eight different ways.

Example 67:

Show that every sphere through the circle $x^2 + y^2 - 2ax + r^2 = 0$, $z = 0$ cuts orthogonally every sphere through the circle $x^2 + z^2 = r^2$, $y = 0$.

Solution:

The given circles may be written as

$$x^2 + y^2 + z^2 - 2ax + r^2 = 0,\ z = 0;\ x^2 + y^2 + z^2 = r^2,\ y = 0.$$

Any two spheres through these circles are respectively

$$x^2 + y^2 + z^2 - 2ax + r^2 + 2k_1z = 0,$$

and $x^2 + y^2 + z^2 - r^2 + 2k_2y = 0$, k_1 and k_2 being parameters.

Now $2u_1u_2 + 2v_1v_2 + 2w_1w_2 = 2(-a)\,0 + 2(0)(k_2) + 2(k_1)(0) = 0$,

and $d_1 + d_2 = r^2 - r^2 = 0.$

Hence the two spheres cut orthogonally.

Example 68:

Find the equation of the sphere that passes through the circle $x^2 + y^2 + z^2 - 2x + 3y - 4z + 6 = 0$, $3x - 4y + 5z - 15 = 0$ and cuts the sphere $x^2 + y^2 + z^2 + 2x + 4y - 6z + 11 = 0$ orthogonally.

Solution:

Any sphere through the given circle is

$$x^2 + y^2 + z^2 - 2x + 3y - 4z + 6 + k(3x - 4y + 5z - 13) = 0$$

$$\Rightarrow x^2 + y^2 + z^2 + (3k - 2)x + (3 - 4k)y + (5k - 4)z + 6 - 15k = 0 \quad ...(i)$$

Since (i) cuts $x^2 + y^2 + z^2 + 2x + 4y - 6z + 11 = 0$ orthogonally,

$$\therefore \quad 2\left(\frac{3k-2}{2}\right).1 + 2\left(\frac{3-4k}{2}\right).2 + 2.\left(\frac{5k-4}{2}\right)(-3) = 6 - 15k + 11.$$

This gives $k = -1/5$. Putting in (i), the required sphere is

$$5(x^2 + y^2 + z^2) - 13x + 19y - 25z - 45 = 0.$$

Example 69:

Two spheres of radii r_1 and r_2 cut orthogonally. Prove that the radius of the common circle is $r_1 r_2/\sqrt{(r_1^2 + r_2^2)}$.

Solution:

Let the common circle be $x^2 + y^2 = a^2$, $z = 0$. Its radius is a. Any sphere through this circle is

$$x^2 + y^2 + z^2 - a^2 + 2kz = 0.$$

We choose the two spheres through the circle as

$$x^2 + y^2 + z^2 + 2k_1z - a^2 = 0, \; x^2 + y^2 + z^2 + 2k_2z - a^2 = 0.$$

These spheres will cut orthogonally if $2k_1k_2 = -a^2 - a^2$

i.e., $\quad k_1^2 k_2^2 = a^4. \quad ...(i)$

The radii of the above two spheres are given by

$$r_1^2 = k_1^2 + a^2 \text{ and } r_2^2 = k_2^2 + a^2.$$

From (i) and (ii) we obtain

$$(r_1^2 - a^2)(r_2^2 - a^2) = a^4 \Rightarrow r_1^2 r_2^2 - r^2(r_1^2 + r_2^2) = 0.$$

Hence $a = \dfrac{r_1 r_2}{\sqrt{r_1^2 + r_2^2}}$.

Example 70:

Find the equation of the sphere that passes through the two points (0, 3, 0), (–2, –1, –4) and cuts orthogonally the two spheres $x^2 + y^2 + z^2 + x - 3z - 2 = 0$, $2(x^2 + y^2 + z^2) + x + 3y + 4 = 0$.

Solution:

Let the required equation of the sphere be

$$x^2 + y^2 + z^2 + 2ux + 2vy + 2wz + d = 0. \qquad \text{...(i)}$$

It passes through the points (0, 3, 0) and (–2, –1, –4). Therefore

$$6y + d + 9 = 0, \qquad \text{...(ii)}$$

$$-4u - 2v - 8w + d + 21 = 0 \qquad \text{...(iii)}$$

Since (i) cuts the given spheres orthogonally,

$\therefore$ $2u(1/2) + 2v.0 + 2w(-3/2) - d - 2$ *i.e.,* $u - 3w - d + 2 = 0$, ...(iv)

and $2u.(1/4)\ .\ 2v.(3/4) + 2w.0 = d + 2$ *i.e.,* $u + 3v - 2d - 4 = 0$. ...(v)

From (ii), $d = -6v - 9$. Putting in (v) and (iv), we obtain respectively

$$u + 15v + 14 = 0, \qquad \text{...(vi)}$$

and $\qquad u - 3w + 6v + 11 = 0$ *i.e.,* $w = \frac{1}{3}(u + 6v + 11)$.

Putting the values of w and d in (iii), we obtain

$$5u + 18v + 13 = 0, \qquad \text{...(vii)}$$

Solving (vi) and (vii) for u and v, we obtain

$$v = -1,\ u = 1.$$

Thus $d = -6(-1) - 9 = -3$ and $w = \frac{1}{3}(1 - 6 + 11) = 2$.

Substituting the values of u, v, w and d in (i), we get

$$x^2 + y^2 + z^2 + 2x - 2y + 4z - 3 = 0.$$

Example 71:

Find the equation of the sphere which touches the plane $3x + 2y - z + 2 = 0$ at the point (1, –2, 1) and cuts orthogonally the sphere $x^2 + y^2 + z^2 - 4x + 6y + 4 = 0$.

Solution:

Let the equation of the required sphere be

$$x^2 + y^2 + z^2 + 2ux + 2vy + 2wz + d = 0. \qquad \text{...(i)}$$

The sphere (i) cuts the given sphere orthogonally.

$\therefore \quad 2u(-2) + 2v(3) + 2w.0 = d + 4$

$\Rightarrow \quad 4u - 6v + d + 4 = 0. \quad ...(ii)$

The tangent plane to (i) at the point (1, –2, 1) is

$$1.x - 2y + 1.z + u(x + 1) + v(y - 2) + w(z + 1) + d = 0$$

$\Rightarrow \quad x(1 + u) + y(v - 2) + z(1 + w) + (u - 2v + 2w + d) = 0. \quad ...(iii)$

The equation (iii) must be identical with $3x + 2y - z + 2 = 0$.

$$\therefore \quad \underset{(i)}{\frac{1+u}{3}} = \underset{(ii)}{\frac{v-2}{2}} = \underset{(iii)}{\frac{1+w}{-1}} = \underset{(iv)}{\frac{u-2v+2w+d}{2}}.$$

Now (i) = (ii) $\Rightarrow$ $2u - 3v + 8 = 0,$ (iv)

(ii) = (iii) $\Rightarrow$ $v + 2w = 0, \Rightarrow w = -v/2,$ ...(v)

(iii) = (iv) $\Rightarrow$ $u - 2v + 3w + d + 2 = 0.$ ...(vi)

Subtracting (vi) from (ii), we get $3u - 4v - 3w + 2 = 0$

$\Rightarrow \quad 6u - 5v + 4 = 0$, by (v) ...(vii)

Solving (iv) and (vii), we get $u = 7/2$, $v = 5$.

Now from (v), (vi), we obtain $w = -5/2$ and $d = 12$.

Substituting the values of u, v, w and d in (i),

$x^2 + y^2 + z^2 + 7x + 10y - 5z + 12 = 0$ is the required sphere.

Example 72:

Show that all the spheres that can be drawn through the origin and each set of points where planes parallel to the plane x/a + y/b + z/c = 0 cut the coordinate axes form a system of spheres which are cut orthogonally by the spheres.

$$x^2 + y^2 + z^2 + 2fx + 2gy + 2hz = 0 \text{ if } af + bg + ch = 0$$

Solution:

Any plane parallel to the plane

$$x/a + y/b + z/c = 0 \text{ is}$$

$$\frac{x}{a} + \frac{y}{b} + \frac{z}{c} = k.$$

This plane meets the coordinate axes at

$$A\ (ka, 0, 0),\ B(0, kb, 0),\ C(0, 0, kc).$$

Now the equation of any sphere through O, A, B and C is

$$x^2 + y^2 + z^2 - k\,(ax + by + cz) = 0. \qquad \ldots(1)$$

The equation (1) represents a system of spheres for different values of k. The sphere (1) cuts orthogonally the sphere

$$x^2 + y^2 + z^2 + 2fx + 2gy + 2hz = 0 \text{ if}$$

$$2\left(-\frac{ak}{2}\right)f + 2\left(-\frac{bk}{2}\right)g + 2\left(-\frac{ck}{2}\right)h = 0$$

i.e., if $\quad af + bg + ch = 0.$

Example 73:

Prove that the spheres, that can be drawn through the origin and each sets of points where the planes parallel to the plane $x/a + y/b + z/c = 0$ cut the co-ordinates axes, form a system of spheres which are cut orthogonally by the sphere $x^2 + y^2 + z^2 + 2ux + 2vy + 2wz = 0$ if $au + bv + cw = 0$.

Solution:

The equation of any plane parallel to the plane

$$(x/a) + (y/b) + (z/c) = 0 \text{ is } (x/a) + (y/b) + (z/c) = k. \ldots(i)$$

This plane cuts the axes at A (ak, 0, 0), B(0, bk, 0) and C(0, 0, ck).

The equation of sphere through O(0, 0, 0), A, B and C is

$$x^2 + y^2 + z^2 - akx - bky - ckz = 0$$

If this sphere cuts the given sphere orthogonally, then we have

$$2u\left(-\frac{1}{2}ak\right) + 2v\left(-\frac{1}{2}bk\right) + 2w\left(-\frac{1}{2}ck\right) = 0 + 0$$

$\Rightarrow \quad au + bv + cw = 0$, for all values of k. **Hence proved.**

Example 74:

Find the equation of the sphere that passes through the circle $x^2 + y^2 + z^2 - 2x + 3y - 4z + 9 = 0$, $3x - 4y + 5z - 15 = 0$ and cuts the sphere $x^2 + y^2 + z^2 + 2x + 4y - 6z + 11 = 0$ orthogonally.

Solution:

The equation of the sphere through the given circle is

$$(x^2 + y^2 + z^2 - 2x + 3y - 4z + 9) + \lambda\,(3x - 4y + 5z - 15) = 0$$

$$\Rightarrow x^2 + y^2 + z^2 + (3\lambda - 2)\,x + (3 - 4\lambda)\,y + (5\lambda - 4)z + (9 - 15\lambda) = 0 \qquad \ldots(i)$$

The other given sphere is $x^2 + y^2 + z^2 + 2x + 4y - 6z + 11 = 0 \ldots(ii)$

The condition of orthogonal intersection of the spheres (i) and (ii) is

$$\text{"}2uu' + 2vv' + 2ww' = d + d'\text{"}$$

$$\Rightarrow 2\frac{1}{2}(3\lambda - 12).1 + 2.\frac{1}{2}(3 - 4\lambda)2 + 2.\frac{1}{2}(5\lambda - 4)(-3) = (6 - 15\lambda) + 11$$

$$\Rightarrow (3\lambda - 2) + 2(3 - 4\lambda) - 3(5\lambda - 4) = 17 - 15\lambda \text{ or } \lambda = -1/5.$$

$\therefore$ From (i) the required sphere is given as

$$5(x^2 + y^2 + z^2) - 13x + 19y - 25z + 45 = 0. \qquad \textbf{Ans.}$$

Example 75:

Find the condition that the spheres $x^2 + y^2 + z^2 = a^2$ and $x^2 + y^2 + z^2 + 2ux + 2vy + 2wz + d = 0$ may cut orthogonally.

Solution:

Here we have u = 0, v = 0, w = 0, d = $-a^2$ and u' = u, v' = v, w' = w, d' = d.

Now the condition for orthogonality is 2uu' + 2vv' + 2ww' = d + d'

$\Rightarrow$ 2.0u + 2.0v + 2.0w = $-a^2 + d$ or $d - a^2 = 0$ or $d = a^2$. **Ans.**

Example 76:

Two points P and Q are conjugate with respect to a sphere S; prove that the sphere on PQ as diameter cuts S orthogonally.

Solution:

Let the points P and Q be (x_1, y_1, z_1) and (x_2, y_2, z_2) and the equation of the sphere by $x^2 + y^2 + z^2 = r^2$. ...(i)

The polar plane of P (x_1, y_1, z_1) with respect to the sphere (i) is $xx_1 + yy_1 + zz_1 = r^2$ and if Q (x_2, y_2, z_2) lies on this plane then

$$x_1x_2 + y_1y_2 + z_1z_2 = r^2 \qquad \text{...(ii)}$$

Now the equation of the sphere on PQ as diameter is

$$(x - x_1)(x - x_2) + (y - y_1)(y - y_2) + (z - z_1)(z - z_2) = 0$$

$$\Rightarrow x^2 + y^2 + z^2 - (x_1 + x_2)x - (y_1 + y_2) + (z_1 + z_2)z$$
$$+ (x_1x_2 + y_1y_2 + z_1z_2) = 0 \qquad \text{...(iii)}$$

If this sphere (iii) cuts the sphere (i) orthogonally, then

$$\text{"}2uu' + 2vv' + 2ww' = d + d'\text{"}$$

i.e., $0 = (-r^2) + (x_1x_2 + y_1y_2 + z_1z_2)$, $\because$ u' = 0 = v' = w', from (i)

$\Rightarrow$ $x_1x_2 + y_1y_2 + z_1z_2 = r^2$. which is true by virtue of (ii)

Hence the spheres (i) and (iii) cut orthogonally.

Example 77:

Find the pole of the plane $lx + my + nz = p$ with respect to the sphere

$$x^2 + y^2 + z^2 = a^2.$$

Solution:

Let the pole of the plane $lx + my + nz = p$ ...(i)

with respect to the sphere $x^2 + y^2 + z^2 = a^2$ ...(ii)

be (x_1, y_1, z_1).

Then the polar of (x_1, y_1, z_1) with respect to the sphere (ii) is

$$x_1x + y_1y + z_1z = a^2 \quad ...(iii)$$

Now (i) and (iii) represent the same plane, so comparing them we have

$$\frac{x_1}{l} = \frac{y_1}{m} = \frac{z_1}{n} = \frac{a^2}{p}$$

whence $x_1 = a^2l/p;\ y_1 = a^2\ m/p;\ z_1 = a^2n/p.$

$\therefore$ The required pole is $\left(\frac{a^2l}{p}, \frac{a^2m}{p}, \frac{a^2n}{p}\right)$. **Ans.**

Example 78:

Prove that the polar plane of any point on the line $\frac{1}{2}x = \frac{1}{3}(y - 1)$ $= \frac{1}{4}(z + 3)$ *with respect to the sphere* $x^2 + y^2 + z^2 = 1$ *passes through the line* $(1/13)\ (2x + 3) = -\frac{1}{3}(y - 1) = -z.$

Solution:

Any point of the line $\frac{x}{2} = \frac{y-1}{3} = \frac{z+3}{4}$ is P $(2r, 1 + 3r, -3 + 4r)$

$\therefore$ The polar plane of the point P with respect to the sphere $x^2 + y^2 + y^2 = 1$ is $x\ .\ (2r) + y\ .\ (1 + 3r) + z\ .\ (-3 + 4r) = 1$ ...(i)

If this plane (i) passes through the given line

$$\frac{2x+3}{13} = \frac{y-1}{-3} = \frac{z}{-1} \text{ or } \frac{x+\frac{3}{2}}{13} = \frac{y-1}{-6} = \frac{z-0}{-2} \quad ...(ii)$$

then the point (–3/2, 1, 0) on this line must lie on the plane (i) and so we have $-\frac{3}{2}(2r) + 1(1 + 3r) + 0(-3 + 4r) = 1$, which being true for all values of r the point $(-\frac{3}{2}, 1, 0)$ lies on the lane (i).

Example 79:

Find the radius and the coordinates of the centre of the sphere $x^2 + y^2 + z^2 - 4x + 6y + 9 = 0$. Determine whether the point P (1, –2, 1) lies inside or outside the sphere. Find the values of k for which the plane $x + y + kz = 3$ is tangent to this sphere.

Solution:

Given sphere is $x^2 + y^2 + z^2 - 4x + 6y + 9 = 0$ ...(i)

Its centre is (–u, –v, –w) *i.e.,* (2, –3, 0),

and radius $= \sqrt{(u^2 + v^2 + w^2 - d)} = \sqrt{(2^2 + 3^2 + 0^2 - 9)} = 2$.

Also substituting the coordinates of P (1, –2, 1) on the left hand side expression of the equation (i), we get

$$1^2 + (-2)^2 + 1^2 - 4(1) + 6(-2) + 9$$

$$= 1 + 4 + 1 - 4 - 12 + 9 = -1 = \text{negative}$$

∴ The point P (1, –2, 1) is inside the sphere given by (i).

Again if the plane $x + y + kz = 3$ ...(ii)

is a tangent plane to the sphere (i), then the length of perpendicular from the centre (2, –3, 0) of the sphere (i) to the plane (ii) must be equal to the radius of the sphere (i)

i.e., $$\frac{(2) + (-3) + k(0) - 3}{\pm\sqrt{(1^2 + k^2 + 1^2)}} = 2. \quad \text{or} \quad k = \pm\sqrt{2}$$ **Ans.**

Example 80:

Two spheres of radii r_1 and r_2 cut orthogonally. Prove that the radius of the common circle is $r_1 r_2 / \sqrt{(r_1^2 + r_2^2)}$.

Solution:

Let the equation of the common circle be $x^2 + y^2 = a^2 \,.\, z = 0$ ...(i)

Its radius is evidently a and we are to evaluate it.

Now let the equations of the two given spheres through this circle be

$$(x^2 + y^2 - a^2) + 2\lambda z + z^2 = 0 \quad \text{...(ii)}$$

and $$(x^2 + y^2 - a^2) + 2\mu z + z^2 = 0 \quad \text{...(iii)}$$

From (ii) the radius of the sphere is given as

$$= \sqrt{[(-\lambda)^2 - (-a^2)]} = \sqrt{(\lambda^2 + a^2)} = r_1 \text{ (given)}$$

and similarly from (iii) the radius of the sphere $= \sqrt{(\mu^2 + a^2)} = r_2$ (given).

Also as the sphere (ii) and (iii) cut each other orthogonally, so we have

$$2\lambda u = (-a^2) + (-a^2) \quad \text{or} \quad \lambda^2\mu^2 = a^4, \text{ squaring both sides}$$

$$\Rightarrow \quad (r_1^2 - a^2)(r_2^2 - a^2) = a^4 \quad \because \quad \lambda^2 + a^2 = r_1^2 . \mu^2 + a^2 = r_2^2$$

$$\Rightarrow \quad r_1^2 r_2^2 = a^2 (r_1^2 + r_2^2)$$

$$\Rightarrow \quad a = r_1 r_2 / \sqrt{(r_1^2 + r_2^2)}.$$

Example 81:

Find the equation of the sphere passing through the circles $y^2 + z^2 = 9$, $x = 4$ and $y^2 + z^2 = 36$, $x = 1$.

Solution:

The equation of any sphere through the circle $y^2 + z^2 = 9$, $x = 4$ is

$$(x^2 + y^2 + z^2 - 9 - 16) + \lambda (x - 4) = 0 \quad \text{...(i)}$$

(Note the introduction of x^2)

Similarly the equation of any sphere through the circle $x^2 + z^2 = 36$, $x = 1$ is $$(x^2 + y^2 + z^2 - 36 - 1) + \mu (x - 1) = 0 \quad \text{...(ii)}$$

If the two given circles lie on the same sphere then (i) and (ii) should represent that sphere on which the given circles lie *i.e.,* (i) and (ii) represent the same sphere.

$\therefore$ Comparing the coefficients of x and constant terms in (i) and (ii) we get $\lambda = \mu$ and $25 + 4\lambda = 37 + m$

Solving these we get $\lambda = \mu = 4$

$\therefore$ From (i) and (ii) we get the required equation as

$(x^2 + y^2 + z^2 - 25) + 4 (x - 4) = 0$

or $x^2 + y^2 + z^2 + 4x = 42$. **Ans.**

Example 82:

Find the equations of the spheres through the circle $x^2 + y^2 + z^2 = 1$, $2x + 4y + 5z = 6$ and touching the plane $z = 0$.

Solution:

The equation of any sphere through the given circle is given by

$$(x^2 + y^2 + z^2 - 1) + \lambda\,(2x + 4y + 5z - 6) = 0$$

$$\Rightarrow \quad x^2 + y^2 + z^2 + 2\lambda y + 4\lambda y + 5\lambda z - (1 + 6\lambda) = 0 \quad \ldots(i)$$

Its centre is $(-\lambda, -2\lambda, -\frac{5}{2}\lambda)$ and its radius

$$= \sqrt{[(-\lambda)^2 + (-2\lambda)^2 + (-\frac{5}{2}\lambda)^2 + (1 + 6\lambda)} = \sqrt{(\frac{45}{4}\lambda^2 + 6\lambda + 1)]}$$

If the sphere (i) touches the plane $z = 0$ ...(ii)

then the length of the perpendicular from he centre of (i) to (ii) must be equal to the radius of (i).

$$\therefore \quad \frac{-\frac{5}{2}\lambda}{\sqrt{(1)^2}} = \sqrt{[(45/4)\,\lambda^2 + 6\lambda + 1]} \quad \textbf{(Note)}$$

$$\Rightarrow \quad (25/4)\,\lambda^2 = (45/4)\,\lambda^2 + 6\lambda + 1$$

$$\Rightarrow \quad 5\lambda^2 + 6\lambda + 1 = 0 \Rightarrow (5\lambda + 1)(\lambda + 1) = 0 \Rightarrow \lambda = -1, -\frac{5}{2}.$$

Substituting these values of λ in (i) we get the required equations as

$$x^2 + y^2 + z^2 - 2x - 4y - 5z + 5 = 0$$

and $$x^2 + y^2 + z^2 - \frac{2}{5}x - \frac{4}{5}y - z + \frac{1}{5} = 0.$$ **Ans.**

Example 83:

Find the condition that the circles.

$$x^2 + y^2 + z^2 + 2ux + 2vy + 2wz + d = 0,\ lx + my + nz = p;$$

$$x^2 + y^2 + z^2 + 2u'x + 2v'y + 2w'z + d' = 0,\ l'x + m'y + n'z = p'$$

should lie on the same sphere.

Solution:

The equation of any sphere through the first circle is given by

$$(x^2 + y^2 + z^2 + 2ux + 2vy + 2wz + d) + \lambda\,(lx + my + nz - p) = 0 \ldots(i)$$

Similarly the equation of any sphere through second circle is given as

$$(x^2 + y^2 + z^2 + 2u'x + 2v'y + 2w'z + d') + \mu(l'x + m'y + n'z - p') = 0 \quad \ldots(ii)$$

If the given circles lie on the same sphere, then for some values of λ and μ the equations (i) and (ii) should represent the same sphere. Comparing the coefficients of x, y, z and constant terms in (i) and (ii) we get

$$2u + \lambda l = 2u' + \mu l' \quad \text{or} \quad 2(u - u') + \lambda l - \mu l' = 0 \qquad \text{...(iii)}$$

$$2v + \lambda m = 2v' + \mu m' \quad \text{or} \quad 2(v - v') + \lambda m - \mu m' = 0 \qquad \text{...(iv)}$$

$$2w + \lambda n = 2w' + \mu n' \quad \text{or} \quad 2(w - w') + \lambda n - \mu n' = 0 \qquad \text{...(v)}$$

$$\Rightarrow \quad d - \lambda p = d' - \mu p' \quad \text{or} \quad -(d - d') + \lambda p - \mu p' = 0 \qquad \text{...(iv)}$$

Eliminating λ and $-\mu$ from (iii), (iv), (v) and (vi) we get required condition as

$$\begin{vmatrix} 2(u-u') & l & l' \\ 2(v-v') & m & m' \\ 2(w-w') & n & n' \\ -(d-d') & p & p' \end{vmatrix} = 0 \qquad \textbf{Ans.}$$

Example 84:

Obtain the equations of the tangent planes to the sphere $x^2 + y^2 + z^2 + 6x - 2z + 1 = 0$, which pass through the line

$$3(16 - x) = 3z = 2y + 30.$$

Solution:

The given line may be written as

$$16 - x = z, \; 2y - 3z + 30 = 0.$$

The equation of any plane through this line is

$$(x + z - 16) + k(2y - 3z + 30) = 0$$

$$\Rightarrow \quad x + 2ky + (1 - 3k)z + (30k - 16) = 0. \qquad \text{...(i)}$$

The centre of the given sphere is (–3, 0, 1) and

$$\text{radius} = \sqrt{9 + 0 + 1 - 1} = 3.$$

The plane (i) will touch the sphere if

$$\frac{-3 + 0 + (1 - 3k) \cdot 1 + (30k - 16)}{\sqrt{1 + 4k^2 + (1 - 3k)^2}} = 3$$

$$\Rightarrow \quad 9k - 6 = \sqrt{13k^2 - 6k^2 + 2}.$$

$$\Rightarrow \quad 2k^2 - 3k + 1 = 0 \Rightarrow (2k - 1)(k - 1) = 0 \Rightarrow k = 1, \frac{1}{2}.$$

Putting these values of k in (i), the two tangent planes are

$$x + 2y - 2z + 14 = 0,\ 2x + 2y - z = 2.$$

Example 85:

Show that the plane $2x - 2y + z + 12 = 0$ touches the sphere $x^2 + y^2 + z^2 - 2x - 4y + 2z = 3$ and find the point of contact.

Solution:

The centre of the given sphere is (1, 2, –1) and its radius is 3. The plane $2x - 2y + z + 12 = 0$ touches the sphere if

$$\frac{2.1 - 2.2 - 1 + 12}{\sqrt{4 + 4 + 1}} = 3 \text{ or } 3 = 3.$$

Hence the given plane touches the given sphere.

The equations of the normal to the given plane drawn through (1, 2, –1) are

$$\frac{x - 1}{2} = \frac{y - 2}{-2} = \frac{z + 1}{1} = r \text{ (say)}.$$

Any point on this line is P(2r + 1, –2r + 2, r – 1).

The point P becomes the point of contact if it lies on the plane

$$2x - 2y + z + 12 = 0.$$

$$\therefore \quad 2(2r + 1) - 2(-2r + 2) + (r - 1) + 12 = 0 \Rightarrow r = -1.$$

Hence P (–1, 4, –2) is the required point of contact.

Example 86:

Find the equations of the tangent line to the circle

$$x^2 + y^2 + z^2 + 5x - 7y + 2z - 8 = 0,\ 3x - 2y + 4z + 3 = 0$$

at the point (–3, 5,4).

Solution:

Let $\frac{x + 3}{l} = \frac{y - 5}{m} = \frac{z - 1}{n}$ be the required tangent line.

Since this line is tangent to the given circle, it lies in the plane

$$3x - 2y + 4z + 3 = 0.$$

$$\therefore \quad 3l - 2m + 4n = 0. \quad \text{...(i)}$$

The centre of the given sphere is C(–5/2, 7/2, –1) and the given point is P(–3, 5, 4). The direction ratios of CP are

$$-\frac{5}{2}+3, \frac{7}{2}-5, -1-4 \text{ or } \frac{1}{2}, -\frac{3}{2}, -5.$$

Since CP is perpendicular to the tangent line,

$$\therefore \quad \frac{1}{2}l - \frac{3}{2}m - 5n = 0 \text{ or } l - 3m - 10n = 0. \qquad \text{...(ii)}$$

From (i) and (ii), we obtain

$$\frac{l}{32} = \frac{m}{34} = \frac{n}{-7}.$$

Hence the equations of the tangent line are

$$\frac{x+3}{32} = \frac{y-5}{34} = \frac{z-4}{-7}.$$

Example 87:

Find the equation of the spheres through the circle $x^2 + y^2 + z^2 = 1$, $2x + 4y + 5z = 6$ and touching the plane $z = 0$.

Solution:

Do your self. **Ans.** $5(x^2 + y^2 + z^2) - 2x - 4y - 5z + 1 = 0$

Example 88:

Find the equation of a sphere inscribed in the tetrahedron whose faces are $x = 0$, $y = 0$, $z = 0$, $2x - 6y + 3z + 6 = 0$.

Solution:

The equation of the sphere touching the coordinate planes is given by

$$x^2 + y^2 + z^2 - 2\lambda x + 2\lambda y - 2\lambda z + 2\lambda^2 = 0$$

nothing that the fourth plane meets the x, y and z axes in the –ve, +ve and –ve directions.

Its centre is $(\lambda, -\lambda, \lambda)$ and radius $= \sqrt{(\lambda^2 + \lambda^2 - 2\lambda^2)} = \lambda$.

If this sphere touches the plane $2x - 6y + 3z + 6 = 0$, then the length of the perpendicular from the centre $(\lambda, -\lambda, \lambda)$ of the sphere to this plane must be equal to its radius λ.

i.e., $$\frac{2\lambda + 6\lambda + 3\lambda + 6}{\sqrt{(2^2 + 6^2 + 3^2)}} = \lambda \text{ or } \frac{6 + 11\lambda}{7} = -\lambda \text{ or } \lambda = \frac{-1}{3} \qquad \textbf{(Note)}$$

The required equation is $x^2 + y^2 + z^2 + (2/3)(x - y + z) + 2(1/9) = 0$

$\Rightarrow \; 9(x^2 + y^2 + z^2) + 6(x - y + z) + 2 = 0$ **Ans.**

Example 89:

Find the equation of the sphere in the positive octant touching the coordinates planes and the plane $2x + 3y + 6z - 24 = 0$.

Solution:

Do your self.

Example 90:

Prove that the centres of the spheres which touch the lines $y = mx$, $z = c$; $y = -mx$, $z = -c$ lie upon the conicoid $mxy + cz (l + m^2) = 0$.

Solution:

Let the equation of the sphere which touches the lines be given as

$$x^2 + y^2 + z^2 + 2ux + 2vy + 2wz + d = 0 \quad \text{...(i)}$$

If this sphere touches the line $y = mx$, $z = c$, then we get

$$x^2 + m^2x^2 + c^2 + 2ux + 2vmx + 2wc + d = 0.$$

putting $y = mx$, $z = c$ in (i)

$$\Rightarrow \quad x^2 (1 + m^2) + 2 (u + vm) x + (c^2 + 2wc + d) = 0 \quad \text{...(ii)}$$

If the line $y = mx$, $z = c$ touches the sphere (i) then the roots of (ii) must be coincident and the condition for the same is

$$[2(u + vm)]^2 = 4 (1 + m^2) (c^2 + 2wc + d) \quad \text{... “}b^2 = 4ac\text{”}$$

$$\Rightarrow \quad (u + vm)^2 = (1 + m^2) (c^2 + 2wc + d). \quad \text{...(iii)}$$

Similarly if the line $y = -mx$, $z = -c$ touches the sphere (i), then we shall have

$$(u - vm)^2 = (1 + m^2) (c^2 - 2wc + d). \quad \text{...(iv)}$$

putting $-m$ for m and $-c$ for c in (iii)

Subtracting (iv) from (iii), we get $4uvm = (l + m^2) (4wc)$

$$(-u) (-v) m + (1 + m^2) (-w) c = 0. \quad \textbf{(Note)}$$

$\therefore$ The locus of the centre $(-u, -v, -w)$ of the sphere is

$$xym + (1 + m^2) zc = 0$$ **Hence proved.**

Example 91:

Find the equation of the tangent planes to the sphere $x^2 + y^2 + z^2 - 4x + 2y - 6z + 5 = 0$, which are parallel to the plane $2x + y - z = 0$.

Solution:

Let the tangent plane to the given sphere parallel to the plane be given as

$$2x + y - z + k = 0. \quad \text{...(i)}$$

The centre of the sphere is (2, –1, 3)

and radius = $\sqrt{(2^2 + 1^2 + 3^2 - 5)}$ or $\sqrt{(4 + 1 + 9 - 5)}$ or 3.

If the plane (i) touches the given sphere, then the length of perpendicular from the centre (2, –1, 3) to (i) must be equal to the radius 3.

i.e., $$\frac{2(2) + (-1) - (3) + k}{\sqrt{(2^2 + 1^2 + 1^2)}} = 3 \quad \text{or} \quad k = \pm 3\sqrt{6} \qquad \textbf{(Note)}$$

from (i) the required plane are $2x + y - z \pm 3\sqrt{6} = 0$

Example 92:

In one end of a diameter of the sphere

$x^2 + y^2 + z^2 - 2x + 4y - 6z - 11 = 0$ be (1, 2, 4) find the coordinates of the other end.

Solution:

If C be the centre of the given sphere then C is (1, –2, 3)

Also let A be the point (–1, 2, 4). Then the direction ratios of the line AC are (1 + 1, –2 –2, 3 – 4) or (2, –4, –1).

∴ The equation of the line AC is given as $\frac{x+1}{2} = \frac{y-2}{-4} = \frac{z-4}{-1}$

Any point on this line is B (–1 + 2r, 2 –4r, 4 – r). ...(i)

If the point B is the other end of the diameter through A, then B must lie on the sphere, and so we have

$$(-1 + 2r)^2 + (2 - 4r)^2 + (4 - r)^2 - 2(-1 + 2r) + 4(2 - 4r) - 6(4 - r) - 11 = 0$$

⇒ $21r^2 - 42r - 4 = 0 \Rightarrow r = 1 + [5/\sqrt{(21)}]$, taking +ve value.

∴ The required point B is

$$\left[1 + \frac{10}{\sqrt{(21)}}, -2 - \frac{20}{\sqrt{(21)}}, 3 - \frac{25}{\sqrt{(21)}}\right] \qquad \textbf{Ans.}$$

Example 93:

Find the equation of the sphere which has its sphere at the origin and which touches the line $2(x + 1) = 2 - y = z + 3$.

Solution:

The equation of given line may be written as

$$\frac{x+1}{1/2} = \frac{y-2}{-1} = \frac{z+3}{1} = r. \qquad \text{...(i)}$$

Let the sphere with its centre at the origin O touch the line (i) at the point P. Then the radius of this sphere is OP. Any point on the line (i) is $P\left(\frac{1}{2}.r-1, -r+2, r-3\right)$.

The directions ratios of OP are $\frac{1}{2}r-1, -r+2, r-3$.

Since OP is perpendicular to the line (i),

$$\therefore \left(\frac{1}{2}.r-1\right).\frac{1}{2}-(-r+2).1+(r-3).1=0 \Rightarrow r=22/9$$

$$\Rightarrow P(2/9, -4/9, -5/9)$$

and $OP = \sqrt{(2/9)^2+(-4/9)^2+(-5/9)^2} = \sqrt{5}/3$.

Hence the equation of the sphere with centre (0, 0, 0) and radius $\sqrt{5}/3$ is $x^8 + y^3 + z^2 = 5/9$ or $9(x^2 + y^2 + z^2) = 5$.

Example 94:

Prove that the centres of spheres which touch the lines $y = mx$, $z = c$ and $y = -mx$, $z = -c$, lie on the conicoid

$$mxy + c(1 + m^2)z = 0.$$

Solution:

Let the centre of the sphere touching the given lines be (α, β, γ) and radius a.

The given lines may be written as

$$\frac{x}{1}=\frac{y}{m}=\frac{z-c}{0}, \qquad \text{...(i)}$$

and

$$\frac{x}{1}=\frac{y}{-m}=\frac{z+c}{0}, \qquad \text{...(ii)}$$

Since the sphere touches the line (i), the distance of the point (α, β, γ) from (i) is a.

$$\therefore \alpha^2+\beta^2+(\gamma-c)^2-\left[\frac{1.\alpha+m.\beta+0(\gamma-c)}{\sqrt{1+m^2+0}}\right]^2=a^2. \qquad \text{...(iii)}$$

Similarly

$$\alpha^2+\beta^2+(\gamma+c)^2-\left[\frac{1.\alpha-m\beta+0(\gamma+c)}{\sqrt{1+m^2+0}}\right]^2=a^2. \qquad \text{...(iv)}$$

From (iii) and (iv), we obtain

$$\alpha^2 + \beta^2 + (\gamma - c)^2 - \frac{(\alpha + m\beta)^2}{1 + m^2} = \alpha^2 + \beta^2 + (\gamma + c)^2 - \frac{(\alpha - m\beta)^2}{1 + m^2}$$

$$\Rightarrow \qquad m\alpha\beta + c\gamma\,(1 + m^2) = 0$$

Hence the locus of (α, β, γ) is

$$mxy + c\,(l + m^2)\,z = 0.$$

Example 95:

Obtain the equations of the planes which pass through the point (3, 0, 3), touch the sphere $x^2 + y^2 + z^2 = 9$ and are parallel to the line $x = 2y = -z$.

Solution:

Any plane through the given line $x = 2y = -z$ is

$$(x - 2y) + \lambda\,(2y + z) = 0, \ \lambda \text{ being a parameter}$$

$$\Rightarrow \qquad x + 2\,(\lambda - 1)\,y + \lambda z = 0. \qquad \text{...(i)}$$

The equation of any plane through the point (3, 0, 3) and parallel to the plane (i) is

$$1\,(x - 3) + 2\,(\lambda - 1)\,(y - 0) + \lambda\,(z - 3) = 0$$

$$\Rightarrow \qquad x + 2\,(\lambda - 1)\,y + \lambda z - 3\,(1 + \lambda) = 0. \qquad \text{...(ii)}$$

The centre of the given sphere $x^2 + y^2 + z^2 = 9$ is (0, 0, 0) and its radius is 3.

The plane (ii) touches the sphere if

$$\frac{3(1 + \lambda)}{\sqrt{1 + 4(\lambda - 1)^2 + \lambda^2}} = 3$$

$$\Rightarrow \qquad (1 + \lambda)^2 = 1 + 4(\lambda - 1)^2 + 1^2 \ \Rightarrow \ 2\lambda^2 - 5\lambda + 2 = 0.$$

$\therefore\ \lambda = 2, \dfrac{1}{2}$. Putting these values of λ in (ii)

$$x + 2y + 2z = 9; \ 2x - 2y + z = 9.$$

These are the required planes.

Example 96:

Find the equations of the spheres which pass through the circle $x^2 + y^2 + z^2 = 5$, $2x + 3y + z = 3$ and touch the plane $3x + 4y = 15$.

Solution:

The equation of any sphere through the given circle is

$$(x^2 + y^2 + z^2 - 5) + \lambda(2x + 3y + z - 3) = 0$$

$$\Rightarrow \quad x^2 + y^2 + z^2 + 2\lambda x + 3\lambda y + \lambda z - (3\lambda + 5) = 0 \quad \text{...(i)}$$

If this sphere touches the plane $3x + 4y - 15 = 0$ then the length of perpendicular from its centre $(-\lambda, -3\lambda/2, -\lambda/2)$ to this plane must be equal to its radius $\sqrt{[(-\lambda)^2 + (-3\lambda/2)^2 + (-\lambda/2)^2 + (3\lambda + 5)]}$

i.e., $\sqrt{[(7\lambda^2/2) + (3\lambda + 5)]}$ **(Note)**

i.e.,
$$\frac{3(-\lambda) + 4(-3\lambda/2) - 15}{\sqrt{(3^2 + 4^2)}} = \sqrt{\left[\left(\frac{7\lambda}{2}\right)^2 + (3\lambda + 5)\right]}$$

$$\Rightarrow \quad \frac{-(9\lambda + 15)}{5} = \sqrt{\left(\frac{7\lambda^2 + 6\lambda + 10}{2}\right)}$$

$$\Rightarrow \quad 18(3\lambda + 5)^2 = 25(7\lambda^2 + 6\lambda + 10), \text{ squaring both sides}$$

$$\Rightarrow \quad 13\lambda^2 - 390\lambda - 200 = 0, \text{ on simplifying.}$$

Substituting the two values of λ obtained from here in (i) by turn, we get the required equations of sphere is given as. **Ans.**

Example 97:

A sphere touches the three coordinate planes and passes through the point (2, 1, 5). Find its equation.

Solution:

Let the sphere be

$$x^2 + y^2 + z^2 + 2ux + 2vy + 2wz + d = 0. \quad \text{...(i)}$$

Its centre is $(-u, -v, -w)$ and radius $\sqrt{(u^2 + v^2 + w^2 - d)}$.

If the sphere touches the yz-plane *i.e.*, $x = 0$, then the length of the perpendicular drawn from the centre $(-u, -v, -w)$ of the sphere to the plane $x = 0$ must be equal to the radius $\sqrt{(u^2 + v^2 + w^2 - d)}$ of the sphere. **(Note)**

$$\therefore \quad \frac{-u}{\sqrt{(1^2 + 0^2 + 0^2)}} = \sqrt{(u^2 + v^2 + w^2 - d)} \text{ or } u^2 = u^2 + v^2 + w^2 - d$$

$$\Rightarrow \quad v^2 + w^2 = d \quad \text{...(ii)}$$

Similarly, if the sphere (i) touches planes $y = 0$ and $z = 0$ (the other coordinate planes) then we shall get

$$u^2 + w^2 = d \quad \text{...(iii)}$$

and $$u^2 + v^2 = d \quad \text{...(iv)}$$

Adding (ii), (iii) and (iv) we get $u^2 + v^2 + w^2 = \frac{3}{2}d$ ...(v)

∴ From (v) we get $u^2 = \frac{1}{2}d$.

Similarly from (iii), (iv) and (v) we get $v^2 = \frac{1}{2}d = w^2$.

∴ From (i) the equation of the sphere reduces to

$$x^2 + y^2 + z^2 + 2\sqrt{(\frac{1}{2}d)}\,(x + y + z) + d = 0, \quad \text{...(vi)}$$

on substituting values of u, v and w.

If this sphere passes through (2, 1, 5) then we get

$$4 + 1 + 25 + \sqrt{(2d)}\,(2 + 1 + 5) + d = 0$$

$$\Rightarrow \quad d + 8\sqrt{2}\,\sqrt{d} + 30 = 0 \text{ or } \sqrt{d} = \frac{1}{2}[-8/\sqrt{2} \pm \sqrt{(128 - 120)}]$$

$$\Rightarrow \quad \sqrt{d} = \frac{1}{2}[-8\sqrt{2} \pm 2\sqrt{2}] = -3\sqrt{2}, -5\sqrt{2} \Rightarrow d = 18, 50$$

∴ From (vi), the required equations of the sphere are

$$x^2 + y^2 + z^2 - 6\,(x + y + z) + 18 = 0$$

and $$x^2 + y^2 + z^2 - 10\,(x + y + z) + 30 = 0$$ **Ans.**

EXERCISES

1. Find the length of the chord intercepted by the line $(x - \alpha)/1 = (y - \beta)/m = (z - \gamma)/n$ on the sphere $x^2 + y^2 + z^2 = a^2$.

   Ans. $d = 2[(l\alpha + m\beta + n\gamma)^2 - (\alpha^2 + \beta^2 + \gamma^2 - a^2)]^{1/2}$

2. Find the equations of the sphere passing through the circle $x^2 + y^2 + z^2 - 4x - y + 3z + 12 = 0$; $2x + 3y + 7z = 10$ and touching the plane $x - 2y + 2z = 1$.

   Ans. $x^2 + y^2 + z^2 - 2x + 2y - 4z + 2 = 0$,
   $x^2 + y^2 + z^2 - 6x - 4y + 10z + 22 = 0$.

3. Find the equations of the spheres passing through $x^2 + y^2 + z^2 = 5$, $x + 2y + 3z = 3$ and touch the plane $4x + 3y = 15$.

   Ans. $x^2 + y^2 + z^2 + 2x + 4y + 6z = 11$,
   $5(x^2 + y^2 + z^2) - 4x - 8y - 12z = 13$.

4. Obtain the condition that the spheres $a(x^2 + y^2 + z^2) + 2lx + 2my + 2nz + p = 0$ and $b(x^2 + y^2 + z^2) = k^2$ may cut orthogonally.

   Ans. $ak^2 = bp$.

5. Obtain the condition that the spheres $a(x^2 + y^2 + z^2) + 2ux + 2vy + 2wz + d = 0$ and $a'(x^2 + y^2 + z^2) + 2u'x + 2v'y + 2w'z + d' = 0$ may cut orthogonally. **Ans.** $2(uu' + vv' + ww') = da' + d'a$.
6. Show the equation of the tangent plane to the sphere $x^2 + y^2 + z^2 = 9$ at $(1, -2, 2)$ is $x - 2y + 2z = 9$.
7. If a tangent plane to the sphere $k^2(x^2 + y^2 + z^2) = r^2$ makes intercepts a, b, c on the axes of coordinates, prove that
$$k^{-2} (a^{-2} + b^{-2} + c^{-2}) = r^{-2}.$$
8. Obtain the equations of the tangent planes to the sphere $x^2 + y^2 + z^2 + 6x - 2z + 1 = 0$ which passes through the line
$$3 (16 - x) = 3z = 2y + 30$$
Ans. $x + 2y - 2z + 14 = 0$, $2x + 2y - z - 2 = 0$.
9. A sphere is inscribed in the tetrahedron whose faces are $x = 0$, $y = 0$, $z = 0$ and $2x + 6y + 3z = 14$. Find its centre, radius and equation.

Ans. $(7/9, 7/9, 7/9)$; $7/9$; $81 (x^2 + y^2 + z^2) = 128 (x + y + z) - 98$.

10. Find the equation of the sphere which touches the sphere $x^2 + y^2 + z^2 + 2x - 6y + 1 = 0$ at $(1, 2, -2)$ and passes through $(0, 0, 0)$.
Ans. $x^2 + y^2 + z^2 + (5/2)x - (25/4)y - (1/2) z = 0$.
11. Find the equation of the sphere which touches the sphere $x^2 + y^2 + z^2 - x + 3y + 2z - 3 = 0$ at $(1, 1, -1)$ and passes through $(2, 0, 1)$.
Ans. $2 (x^2 + y^2 + z^2) - x + 11y + 4z - 12 = 0$.
12. Find the centre and radius of the sphere given by
$$2(x^2 + y^2 + z^2) - 2x + 4y - 6z = 1$$
Ans. $(1/2, -1, 3/2)$; $\sqrt{(11)}$.
13. Show that the spheres $x^2 + y^2 + z^2 = 100$ and $x^2 + y^2 + z^2 - 24x - 30y - 32z + 400 = 0$ touch externally and find their point of contact. **Ans.** Point of contact is $(24/5, 6, 32/5)$.
14. State which of the following statements are true and which are false. Justify your answer:
    (i) One and only one sphere of given radius R can be made to pass through a circle of radius $r < R$.
    (ii) A circle on sphere is uniqueluy determined if we know its radius and the coordinates of its centre.
15. Find the co-ordinates of the centre and the radius of the circle $x + 2y + 2z = 15$, $x^2 + y^2 + z^2 - 2y - 4z = 11$. **Ans.** $(1, 2, 3)$ and $\sqrt{7}$.

16. Find the equation to the sphere which passes through the circle $x^2 + y^2 + z^2 = 4$, $z = 0$ and is cut by the plane $x + 2y + 2z = 0$ in a circle of radius 3. **Ans.** $x^2 + y^2 + z^2 \pm 6z = 4$.

17. Prove that the radical planes of three spheres taken two by two pass through one line.

18. Show that the limiting points of the co-axial system of spheres determined by $x^2 + y^2 + z^2 + 3x - 3y + 6 = 0$ and $x^2 + y^2 + z^2 - 6y - 6z + 6 = 0$ are $(-1, 2, 1)$ and $(-2, 1, -1)$.

19. Find the equations of the tangent planes to the sphere $x^2 + y^2 + z^2 - 2(x + y + z) = 0$ perpendicular to the line with direction numbers 1, 1, 1.

20. Prove that in general two spheres can be drawn through a given point to touch the coordinate planes and find for what positions of the point the spheres are :

    (i) real; (ii) coincident.

21. Find the equation of the sphere circumscribing the tetrahedron formed by the planes $4y + 3z = 0$, $3x + 2y = 0$, $z + 2x = 0$ and $6x + 4y + 3z = 12$.

    Ans. $12(x^2 + y^2 + z^2) = 174x + 116y + 87z$.

22. Find the equation of the sphere circumscribing the tetrahedron whose faces are $y + z = 0$, $z + x = 0$, $x + y = 0$ and $x + y + z = 1$.

    Ans. $x^2 + y^2 + z^2 = 3(x + y + z)$

23. Find the equation of the sphere the extremities of whose diameter are the points $(3, 4, -2)$ and $(-2, -1, 0)$.

    Ans. $x^2 + y^2 + z^2 - x - 3y + 2z = 10$.

24. A sphere has its centre at $(1, -3, 4)$ and passes through $(3, -1, 3)$. Find its equation. Find also the equation of the concentric sphere which touches the plane $2x + 5y - 5z = 21$.

    Ans. $x^2 + y^2 + z^2 - 2x + 6y - 8z + 17 = 0$;

    $x^2 + y^2 + z^2 - 2x + 6y - 8z - 28 = 0$.

25. Find the centre of the sphere through the four points $(4, -1, 2)$, $(0, -2, 3)$, $(1, -5, 1)$ and $(2, 0, 1)$. **Ans.** $(2, -13/15, 7/5)$

26. Find the equation of the sphere passing through the points $(3, 0, 2)$, $(-1, 1, 1)$, $(2, -5, 4)$ and having its centre on $2x + 3y + 4z = 6$.

    Ans. $x^2 + y^2 + z^2 + 4y - 6z - 1 = 0$.

27. Find the equation of the sphere through the origin cutting off intercepts 3, 4 and 5 from the coordinate axes.

Ans. $x^2 + y^2 + z^2 - 3x - 4y - 5z = 0$.

28. A plane through a fixed point (1, 1, 1) cuts the axes in A, B, C. Find the locus of the centre of the sphere OABC, where O is the origin. **Ans.** $x^{-1} + y^{-1} + z^{-1} = 2$.

29. Find the equations of the spheres through the following points and find their centres and radii:-

(i) (0, 0, 0); (2, 0, 0); (0, 4, 0); (0, 0, –1).

(ii) (0, 0, 0); (–4, 3, 1); (4, –3, 1); (4, 3, –1).

(iii) (0, 0, 0); (–a, b, c); (a, –b, c); (a, b, –c).

(iv) (0, 0, 0); (–1, 1, 1); (1, –1, 1); (1, 1, –1).

Ans. (i) $x^2 + y^2 + z^2 = 2x + 4y - z$; $\left(1, 2, -\frac{1}{2}\right)$; $\frac{1}{2}\sqrt{(21)}$

(ii) $6(x^2 + y^2 + z^2) = 13(3x + 4y + 12z)$; $\left(\frac{13}{4}, \frac{13}{2}, 13\right)$, $14\frac{1}{13}$.

(iii) $\frac{x^2 + y^2 + z^2}{a^2 + b^2 + z^2} - \frac{x}{a} - \frac{y}{b} - \frac{z}{c} = 0$;

$$\left(\frac{a^2 + b^2 + c^2}{2a}, \frac{a^2 + b^2 + c^2}{2b}, \frac{a^2 + b^2 + c^2}{2c}\right),$$

$$\frac{(a^2 + b^2 + c^2)}{2abc}\sqrt{(a^2b^2 + b^2c^2 + c^2a^2)}.$$

(iv) $x^2 + y^2 + z^2 = 3(x + y + z)$, (3/2, 3/2, 3/2), $3\sqrt{3}/2$

30. Find the equation of the sphere passing through the circle $x^2 + y^2 = 25$, $y = 2$ and $x^2 + z^2 = 16$, $y = 3$.

31. Find the equations of the spheres which passes through the circle $x^2 + y^2 + z^2 - 2x + 2y + 4z = 3$; $2x + y + z + 4 = 0$ and touch the plane $3x + 4y = 14$.

Ans. $x^2 + y^2 + z^2 - 2x + 2y + 4z = 3$;
$x^2 + y^2 + z^2 + 34x + 20y - 22z + 69 = 0$.

32. Find the equation of the spheres which pass through the circle $x^2 + y^2 + z^2 - 2x + 2y + 2z = 2$, $y = 0$ and touch the plane $y + z = 2$.

Ans. $x^2 + y^2 + z^2 - 2x - 4y + 2z = 2$;
$x^2 + y^2 + z^2 - 2x + 28y + 2z = 2$.

33. Find the radius of the circle in which the sphere $x^2 + y^2 + z^2 = 4$ is cut by the plane $x + y + z = 1$. **Ans.** $\sqrt{(11/3)}$

34. Find the radius and centre of the circle of intersection of the sphere $x^2 + y^2 + z^2 - 2y - 4z - 11 = 0$ and the plane $x + 2y + 2z = 15$.
Ans. $\sqrt{7}$ and (1, 3, 4)

35. Obtain the radius and centre of the circle $x^2 + y^2 + z^2 + x + y + z - 4 = 0$, $x + y + z = 0$. **Ans.** 2; (0, 0, 0)

36. Find the equation of the sphere through the circle $x^2 + y^2 + z^2 = 9$; $2x + 3y + 4z = 5$ and the point (1, 2, 3).
Ans. $3(x^2 + y^2 + z^2) - 2x - 3y - 4z = 22$.

37. Show that the two circles $x^2 + y^2 + z^2 + 3x - 4y - 3z = 0$, $x - y + 2z = 4$ and $2(x^2 + y^2 + z^2) + 8x - 18y + 17z = 17$, $2x + y - 3z + 1 = 0$ lie on the same sphere. Find its equation.
Ans. $x^2 + y^2 + z^2 + 5x - 6y + 7x = 8$.

38. A circle with centre (2, 3, 0) and radius 1, is drawn in the plane $z = 0$. Find the equation of the sphere which passes through this circle and through the point (1, 1, 1).

39. A variable sphere passes through the point $(0, 0, \pm c)$ and cuts the line $y = x \tan \alpha$, $z = c$; $y = -x \tan \alpha$, $z = -c$ in the points P, P'. If PP' has constant length 2a, show that the centre of the sphere lies on the circle $z = 0$, $x^2 + y^2 = (a^2 - c^2) \operatorname{cosec}^2 2\alpha$.

40. Find the radius of the circle given by the equation
$3x^2 + 3y^2 + 3z^2 + x - 5y - 2 = 0$, $x + y = 2$ **Ans.** $1/\sqrt{2}$.

41. Find out the equation of the sphere having the circle given by $x^2 + y^2 + z^2 = 9$, $x - 2y - 2z = 5$ for a great circle. Also find out its centre and radius. **Ans.** $9(x^2 + y^2 + z^2) - 10x + 20y - 20z = 31$;
$(5/9, -10/9, 10/9)$; $\frac{2}{3}\sqrt{(14)}$.

42. Find the equation of the sphere having the circle $x^2 + y^2 + z^2 - 2x + 4y - 6z + 7 = 0$, $2x - y + 2z = 5$ for great circle.
Ans. $9(x^2 + y^2 + z^2) + 2x + 26y - 34z + 13 = 0$.

43. Find the equation of the plane which cuts the spehre $x^2 + y^2 + z^2 - 2x + 4y - 6z + 3 = 0$ in a circle whose centre is (2, 3, –4). Also find the radius of this circle.

44. Show that the centres of the spheres which cut both the sp[illegible] $x^2 + y^2 + z^2 + 2\lambda x + d = 0$ and $x^2 + y^2 + z^2 + 2\mu x + d = 0$ [illegible] circles, lies on the plane $x + \lambda + \mu = 0$.

45. Show that the circles, $x^2 + y^2 + z^2 - y + 2z = 0$, $x - y +$ [illegible] $x^2 + y^2 + z^2 + x - 3y + z = 5$, $2x - y + 4z = 1$ lie on the sa[illegible] [illegible]ere. Find the equation of the sphere.

46. Show that the equations of any two spheres can be put in the form $x^2 + y^2 + z^2 + 2\lambda_1 x + d = 0$, $x^2 + y^2 + z^2 + 2\lambda_2 x + d = 0$.

47. Prove that only one tangent plane to $x^2 + y^2 + z^2 - 2x + 6y + 2z + 8 = 0$ can pass through the straight line $x = 4t + 4$, $y = 3t + 1$, $z = t + 1$.

48. Rectify the mistakes if any, otherwise assert correctness of the following statement giving adequate reasons : The plane $3x + 4y + 5z = 49$ intersects the sphere $x^2 + y^2 + z^2 = 50$ in a point circle.

49 The equation of a sphere with the join of (x_1, y_1, z_1) and (x_2, y_2, z_2) as diameter is

(i) $\Sigma(x - x_1)(x - x_2) = 0$;

(ii) $\Sigma(x - x_1)(y - y_1) = 0$;

(iii) $\Sigma(x + x_1)^2 (Y - y_1) = 0$;

(iv) $\Sigma(x - x_2)(y - y_1) = 0$. **Ans.** (i)

50. Find the smallest sphere (i.e., the sphere of smallest radius) which touches the lines

$$\frac{1}{2}(x-5) = -(y - 2) = -(z - 5), -\frac{1}{3}(x + 4) = -\frac{1}{6}(y + 5) = \frac{1}{4}(z-4)$$

51. Find the equation of the sphere which passes through the points $(3, 0, 2)$, $(-1, 1, 1)$, $(2, -5, 4)$ and whose centre lies on the plane

$$x + y + 2z = 3.$$

Ans. $8(x^2 + y^2 + z^2) - x + 35y - 41z - 19 = 0$

52. Show that the equation of the sphere through three points $(3, 0, 2)$, $(-1, 1, 1)$, $(2, -5, 4)$ and having its centre on the plane $2x + 3y + 4z = 6$ is $x^2 + y^2 + z^2 + 4y - 6z = 1$.

53. Find the centre and radius of the circle $x^2 + y^2 + z^2 = ax + by + cz$, $(x/a) + (y/b) + (z/c) = 1$.

54. Find the equation of the sphere through the points $(0, 0, 0)$; $(0, 1, -1)$ $(-1, 2, 0)$ and $(1, 2, 1)$.

Ans. $10\ (x^2 + y^2 + z^2) - 2x - 26y - 6z = 0$.

55. Find the angle of intersection of the spheres

$$x^2 + y^2 + z^2 - 2x - 4y - 6z + 10 = 0$$

and $x^2 + y^2 + z^2 - 6x - 2y - 2z + 2 = 0$. **Ans.** $\cos^{-1}(-2/3)$

56. Obtain the equation of the sphere with centre at $(1, -2, 3)$ and touching the plane $6x - 3y + 2z = 4$.

2

THE CONE AND CYLINDER

Definition : *A **cone** is a surface generated by a variable straight line which passes a fixed point and satisfies one more condition i.e., intersecting a given curve or touching a given surface.*

The fixed point is called the *vertex* and the given curve (or surface) is called the *guiding curve* (or *guiding surface)* of the cone. The variable straight line is known as the *generator* of the cone.

A cone whose equation is of second degree is know as *quartic cone* or equation of a cone with second degree equations.

CONE WITH VERTEX AT THE ORIGIN

To prove that the equation of the cone with vertex at the origin is a homogeneous second degree equation in x, y, and z.

Let the general equation of the second degree in x, y, z is given as

$$ax^2 + by^2 + cz^2 + 2fyz + 2gzx + 2hxy + 2ux + 2vy + 2wz + d = 0 \quad ...(i)$$

represent a cone with vertex at the origin O.

Let P (x_1, y_1, z_1) be any point on the cone. The equation of the generator OP is given by

$$\frac{x-0}{x_1} = \frac{y-0}{y_1} = \frac{z-0}{z_1} \quad ...(ii)$$

Any point Q on this generator is (rx_1, ry_1, rz_1). As OP is a generator of the cone (i), so every point on it like Q must lie on cone (i) for all values of r, which means that $r^2\left(ax_1^2 + by_1^2 + cz_1^2 + 2fy_1z_1 + 2gz_1x_1 + 2hx_1y_1\right)$

$$+ 2r\,(ux_1 + vy_1 + wz_1) + d = 0,$$

must be an identity and the conditions for the same given as

$$ax_1^2 + by_1^2 + cz_1^2 + 2fy_1z_1 + 2gz_1x_1 + 2hx_1y_1 = 0 \quad ...(iii)$$

$$ux_1 + vy_1 + wz_1 = 0 \qquad ...(iv)$$

and $$d = 0. \qquad ...(v)$$

From (iv) we conclude that P (x_1, y_1, z_1) is a point on a plane ux + vy + wz = 0 if u, v and w are not all zero and this is against hypothesis. So u, v and w must be all zero. Also (v) is obvious as the cone passes through the origin.

Hence, from (i) the equation of the cone with vertex at the origin is given by

$$ax^2 + by^2 + cz^2 + 2fyz + 2gzx + 2hxy = 0, \qquad ...(vi)$$

which is a homogeneous equation of second degree in x, y and z.

Conversely : *Every homogeneous equation of second degree in x, y, and z represents a cone with its vertex at the origin.*

We know that if P (x_1, y_1, z_1) satisfies the equation (vi) then for all values of r, the point (rx_1, ry_1, rz_1) also satisfies (vi) *i.e.,* all points on the generator OP lies on the surface of the cone *i.e.,* the line OP lies entirely on the cone.

Thus, the surface is generated by straight lines through the origin and hence it is cone with its vertex at the origin.

Corollary : *If the line $x/l = y/m = z/n$ is a generator of the cone given by*

$$ax^2 + by^2 + cz^2 + 2fyz + 2gzx + 2hxy = 0, \qquad ...(i)$$

then its direction cosines l, m, n satisfy the equation of the cone.

Let any point on the given generator is (*l*r, mr, nr). If this lies on the cone (i), then we have

$$al^2 + bm^2 + cn^2 + 2fmn + 2gnl + 2hlm = 0 \qquad ...(ii)$$

Conversely : if the relation (ii) holds good *i.e.,* if the direction ratios of a straight line which always passes through a fixed point satisfy a homogeneous equation, then this line is a generator of a cone whose vertex is at the fixed point.

GENERAL EQUATION OF A CONE OF SECOND DEGREE WHICH PASSES THROUGH THE COORDINATE AXES

If the cone passes through the coordinates axes, then its vertex must be at the origin and as such its equation must be of the form

$$ax^2 + by^2 + cz^2 + 2fyz + 2gzx + 2hyx = 0 \qquad ...(i)$$

Also the d.c.'s of the coordinate axes are 1, 0, 0 : 0, 1, 0 and 0, 0, 1 which should satisfy (i).

So, we get a = 0, b = 0 and c = 0. Hence, the required equation of the cone through the coordinate axes is given as fyz + gzx + hxy = 0.

EQUATION OF THE CONE WITH A GIVEN VERTEX AND A GIVEN CONIC FOR ITS BASE

To find the equation to the cone whose vertex is (α, β, γ) *and the base conic* $ax^2 + 2hxy + by^2 + 2gx + 2fy + c = 0,\ z = 0.$

Proof :

Equation of any line through (α, β, γ) is given by

$$\frac{x-\alpha}{l} = \frac{y-\beta}{m} = \frac{z-\gamma}{n} \qquad ...(i)$$

It meets the plane z = 0 at the point given as

$$\frac{x-\alpha}{l} = \frac{y-\beta}{m} = \frac{z-\gamma}{n} \quad i.e., \left[\alpha - \frac{l\gamma}{n}, \beta - \frac{m\gamma}{n}, 0\right]$$

If this point lies on the given conic, then we have

$$a\left(\alpha - \frac{l\gamma}{n}\right)^2 + 2h\left(\alpha - \frac{l\gamma}{n}\right)\left(\beta - \frac{m\gamma}{n}\right) + b\left(\beta - \frac{m\gamma}{n}\right)^2$$
$$+2g\left(\alpha - \frac{l\gamma}{n}\right) + 2f\left(\beta - \frac{m\gamma}{n}\right) + c = 0. \qquad ...(ii)$$

The equation (ii) is the equation for the line (i) to intersect the given conic and hence the locus of the line (i) is obtained by eliminating l, m, n between (i) and (ii) by writing $\frac{x-\alpha}{z-\gamma}$ and $\frac{y-\beta}{z-\gamma}$ for l/n and m/n respectively.

Hence the required equation of the cone is given by

$$a\left[\alpha - \left(\frac{x-\alpha}{z-\gamma}\right)\gamma\right]^2 + 2h\left[\alpha - \left(\frac{x-\alpha}{z-\gamma}\right)\gamma\right]\left[\beta - \left(\frac{y-\beta}{z-\gamma}\right)\gamma\right]$$
$$+b\left[\beta - \left(\frac{y-\beta}{z-\gamma}\right)\gamma\right]^2 + 2g\left[\alpha - \left(\frac{x-\alpha}{z-\gamma}\right)\gamma\right] + 2f\left[\beta - \frac{y-\beta}{z-\gamma}\gamma\right] + c = 0$$

Simplifying [multiplying each term by $(z - \gamma)^2$], we get

$$a(\alpha z - \gamma x)^2 - 2h(\alpha z - \gamma x)(\beta z - \gamma y) + b(\beta z - \gamma y)^2 + 2g(\alpha x - \gamma x)(z - \gamma)$$
$$+ 2f(\beta z - \gamma y)(z - \gamma) + c(z - \gamma)^2 = 0.$$

CONDITION FOR THE GENERAL EQUATION OF THE SECOND DEGREE TO REPRESENT A CONE AND TO FIND ITS VERTEX

Let the general equation of second degree is given as

$$ax^2 + by^2 + cz^2 + 2fyz + 2gzx + 2hxy + 2ux + 2vy + 2wz + d = 0 \quad \text{...(i)}$$

represent a cone with its vertex at (α, β, γ).

Shifting the origin to (α, β, γ), the equation (i) transforms into

$$a(x + \alpha)^2 + b(y + \beta)^2 + c(z + \gamma)^2 + 2f(y + \beta)(z + \gamma) + 2g(z + \gamma)(x + \alpha) + 2h(x + \alpha)(y + \beta) + 2u(x + \alpha) + 2v(y + \beta) + 2w(z + \gamma) + d = 0$$

$$\Rightarrow \quad (ax^2 + by^2 + cz^2 + 2fyz + 2gzx + 2hxy) + 2x(a\alpha + h\beta + g\gamma + u) + 2y(h\alpha + b\beta + f\gamma + v) + 2z(g\alpha + f\beta + c\gamma + w) + (a\alpha^2 + b\beta^2 + c\gamma^2) + 2f\beta\gamma + 2g\gamma\alpha + 2h\alpha\beta + 2u\alpha + 2v\beta + 2w\gamma + d) = 0 \quad \text{...(ii)}$$

Since equation (ii) represents a cone with vertex at the origin, so it must be homogeneous and therefore the coefficients of x, y, z and absolute term in (ii) must vanish separately hence

i.e., $a\alpha + h\beta + g\gamma + u = 0,$ ...(iii)

$h\alpha + b\beta + f\gamma + v = 0$ ...(iv);

$g\alpha + f\beta + c\gamma + w = 0$ ...(v)

and $a\alpha^2+b\beta^2+c\gamma^2 + 2g\gamma\alpha + 2g\gamma\alpha + 2h\alpha\beta + 2u\alpha + 2v\beta + 2w\gamma + d = 0$...(vi)

Now (vi) can be written as

$$a(a\alpha + h\beta + g\gamma + u) + \beta(h\alpha + b\beta + f\gamma + v) + \gamma(g\alpha + f\beta + c\gamma + w) + (u\alpha + v\beta + w\gamma + d) = 0$$

or with the help of (ii), (iv) and (v) this reduces to

$$u\alpha + v\beta + w\gamma + d = 0. \quad \text{...(vii)}$$

The required condition obtained by eliminating α, β, γ between (iii), (iv), (v) and (vii) is

$$\begin{vmatrix} a & h & g & u \\ h & b & f & v \\ g & f & c & w \\ u & v & w & d \end{vmatrix} = 0. \quad \text{...(viii)}$$

If this condition holds good, the coordinates (α, β, γ) of the vertex are found by solving any three of equations (iii), (iv), (v) and (viii).

Rule : We denote the given equations of the second degree by F (x, y, z) = 0, and introduce a variable t so as to make F (x, y, z) homogeneous in x, y, z, t. Then the equations (iii), (iv), (v) and (vii) of the above article

are given by $\frac{\partial F}{\partial x} = 0, \frac{\partial F}{\partial y} = 0, \frac{\partial F}{\partial z} = 0, \frac{\partial F}{\partial t} = 0,$

where t is replaced by unity after differentiation.

Solve any three of these equations and if the fourth is satisfied by the values of x, y and z so obtained, then the given equation represents a cone with its vertex at (x, y, z).

Example 1:

Find the angle between the lines of intersection of the plane $10x + 7y - 6z = 0$ and the cone $20x^2 + 7y^2 - 108z^2 = 0$.

Solution:

Let $$\frac{x}{l} = \frac{y}{m} = \frac{z}{n} \qquad ...(1)$$

be one of the two lines in which the given plane cuts the given cone so that we have

$$10l + 7m - 6n = 0, \qquad ...(2)$$

and $$20l^2 + 7m^2 - 108n^2 = 0. \qquad ...(3)$$

From (2), $n = \frac{1}{6}(10l + 7m)$. Putting in (3), we get

$$20l^2 + 7m^2 - 3(10l + 4m)^2 = 0 \text{ or } 2l^3 + 3lm + m^2 = 0$$

$$\Rightarrow \quad \frac{2l^2}{m^2} + \frac{3l}{m} + 1 = 0 \Rightarrow \left(\frac{2l}{m} +\right)\left(\frac{l}{m} + 1\right) = 0$$

$$\Rightarrow \quad \frac{l}{m} = -1 \text{ or } -\frac{1}{2}.$$

From (2), $\frac{10l}{m} + 7 = 6\frac{n}{m}$.

Now $\frac{l}{m} = -1 \Rightarrow \frac{n}{m} = -\frac{1}{2}$ and so $\frac{l}{-2} = \frac{m}{2} = \frac{n}{-1}$, and $\frac{l}{m} = -\frac{1}{2} \Rightarrow \frac{n}{m} = \frac{1}{3}$ and so $\frac{l}{-3} = \frac{m}{6} = \frac{n}{2}$.

Hence the two lines of intersection are

$$\frac{x}{-2} = \frac{y}{2} = \frac{z}{-1} \text{ and } \frac{x}{-3} = \frac{y}{6} = \frac{z}{2}.$$

The angle between these lines is

$$\cos\beta = \frac{(-2)(-3) + 2.6 + (-1).2}{\sqrt{4+4+1}\sqrt{9+36+4}} = \frac{16}{3 \times 7}$$

Hence $\theta = \cos^{-1}\left(\frac{16}{21}\right)$ is the required angle.

Example 2:

Show that the equation of the quadric cone which contains the three coordinate axes and the lines in which the plane

$x - 5y - 3z = 0$ *cuts the cone* $7x^2 + 5y^2 - 3z^2 = 0$ *is*

$yz + 10zx + 18xy = 0.$

Solution:

Let $\frac{x}{l} = \frac{y}{m} = \frac{z}{n}$ ...(1)

be one of the two lines in which the given plane cuts the given cone so that we have

$l - 5m - 3n = 0$ and $7l^2 + 5m^2 - 3n^2 = 0.$ ...(2)

From these equations, we obtain

$7(5m + 3n)^2 + 5m^2 - 3n^2 = 0$ or $6m^2 + 7mn + 2n^2 = 0$

$$\Rightarrow \frac{6m^2}{n^2} + \frac{7m}{n} + 2 = 0. \therefore \frac{m}{n} = \frac{-7 \pm \sqrt{49-48}}{12} = -\frac{1}{2} \text{ or } \frac{-2}{3}.$$

From (2), $l - 5m - 3n = 0 \Rightarrow \frac{l}{n} = 5\frac{m}{n} + 3.$

Now $\frac{m}{n} = -\frac{1}{2} \Rightarrow \frac{l}{n} = \frac{1}{2}$ and so $\frac{l}{1} = \frac{m}{-1} = \frac{n}{2}$,

and $\frac{m}{n} = -\frac{2}{3} \Rightarrow \frac{l}{n} = -\frac{1}{3}$ and so $\frac{l}{1} = \frac{m}{2} = \frac{n}{-3}$.

Hence by (1), the two lines of intersection are

$$\frac{x}{1}=\frac{y}{-1}=\frac{z}{2} \text{ and } \frac{x}{1}=\frac{y}{2}=\frac{z}{-3} \qquad ...(3)$$

We know that the equation of the cone through the axes is

$$fyz + gzx + hxy = 0. \qquad ...(4)$$

It will contain the lines given in (3), if

$$f(-1)(2) + g(2)(1) + h(1)(-1) = 0 \Rightarrow 2f - 2g + h = 0,$$

and $\quad f(2)(-3) + g(-3)(1) + h(1)(2) = 0 \Rightarrow 6f + 3g - 2h = 0.$

Solving these equations for f, g, h; we obtain

$$\frac{f}{4-3}=\frac{g}{6+4}=\frac{h}{6+12}=k \text{, say.}$$

$\therefore$ f = k, g = 10 k, h = 18 k. Putting these values in (4), we get $yz + 10zx + 18xy = 0$ as the required cone.

Example 3:

Find the equation of the cone which passes through the common generators of the cones $-2x^2 + 4y^2 + z^2 = 0$ and $10xy - 2yz + 5zx = 0$ and the line with direction cosines proportional to 1, 2, 3.

Solution:

Clearly the given cones have the obtain as the common vertex. Therefore the equation the equation of the cone with its vertex at the origin and passing through the common generators of the given cones is

$$-2x^2 + 4y^2 + z^2 + k(10xy - 2yz + 5zx) = 0, \qquad ...(1)$$

where k is a parameter.

Since the cone (1) contains the line with direction cosines proportional to 1, 2, 3; therefore

$$-2(1)^2 + 4(2)^2 + (3)^2 + k(10.1.2 - 2.2.3 + 5.3.1) = 0$$

$\Rightarrow \qquad 23 + 23k = 0$ or $k = -1.$

Putting in (1), we get $2x^2 - 4y^2 - z^2 + 10xy - 2yz + 5zx = 0$,

as the required cone.

Example 4:

Find the equation of the quadric cone which passes through the three coordinate axes and the three mutually perpendicular lines,

$$\frac{1}{2}x = y = -z,\ x = \frac{1}{3}y = \frac{1}{5} - z,\ \frac{1}{8}x = -\frac{1}{11}y = \frac{1}{5}z.$$

Solution:

Any cone passing through the three coordinate axes is

$$fyz + gzx + hxy = 0. \qquad ...(1)$$

This cone contains the given lines

$$\frac{x}{2} = \frac{y}{1} = \frac{z}{-1} \text{ and } \frac{x}{1} = \frac{y}{3} = \frac{z}{5}. \text{ Therefore,}$$

$$f(1)(-1) + g(-1)(2) + h.2.1 = 0, \Rightarrow f + 2g - 2h = 0,$$

and $$f.3.5 + g.5.1 + h.1.3 = 0. \quad 15f + 5g + 3h = 0.$$

Solving for f and g, we obtain

$$\frac{f}{6+10} = \frac{g}{-30-3} = \frac{h}{5-30} \text{ or } \frac{f}{16} = \frac{g}{-33} = \frac{h}{-25}.$$

Hence, by (1), the required cone is

$$16yz - 33zx - 25xy = 0.$$

It contains the third line $\frac{x}{8} = -\frac{y}{11} = \frac{z}{5}$, since

$$16(-11)(5) - 33(5)(8) - 25(8)(-11)$$
$$= -880 - 1320 + 2200 = 0.$$

Example 5:

Find the equation of the cone which contains the three coordinate axes and the two lines through the origin with direction cosines (l_1, m_1, n_1) and (l_2, m_2, n_2).

Solution:

The cone passing through the three coordinate axes is

$$fyz + gzx + hxy = 0. \qquad ...(i)$$

It contains the lines $\frac{x}{l_1} = \frac{y}{m_1} = \frac{z}{n_1}$ and $\frac{x}{l_2} = \frac{y}{m_2} = \frac{z}{n_2}$ if

$$fm_1n_1 + gn_1l_1 + hl_1m_1 = 0,$$

and $$fm_2n_2 + gn_2l_2 + hl_2m_2 = 0.$$

Solving these equations, we obtain

$$\frac{f}{l_1l_2(m_2n_1 - m_1n_2)} = \frac{g}{........} = \frac{h}{.........}.$$

Hence, from (1), the required cone is

$$\Sigma l_1 l_2 \, (m_2 n_1 - m_2 n_1) yz = 0.$$

Example 6:

Show that the angle between the lines given by $x + y + z = 0$, $ayz + bzx + cxy = 0$ is $\pi/2$ if $a + b + c = 0$ and $\pi/3$ if

$$a^{l} + b^{l} + c^{l} = 0.$$

Solution:

Let $\frac{x}{l} = \frac{y}{m} = \frac{z}{n}$ be cone of the lines in which the given plane cuts the given cone so that we have

$$l + m + n = 0 \text{ and } amn + bnl + clm = 0.$$

From the above equations, we obtain

$$-\,am(l + m) - bl\,(l + m) + clm = 0.$$

$$\Rightarrow \quad bl^2 + (a + b - c)\, lm + am^2 = 0.$$

$$\Rightarrow \quad b\,\frac{l^2}{m^2} + (a + b - c)\,\frac{l}{m} + a = 0$$

This is a quadratic equation in $\frac{l}{m}$. Let its two roots be $\frac{l_1}{m_1}$ and $\frac{l_2}{m_2}$. Therefore,

$$\frac{l_1 l_2}{m_1 m_2} = \frac{a}{b} \text{ and } \frac{l_1}{m_1} + \frac{l_2}{m_2} = -\frac{a + b - c}{b}. \qquad ...(1)$$

$$\Rightarrow \quad \frac{l_1 l_2}{a} = \frac{m_1 m_2}{b} = \frac{n_1 n_2}{c} \text{ (= k, say)} \qquad \text{[By symmetry]}$$

The angle between the two lines will be $\frac{\pi}{2}$ if

$$l_1 l_2 + m_1 m_2 + n_1 n_2 = 0 \Rightarrow k(a + b + c) = 0 \Rightarrow a + b + c = 0$$

From (1), $\frac{l_1 m_2 + l_2 m_1}{m_1 m_2} = \frac{c - a - b}{b} \Rightarrow l_1 m_1 + l_2 m_1 = k\,(c - a - b).$

Now $\quad (l_1 m_2 - l_2 m_1)^2 = (l_1 m_2 + l_2 m_1)^2 - 4 l_1 l_2 m_1 m_2$

$$= k^2[(c - a - b)^2 - 4ab].$$

We know that the angle between two lines is given by

$$\tan \theta = \frac{[\Sigma\,(l_1 m_2 - l_2 m_1)^2]^{1/2}}{\Sigma l_1 l_2}$$

Squaring both sides and taking $\theta = \frac{\pi}{3}$, we obtain

$$3 = \frac{\Sigma[(c-a-b)^2 - 4ab]}{(a+b+c)^2} = \frac{\Sigma(a^2+b^2+c^2-2ab-2bc-2ca)}{(a+b+c)^2}$$

$\Rightarrow$ $3(a + b + c)^2 = 3\ (a^2 + b^2 + c^2 - 2ab - 2bc - 2ca)$

$\Rightarrow$ $12(ab + bc + ca) = 0$. Dividing by 12abc, we obtain

$$\frac{1}{a}+\frac{1}{b}+\frac{1}{c}=0 \text{ or } a^{-1} + b^{-1} + c^{-1} = 0.$$

Example 7:

Find the angle between the lines given by $x + y + z = 0$ and the cone

$$\frac{yz}{q-r}+\frac{zx}{r-p}+\frac{xy}{p-q}=0.$$

Solution:

Let $a = \dfrac{1}{q-r},\ b = \dfrac{1}{p-r}$

and $c = \dfrac{1}{p-q}$.

The equation of the given cone becomes $ayz + bzx + cxy = 0$.

Now $\dfrac{1}{a}+\dfrac{1}{b}+\dfrac{1}{c} = q + r - p + p - q = 0.$

$\therefore$ $ab + bc + ca = 0$ ...(1)

The angle between the lines given by

$x + y + z = 0$ and $ayz + bzx + cxy = 0$ is

$$\tan^2\theta = \frac{\Sigma\left(a^2+b^2+c^2-2ab-2bc-2ca\right)}{\left(a+b+c\right)^2}$$

$$= \frac{\Sigma\left[\left(a^2+b^2+c^2\right)-2\left(ab+bc+ca\right)\right]}{\left(a^2+b^2+c^2\right)+2\left(ab+bc+ca\right)}$$

$$= \frac{3\left(a^2+b^2+c^2\right)}{\left(a+b^2+c^2\right)}, \text{ using (1)}$$

$\therefore$ $\tan\theta = \sqrt{3}$. Hence $\theta = \dfrac{\pi}{3}$ is the required angle.

Example 8:

Show that the plane $ax + by + cz = 0$ cuts the cone $yz + zx + xy = 0$ in two lines inclined at an angle

$$\tan^{-1}\left[\frac{\left[\left(a^2+b^2+c^2\right)\left(a^2+b^2+c^2-2bc-2ca-2ab\right)\right]^{1/2}}{bc+ca+ab}\right].$$

Solution:

Let $\frac{x}{l}=\frac{y}{m}=\frac{z}{n}$ be one of the lines in which the given plane cuts the given cone so that

$al + bm + cn = 0$ and $mn + nl + lm = 0$.

From these equations, we obtain

$$lm + (l + m)\left(-\frac{al+bm}{c}\right)=0 \text{ or } al^2 + (a + b - c)\, lm + bm^2 = 0$$

$$\therefore \qquad a\frac{l^2}{m^2} + (a + b - c)\frac{l}{m}+b=0.$$

Let the two roots of the above equation be $\frac{l_1}{m_1}$ and $\frac{l_2}{m_2}$.

Then $\quad \frac{l_1 l_2}{m_1 m_2}=\frac{b}{a}$ and $\frac{l_1}{m_1}+\frac{l_2}{m_2}=\frac{c-a-b}{a}$.

$$\Rightarrow \qquad \frac{l_1 l_2}{\frac{1}{a}}=\frac{m_1 m_2}{\frac{1}{b}}=\frac{n_1 n_2}{\frac{1}{c}}=k,\ l_1 m_2 + l_2 m_1 = k\frac{(c-a-b)}{ab}.$$

Now $(l_1 m_2 + l_2 m_1)^2 = (l_1 m_2 + l_2 m_1)^2 - 4l_1 l_2 m_1 m_2$

$$= \frac{k^2}{a^2b^2}(c - a - b)^2 - \frac{4k^2}{ab}=\frac{k^2}{a^2b^2}\left[(c - a - b)^2 - 4ab\right]$$

$$= \frac{k^2}{a^2b^2}(a^2 + b^2 + c^2 - 2ab - 2bc - 2ca) = \frac{k^2P}{a^2b^2},$$

where $P = a^2 + b^2 + c^2 - 2ab - 2bc - 2ca$.

Similarly $\Sigma\,(m_1 n_1 - m_2 n_1)^2 = \frac{k^2P}{b^2c^2},\ \Sigma\,(n_1 l_1 - n_2 l_1)^2 = \frac{k^2P}{c^2a^2}$.

Now $\quad \Sigma l_1 l_2 = k\left(\frac{1}{a}+\frac{1}{b}+\frac{1}{c}\right)=\frac{k}{abc}(bc + ca + ab)$.

The angle between the two lines is given by

$$\tan\theta = \frac{\left[\Sigma\,(l_1 m_2 - l_2 m_1)^2\right]^{1/2}}{\Sigma l_1 l_2}$$

$$= \frac{\left[k^2 P\left(\frac{1}{a^2b^2} + \frac{1}{b^2c^2} + \frac{1}{c^2a^2}\right)\right]^{1/2}}{\frac{k}{abc}(bc + ca + ab)} = \frac{\left[(c^2 + a^2 + b^2)\,P\right]^{1/2}}{bc + ca + ab}$$

Hence the required angle is

$$\tan^{-1}\left[\frac{\left\{\left(a^2 + b^2 + c^2\right)\left(a^2 + b^2 + c^2 - 2ab - 2bc - 2ca\right)\right\}^{1/2}}{bc + ca + ab}\right].$$

Example 9:

Find the equation of the cone generated by straight lines drawn from the origin to cut the circle through the three points (1, 0, 0), (0, 2, 0), (2, 1, 1) and prove that acute angle between the two lines in which the plane $x = 2y$ *cuts the cone is* $\cos^{-1}\sqrt{(5/14)}$.

Solution:

First we shall determine the equation of the plane of the circle through the given three points. Any plane through the point (2, 1, 1) is

$$a(x - 2) + b(y - 1) + c(z - 1) = 0. \qquad ...(1)$$

This plane passes through the points (1, 0, 0) and (0, 2, 0).

$\therefore \quad a + b + c = 0$ and $2a - b + c = 0$.

Solving these, we obtain $\frac{a}{2} = \frac{b}{1} = \frac{c}{-3} = k$.

Putting $a = 2k$, $b = k$ and $c = -3k$ in (1), we obtain

$$2x + y - 3z - 2 = 0. \qquad ...(2)$$

The equation of any sphere passing through the origin is

$$x^2 + y^2 + z^2 + 2ux + 2vy + 2wz = 0. \qquad ...(3)$$

It passes through the points (1, 0, 0), (0, 2, 0) and (2, 1, 1).

$\therefore \quad 1 + 2u = 0,\ 4 + 4v = 0$ and $6 + 4u + 2v + 2w = 0$.

$\Rightarrow \quad u = -1/2,\ v = -1$ and $w = -1$.

Substituting these values in (3), we obtain

$$x^2 + y^2 + z^2 - x - 2yz = 0. \qquad ...(4)$$

The equations (2) and (4) represent the given circle.

The equation of the cone with vertex at the origin and intersecting the given circle is obtained by making the equation (4) homogeneous with the help of (2). Hence the equation of the required cone is

$$(x^2 + y^2 + z^2 - (x + 2y + 2z)\left(\frac{2x+y-3z}{2}\right) = 0$$

$$\Rightarrow \quad 8z^2 - zx - 5xy + 4yz = 0. \quad ...(5)$$

Suppose now $x/l=y/m=z/n$ is one of the two lines in which the given plane $x = 2y$ cuts the cone (5).

$$\therefore \quad l = 2m \text{ and } 8n^2 - nl - 5lm + 5mn = 0.$$

Eliminating l between these equations, we get

$$5m^2 - mn - 4n^2 = 0 \text{ or } \frac{m^2}{n^2} - \frac{m}{n} - 4 = 0.$$

$$\therefore \quad \frac{m}{n} = 1, -\frac{4}{5}. \text{ Also } \frac{l}{m} = 2.$$

Now $\quad \frac{l}{m} = 2$ and $\frac{m}{n} = 1 \Rightarrow \frac{l_1}{2} = \frac{m_1}{1} = \frac{n_1}{1}$

and $\quad \frac{l}{m} = 2$ and $\frac{m}{n} = -\frac{4}{5} \Rightarrow \frac{l_2}{8} = \frac{m_2}{4} = \frac{n_2}{-5}$.

Now the angle between the two lines with direction ratios (2, 1, 1) and (8, 4, – 5) is given by

$$\cos\theta = \frac{l_1l_2 + m_1m_2 + n_1n_2}{\sqrt{l_1^2 + m_1^2 + n_1^2}\ \sqrt{l_2^2 + m_2^2 + n_2^2}} = \frac{2.8 + 1.4 - 1.5}{\sqrt{4+1+1}\ \sqrt{64+16+25}}$$

$$= \frac{15}{\sqrt{6 \times 105}} = \frac{15}{3\sqrt{70}} = \frac{5}{\sqrt{70}} = \sqrt{\frac{5}{14}}$$

Hence $q = \cos^{-1}\left(\sqrt{\frac{5}{14}}\right)$.

Example 10:

Show that the plane $3x + 2y - 4z = 0$ passes through a pair of common generators of the comes

$$27x^2 + 20y^2 - 32z^2 = 0 \text{ and } 2yz + zx - 4xy = 0.$$

Also show that the plane containing the other two generators is

$$9x + 10y + 8z = 0.$$

Solution:

The equation of the cone through the intersection of the given cones is

$$27x^2 + 20y^2 - 32z^2 + k(2yz + zx - 4xy) = 0, \text{ k being a parameter.}$$

This will represent a pair of planes, one of the which is

$$3x + 2y - 4z = 0,$$

if $$27x^2 + 20y^2 - 32z^2 + k(2yz + zx - 4xy)$$
$$= (3x + 2y - 4z)(ax + by + cz). \quad ...(1)$$

Comparing the coefficients of similar terms, we get

$$3a + 27,\ 2b = 20,\ -4c = -32,\ 2c - 4b = 2k,$$
$$3c - 4a = k \text{ and } 3b + 2a = -4k.$$
$$\Rightarrow \quad a = 9,\ b = 10,\ c = 8 \quad \text{and} \quad k = -12.$$

Hence the two planes through the common generators are

$$3x + 2y - 4z = 0 \text{ and } ax + by + cz = 9x + 10y + 8z = 0.$$

Example 11:

Find the conditions that the lines of section of the plane $lx + my + nz = 0$ and the cones $fyz + gzx + hxy = 0$,

$ax^2 + by^2 + cz^2 = 0$ should be coincident.

Solution:

The equation of the cone through the intersection of the given cones is

$$ax^2 + by^2 + cz^2 + k(fyz + gzx + hxy) = 0.$$

This will represent a pair of planes, one of which is

$$lx + my + nz = 0, \text{ if}$$

$$ax^2 + by^2 + cz^2 + k(fyz + gzx + hxy) = (lx + my + n)(l'x + m'y + n').$$

$$\Rightarrow \quad ll' = a,\ mm' = b,\ nn' = c,\ mn' + m'n = kf,\ nl' + n'l\ kg,$$

and $$lm' + l'm = kh.$$

$$\Rightarrow \quad l' = \frac{a}{l},\ m' = \frac{b}{m},\ n' = \frac{c}{n}$$

and $$kh = lm' + l'm = \frac{bl}{m} + \frac{am}{l} \Rightarrow k = \frac{am^2 + bl^2}{h/m}$$

$$\text{Similarly } kf = mn' + m'n \Rightarrow k = \frac{bn^2 + cm^2}{fmn},$$

and $\quad kg = nl' + n'l \Rightarrow k = \dfrac{cl^2 + an^2}{gnl}$.

Hence the required conditions are

$$\frac{am^2 + bl^2}{hlm} = \frac{bn^2 + cm^2}{fmn} = \frac{cl^2 + an^2}{gnl}.$$

Example 12:

The section of a cone whose vertex is P and guiding curve the ellipse $\frac{x^2}{a^2} + \frac{y^2}{b^2} = 1$, $z = 0$ by the plane $x = 0$ is a rectangular hyperbola. Show that the locus of P is $\frac{x^2}{a^2} + \frac{y^2 + z^2}{b^2} = 1$.

Solution:

The equation of the cone with vertex P (α, β, γ) and guiding curve $\frac{x^2}{a^2} + \frac{y^2}{b^2} = 1$, $z = 0$ is

$$\left(\frac{\alpha z - x\gamma}{a}\right)^2 + \left(\frac{\beta z - y\gamma}{b}\right)^2 = (z - \gamma)^2. \quad \text{...(i)}$$

The section of the cone (i) by the plane $x = 0$ is

$$\left(\frac{\alpha z}{a}\right)^2 + \left(\frac{\beta z - y\gamma}{b}\right)^2 = (z - \gamma)^2.$$

$$\Rightarrow \quad \left(\frac{\alpha^2}{a^2} + \frac{\beta^2}{b^2} - 1\right) z^2 + \left(\frac{\gamma^2}{\beta^2}\right) y^2 - \left(\frac{2\beta\gamma yz}{b^2} + \gamma^2 - 2\gamma z\right) = 0. \quad \text{...(ii)}$$

Since (ii) is given to be a rectangular hyperbola, therefore, the sum of the coefficients of y^2 and z^2 in (ii) is zero

i.e., $\quad \frac{\alpha^2}{a^2} + \frac{\beta^2}{b^2} - 1 + \frac{\gamma^2}{b^2} = 0$ or $\frac{\alpha^2}{a^2} + \frac{\beta^2 + \gamma^2}{b^2} = 1$.

Hence the locus of P (α, β, γ) is

$$\frac{x^2}{a^2} + \frac{y^2 + z^2}{b^2} = 1.$$

Example 13:

Find the equation of the cone with vertex (5, 4, 3) and with $3x^2 + 2y^2 = 6$, $y + z = 0$ as base.

Solution:

Any generator of cone through the vertex the (5, 4, 3) is

$$\frac{x-5}{l}=\frac{y-4}{m}=\frac{z-3}{n}. \qquad ...(1)$$

It meets the plane y + z = 0 *i.e.,* z = – y. From (1),

$$\frac{y-4}{m}=\frac{-m-3}{n} \Rightarrow y=\frac{4n-3m}{m+n}.$$

Putting this value of y in (x – 5)/l = (y – 4)/m, we obtain

$$x=\frac{5m+5n-7l}{m+n}.$$

Substituting the values of x and y in $3x^2 + 2y^2 = 6$, we get

$$3\,(5m + 5n - 7l)^2 + 2\,(4n - 3m)^2 = 6\,(m + n)^2. \qquad ...(2)$$

Eliminating l, m, n between (1) and (2), we get

$$3[5\,(y-4) + 5\,(z-3) - 7\,(x-5)]^2 + 2\,[4\,(z-3) - 3\,(y-4)]^2$$
$$= 6\,(y - 4 + z - 3)^2$$

$$\Rightarrow \qquad 3\,(-7x + 5y + 5z)^2 + 2\,(-3y + 4z)^2 = 6\,(y + z - 7)^2.$$

This is the required equation of the cone.

Example 14 (a):

Find the equation of the cone whose generators pass through the point (α, β, γ) *and have their direction cosines satisfying the relation* $al^2 + bm^2 + cn^2 = 0$.

Solution:

Any generator of the cone through (α, β, γ) is

$$\frac{x-\alpha}{l}=\frac{y-\beta}{m}=\frac{z-\gamma}{n} \qquad ...(1)$$

Eliminating *l*, m, n between (1) and the relation

$al^2 + bm^2 + cn^2 = 0$, we obtain

$$a\,(x-\alpha)^2 + b\,(y-\beta)^2 + c\,(z-\gamma)^2 = 0.$$

This is the required equation of the cone.

Example 14 (b):

Find the equation of the cone whose vertex is the point (1, 2, 3) and guiding curve the circle $x^2 + y^2 + z^2 = 4$, $x + y + z = 1$.

Solution:

Any generator of the cone through the vertex (1, 2, 3) is

$$\frac{x-1}{l}=\frac{y-2}{m}=\frac{z-3}{n}. \qquad ...(1)$$

It meets the plane $x + y + z = 1$

i.e., $\quad z = 1 - x - y. \qquad ...(2)$

$$\therefore \quad \frac{x-1}{l}=\frac{y-2}{m}=\frac{-(x+y+2)}{n}.$$

Now $\quad \dfrac{x-1}{l}=\dfrac{y-2}{m}$

$\Rightarrow \quad mx - ly + (2l - m) = 0,$

$$\frac{y-2}{m}=\frac{-(x+y+2)}{n}$$

$\Rightarrow \quad mx + (n + m)\, y + (2m - 2n) = 0.$

Solving these equations for x and y, we obtain

$$x=\frac{m-4l+n}{l+m+n},\ y=\frac{2l+2n-3m}{l+m+n}$$

Putting these values of x and y in (2), we get

$$z=\frac{3l+3m-2n}{l+m+n}.$$

Substituting these values of x, y, z in $x^2 + y^2 + z^2 = 4$, we obtain

$$5l^2 + 3m^2 + n^2 - 2lm - 6mn - 4nl = 0. \qquad ...(3)$$

Eliminating l, m, n between (1) and (3), we obtain

$$5(x-1)^2 + 3(y-2)^2 + (z-3)^2 - 2(x-1)(y-2) - 6(y-2)(z-3) - 4(x-1)(z-3) = 0.$$

Simplifying it, we obtain

$$5x^2\ 3y^2 + z^2 - 2xy - 6y_2 - 4zx + 6x + 8y + 10z = 26.$$

This is the required equation of the cone.

Example 15:

A cone has for its guiding curve the circle

$$x^2 + y^2 + 2ax + 2by = 0,\ z = 0$$

and passes through a fixed point (0, 0, c). If the section of the cone by the plane x = 0 is a rectangular hyperbola, prove that the vertex lies on the fixed circle $x^2 + y^2 + z^2 + 2ax + 2by = 0,\ 2ax + 2by + cz = 0$.

Solution:

Let P (α, β, γ) be the vertex of the cone. Any generator through this vertex is

$$\frac{x-\alpha}{l}=\frac{y-\beta}{m}=\frac{z-\gamma}{n}. \qquad ...(1)$$

(1) meets the plane z = 0 at $\left(\alpha-\frac{l}{n}\gamma, \beta-\frac{m}{n}\gamma, 0\right)$

This point lies on $x^2 + y^2 + 2ax + 2by = 0$.

$$\therefore \quad \left(\alpha-\frac{l}{n}\gamma\right)^2+\left(\beta-\frac{m}{n}\gamma\right)^2+2a\left(\alpha-\frac{l}{n}\gamma\right)$$

$$+\ 2b\left(\beta-\frac{m}{n}\gamma\right)=0. \qquad ...(2)$$

Eliminating *l*, m, n between (1) and (2), we obtain

$$\left(\alpha-\frac{x-\alpha}{z-\gamma}\gamma\right)^2+\left(\beta-\frac{y-\beta}{z-\gamma}\gamma\right)^2+2a\left(\alpha-\frac{x-\alpha}{z-\gamma}\gamma\right)+2b\left(\beta-\frac{y-\beta}{z-\gamma}\gamma\right)=0$$

$$\Rightarrow \quad (\alpha z - x\gamma)^2 + (\beta z - y\gamma)^2 + 2a(\alpha z - x\gamma)(z-\gamma)$$

$$+\ 2b(\beta z - y\gamma)(z-\gamma) = 0 \qquad ...(3)$$

The section of the cone (3) by the plane x = 0 is

$$\alpha^2 z^2 + (\beta z - y\gamma)^2 + 2a\alpha z(z-\gamma) + 2b(\beta z - y\gamma)(z-\gamma) = 0. \qquad ...(4)$$

Since (4) is a rectangular hyperbola, therefore, the sum of the coefficients of y^2 and z^2 in (4) is zero.

i.e., $$\alpha^2 + \beta^2 + \gamma^2 + 2a\alpha + 2b\beta = 0. \qquad ...(5)$$

It is given that the cone (3) passes through the point (0, 0, c).

$$\therefore \quad \alpha^2c^2 + \beta^2c^2 + 2a\alpha c(c-\gamma) + 2b\beta c(c-\gamma) = 0$$

$$\Rightarrow \quad c^2(\alpha^2 + \beta^2 + 2a\alpha + 2b\beta) - 2a\alpha c\gamma - 2b\beta c\gamma = 0$$

$$\Rightarrow \quad -c^2\gamma^2 - 2a\alpha c\gamma - 2b\beta c\gamma = 0, \text{ using (5)}$$

$$\Rightarrow \quad -c\gamma(2a\alpha + 2b\beta + c\gamma) = 0$$

$$\Rightarrow \quad 2a\alpha + 2b\beta + c\gamma = 0. \qquad (\because \gamma \neq 0) \qquad ...(6)$$

From (5) and (6), it follows that the vertex (α, β, γ) lies on the circle

$$x^2 + y^2 + z^2 + 2ax + 2by = 0,$$

$$2ax + 2by + cz = 0.$$

Example 16:

Planes through x-axis and y-axis include an angle a; show that locus of their lines of intersection is the cone

$$z^2 (x^2 + y^2 + z^2) = x^2y^2 \tan^2 \alpha.$$

Solution:

The equations of the x-axis are y = 0 and z = 0.

Any plane through the x-axis is $y = \lambda_1 z = 0$, λ_1 being a parameter.

Similarly $x + \lambda_2 z = 0$ is any plane through the y-axis.

The equations of the two planes may be rewritten as

$$0.x + 1.y + \lambda_1 z = 0 \text{ and } \quad 1.x + 0.y + \lambda_2 z = 0.$$

The angle 'α' between the two planes is given by

$$\cos \alpha = \frac{0.1 + 1.0 + \lambda_1\lambda_2}{\sqrt{0+1+\lambda_1^2}\sqrt{1+0+\lambda_2^2}} = \frac{\lambda_1\lambda_2}{\sqrt{1+\lambda_1^2+\lambda_2^2+\lambda_1^2\lambda_2^2}}$$

$$\text{Now } \tan^2 \alpha = \sec^2 a - 1 = \frac{1+\lambda_1^2+\lambda_2^2+\lambda_1^2\lambda_2^2}{\lambda_1^2\lambda_2^2} - 1$$

$$\therefore \qquad \tan^2 \alpha = \frac{1+\lambda_1^2+\lambda_2^2}{\lambda_1^2\lambda_2^2}. \qquad ...(1)$$

The given planes intersect when $\lambda_1 = -y/z$ and $\lambda_2 = -x/z$.

Substituting these values in (1), we obtain

$$\frac{x^2y^2}{z^4} \tan^2 \alpha = 1 + \frac{y^2}{z^2} + \frac{x^2}{z^2}.$$

Hence $x^2y^2 \tan^2 \alpha = z^2 (x^2 + y^2 + z^2)$ is the required locus.

Example 17:

Find the equation of the cone whose vertex is the point (1, 1, 0) and whose guiding curve is $x^2 + z^2 = 4$, $y = 0$.

Solution:

Any line through (1, 1, 0) is $(x - 1)/l = (y - 1)/m = z/n$ ...(i)

It meets the plane y = 0 at $\left(-\frac{l}{m}+1, 0, -\frac{n}{m}\right)$

and if this point lies on $x^2 + z^2 = 4$, $y = 0$, then we have

$$\left(-\frac{l}{m}+1\right)^2\left(-\frac{n}{m}\right)^2=4 \qquad \text{...(ii)}$$

Also from (i), we have $\frac{l}{m}=\frac{x-1}{y-1}$ and $\frac{n}{m}=\frac{z}{y-1}$ ...(iii)

Eliminating l, m, n between (i) and (ii) with the help of (iii), we obtain

$$\left[-\frac{x-1}{y-1}+1\right]^2+\left[-\frac{z}{y-1}\right]^2=4 \Rightarrow \left[\frac{-x+1+y-1}{y-1}\right]^2+\frac{z^2}{(y-1)^2}=4$$

$\Rightarrow \quad (y-x)^2+z^2=4(y-1)^2$

$\Rightarrow \quad x^2-3y^2+z^2-2xy+8y-4=0.$ **Ans.**

Example 18:

Find the equation of the cone whose vertex is the joying (α, β, γ) and whose guiding curve is

(i) $ax^2+by^2=1,\ z=0.$

(ii) $x^2/a^2+y^2/b^2=1,\ z=0.$

(iii) $ax^2+2hxy+by^2+2gx+2fy+c=0,\ z=0.$

Solution:

Any generator of the cone through the vertex (α, β, γ) is

$$\frac{x-\alpha}{l}=\frac{y-\beta}{m}=\frac{z-\gamma}{n}. \qquad \text{...(1)}$$

It meets the plane $z=0$, so

$$\frac{x-\alpha}{l}=\frac{-\gamma}{n} \Rightarrow x=\alpha-\frac{l}{n}\gamma.$$

Similarly, $\frac{y-\beta}{m}=-\frac{\gamma}{n} \Rightarrow y=\beta-\frac{m}{n}\gamma.$

Thus the generator (1) meets the plane $z=0$ in the point

$$P\left(\alpha-\frac{l}{n}\gamma, \beta-\frac{m}{n}\gamma, 0\right).$$

(i) The point P lies on the conic $ax^2+by^2=1$.

$$\therefore \quad a\left(\alpha-\frac{l}{n}\gamma\right)^2+b\left(\beta-\frac{m}{n}\gamma\right)^2=1. \qquad \text{...(2)}$$

Eliminating l, m, n between (1) and (2), we get

$$a\left(\alpha-\frac{x-\alpha}{z-\gamma}\gamma\right)^2+b\left(\beta-\frac{y-\beta}{z-\gamma}\gamma\right)^2=1$$

$$\Rightarrow \quad a(\alpha z - x\gamma)^2 + b(\beta z - y\gamma)^2 = (z-\gamma)^2,$$

which is the required equation of the cone.

(ii) The point P lies on the conic $x^2/a^2 + y^2/b^2 = 1$.

$$\therefore \quad \frac{1}{a^2}\left(\alpha-\frac{l}{n}\gamma\right)^2+\frac{1}{b^2}\left(\beta-\frac{m}{n}\gamma\right)^2=1 \qquad ...(3)$$

Eliminating l, m, n between (1) and (3), we get

$$\frac{1}{a^2}\left(\alpha-\frac{x-\alpha}{z-\gamma}\gamma\right)^2+\frac{1}{b^2}\left(\beta-\frac{y-\beta}{z-\gamma}\gamma\right)^2=1$$

$$\Rightarrow \quad \left(\frac{\alpha z-x\gamma}{a}\right)^2+\left(\frac{\beta z-y\gamma}{b}\right)^2=(z-\gamma)^2,$$

which is the required equation of the cone.

(iii) The point P lies on $ax^2 + 2hxy + by^2 + 2gx + 2fy + c = 0$.

$$\therefore \quad a\left(\alpha-\frac{l}{n}\gamma\right)^2+2h\left(\alpha-\frac{l}{n}\gamma\right)\left(\beta-\frac{m}{n}\gamma\right)$$

$$+b\left(\beta-\frac{m}{n}\gamma\right)^2+2g\left(\alpha-\frac{l}{n}\gamma\right)+2f\left(\beta-\frac{m}{n}\gamma\right)+c=0. \qquad ...(4)$$

Eliminating l, m, n between (1) and (4), we get

$$a\left(\alpha-\frac{x-\alpha}{z-\gamma}.\gamma\right)^2+2h\left(\alpha-\frac{x-\alpha}{z-\gamma}.\gamma\right)\left(\beta-\frac{y-\beta}{z-\gamma}\gamma\right)$$

$$+b\left(\beta-\frac{y-\beta}{z-\gamma}\gamma\right)^2+2g\left(\alpha-\frac{x-\alpha}{z-\gamma}.\gamma\right)+2f\left(\beta-\frac{y-\beta}{z-\gamma}.\gamma\right)+c=0.$$

Hence $a(\alpha z - x\gamma)^2 + 2h(\alpha z - x\gamma)(\beta z - y\gamma) + b(\beta z - y\gamma)^2$
$+ 2g(\alpha z - x\gamma)(z-\gamma) + 2f(\beta z - y\gamma)(z-\gamma) + c(z-\gamma)^2 = 0$
is the required equation of the cone.

Example 19:

Find the equation of the cone whose vertex is the point (1, 1, 0) and whose curve is $y = 0, x^2 + z^2 = 4$.

Solution:

Any generator through the vertex (1, 1, 0) is

$$\frac{x-1}{l}=\frac{y-1}{m}=\frac{z-0}{n}. \qquad ...(1)$$

It meets the plane y = 0, so

$$\frac{x-1}{l}=-\frac{1}{m} \text{ and } \frac{z}{n}=-\frac{1}{m}.$$

$$\Rightarrow \quad x = 1-\frac{l}{m},\ z=-\frac{n}{m}.$$

Thus (1) meets the plane y = 0 in the point P $\left(1-\frac{l}{m}, 0, -\frac{n}{m}\right)$.

The point P lies on the conic $x^2 + z^2 = 4$. Therefore

$$\left(1-\frac{l}{m}\right)^2+\left(-\frac{n}{m}\right)^2=4. \qquad ...(2)$$

Eliminating *l*, m, n between (1) and (2), we obtain

$$\left(1-\frac{x-1}{y-1}\right)^2+\left(-\frac{z}{y-1}\right)^2=4$$

$$\Rightarrow \quad (y-x)^2 + z^2 = 4(y-1)^2.$$

Hence $x^2 - 3y^2 + z^2 - 2xy + 8y - 4 = 0$ is the required cone.

EQUATION OF THE CONE WITH VERTEX AT THE ORIGIN

In this section we shall show that a *homogeneous equation in x, y, z of second degree of the form*

$$ax^2 + by^2 + cz^2 + 2fyz + 2gzx + 2hxy = 0 \qquad ...(1)$$

represents a cone with vertex at the origin called ***Quadric Cone***.

Any equation f(x, y, z) = 0 is called *homogeneous* if

f(rx, ry, rz) = 0 for all real numbers r.

We may see that in equation (1),

$$f(rx, ry, rz) = r^2(ax^2 + by^2 + cz^2 + 2fyz + 2gzx + 2hxy) = 0$$

and so (1) is homogeneous.

Let f(x, y, z) = 0 be the equation of a cone with vertex at the origin O (0, 0, 0). Let P (α, β, γ) be any point on the cone so that $f(\alpha, \beta, \gamma) = 0$. The direction ratios of the line OP are $\alpha - 0, \beta - 0, \gamma - 0$ *i.e.*, α, β, γ. Thus the equations of the line OP (through the origin) are

$$\frac{x}{\alpha} = \frac{y}{\beta} = \frac{z}{\gamma} = r.$$

For any real number r, $(r\alpha, r\beta, r\gamma)$ is a point on the line OP. Since the line OP lies wholly on the cone $f(x, y, z) = 0$, so

$$f(r\alpha, r\beta, r\gamma) = 0 \text{ for all real numbers } r.$$

and so $f(x, y, z) = 0$ is homogeneous.

Hence a homogeneous equation of second degree

$$f(x, y, z) \equiv ax^2 + by^2 + cz^2 + 2fyz + 2gzx + 2hxy = 0$$

represents a cone with its vertex at the origin.

Corollary 1 : *If $C_1 = 0$ and $C_2 = 0$ be the equations of two cones with the origin as the common vertex, then $C_1 + kC_2 = 0$, k being a parameter, represents a cone with its vertex at the origin and passing through the common generators of the given cones.*

Corollary 2 : *If the homogeneous equation $f(x, y, z) = 0$ represents a cone, one of whose generators is*

$$\frac{x}{l} = \frac{y}{m} = \frac{z}{n}, \text{ then } f(l, m, n) = 0.$$

Corollary 3 : *The general equation of a cone which passes through the three axes is*

$$fyz + gzx + hxy = 0.$$

Since the x-axis is a generator of the required cone, so by Corollart 1 its direction cosines 1, 0, 0 satisfy the equation

$$ax^2 + by^2 + cz^2 + 2fyz + 2gzx + 2hxy = 0. \qquad ...(1)$$

$\therefore$ $a = 0$. Similarly, $b = 0$ and $c = 0$.

Substituting $a = b = c = 0$ in (1), the required cone is

$$fyz + gzx + hxy = 0.$$

Example 1:

Find the equations to the cones with vertex at the origin and which pass through the curves given by the equations :

(i) $x^2 + y^2 = 4,\ z = 2.$

(ii) $ax^2 + by^2 + cz^2 + 1,\ \ lx + my + nz = p.$

(iii) $ax^2 + by^2 = 2z,\ lx + my + nz = p.$

(iv) $ax^2 + by^2 + cz^2 = 1,\ \alpha x^2 + \beta y^2 = 2z.$

(v) $x^2 + y^2 + z^2 + 2ax + b = 0,\ lx + my + nz = p.$

(vi) $x^2 + y^2 + z^2 + x - 2y + 3z = 4,\ x^2 + y^2 + z^2 + 2x - 3y + 4z = 5.$

Solution:

Since the equation of a cone with its vertex at the origin is a second degree homogeneous equation, we shall obtain this equation by making the first expression of the given curve homogeneous with the help of the second expression

(i) We have $x^2 + y^2 = 4 = 2^2 = z^2$ ($\therefore$ z = 2)

Hence $x^2 + y^2 - z^2 = 0$ is the required equation of the cone.

(ii) Consider $lx + my + nz = p \Rightarrow \frac{1}{p}(lx + my + nz) = 1$. Now

$$ax^2 + by^2 + cz^2 = 1 = 1^2 = \frac{1}{p^2}(lx + my + nz)^2$$

$$\Rightarrow \quad p^2 (ax^2 + by^2 + cz^2) - (lx + my + nz)^2 = 0.$$

$(ap^3 - l^2) x^2 + (bp^2 - m^2) y^2 + (cp^2 - n^2) z^2 - 2lmxy - 2mnyz - 2nlzx = 0$ is the required equation of the cone.

(iii) Consider $ax^2 + by^2 = 2z \Rightarrow p\ ((ax^2 + by^2) = 2zp$

Hence $p\ (ax^2 + by^2) = 2z\ (lx + my + nz)$ is the required cone.

(iv) Consider $ax^2 + by^2 + cz^2 = 1 \Rightarrow (ax^2 + by^2 + cz^2)\ (2z)^2 = (2z)^2$

Hence $4z^2(ax^2 + by^2 + cz^2) = (\alpha x^2 + \beta y^2)^2$ is the required cone.

(v) We have $\frac{1}{p}(lx + my + nz) = 1$. Now

$$x^2 + y^2 + z^2 + 2ax + b = 0$$

$$\Rightarrow \quad x^2 + y^2 + z^2 + 2ax.1 + b.1^2 = 0$$

$$\therefore (x^2 + y^2 + z^2) + 2a\frac{x}{p}(lx + my + nz) + \frac{b}{p^2}(lx + my + nz)^2 = 0$$

Hence $(p^2 + 2alp + bl^2)\ x^2 + (p^2 + 2amp + bm^2)y^2 + (p^2 + 2anp + bn^2)z^2 + 2blmxy + 2bmnyz + 2blzx = 0,$

which is the required equation of the cone.

(vi) Subtracting the first equation from the second equation, we get

$$x - y + z = 1.$$

Now $\quad x^2 + y^2 + z^2 + x - 2y + 3z = 4$

$$\Rightarrow \quad x^2 + y^2 + z^2 + (x - 2y + 3z).1 = 4.l^2$$

$\Rightarrow \quad (x^2 + y^2 + z^2) + (x - 2y + 3z)(x - y + z) = 4(x - y + z)^2.$

Hence $\quad 2x^2 + y^2 - 5xy - 3yz + 4zx = 0$ is the required cone.

Example 2:

Show that the equation of the cone whose vertex the origin and whose base is the circle through the three points (a, 0, 0), (0, b, 0), (0, 0, c) is $\Sigma a(b^2 + c^2) yz = 0.$

Solution:

The plane of the circle through the three given points is

$$\frac{x}{a}+\frac{y}{b}+\frac{z}{c}=1. \qquad ...(1)$$

Now the equation of the sphere through 0, A, B and C is

$$x^2 + y^2 + z^2 - (ax + by + cz) = 0. \qquad ...(2)$$

The equations (1) and (2) represent the circle ABC.

The required cone obtained from (1) and (2) is

$$x^2 + y^2 + z^2 - \left(\frac{x}{a}+\frac{y}{b}+\frac{z}{c}\right)(ax + by + cz) = 0$$

$$\Rightarrow \quad -\frac{x}{a}(by + cz) - \frac{y}{b}(ax + cz) - \frac{z}{c}(ax + by) = 0$$

$$\Rightarrow \quad yz\left(\frac{c}{b}+\frac{b}{c}\right)+zx\left(\frac{c}{a}+\frac{a}{c}\right)+xy\left(\frac{a}{b}+\frac{b}{a}\right)=0.$$

Hence $a(b^2 + c^2) yz + b(c^2 + a^2) zx + c(a^2 + b^2)xy = 0.$

Example 3:

Find the equation of the cone with vertex at origin and be the circle $x = a, y^2 + z^2 = b^2$. *Show that the section of the cone by a plane parallel to the plane XOY is a hyperbola.*

Solution:

The equation of the required cone is given by making $y^2 + z^2 = b^2$ homogeneous with the help of $x = a$ or $\frac{x}{a}=1$.

We have $y^2 + z^2 = b^2.1 = b^2\left(\frac{x}{a}\right)^2$

$$\Rightarrow \quad a^2(y^2 + z^2) = b^2x^2.$$

The section of this cone by a plane parallel to the XOY plane viz., z = k, is

$$a^2 (y^2 + k^2) = b^2x^2 \text{ or } b^2 x^2 - a^2y^2 = a^2K^2$$

$$\Rightarrow \quad \frac{x^2}{a^2k^2 / b^2} - \frac{y^2}{k^2} = 1, \text{ which is a hyperbola.}$$

Example 4:

Show $ax^2 + by^2 + cz^2 + 2ux + 2vy + 2wz + d=0$ represents a cone if

$$\frac{u^2}{a} + \frac{v^2}{b} + \frac{w^2}{c} = d.$$

Solution:

Let P (α, β, γ) be the vertex of the given cone. Shifting the origin at P by means of the transformations

$$x = X + \alpha,\ y = Y + \beta,\ z = Z + \gamma,$$

the given equation becomes

$$a(X + \alpha)^2 + b(Y + \beta)^2 + c(Z + \gamma)^2 + 2u(X + \alpha) + 2v(Y + \beta) + 2w(Z + \gamma) + d = 0$$

$$\Rightarrow \quad aX^2 + bY^2 + cZ^2 + 2X(c\alpha + u) + 2Y(b\beta + v) + 2z(c\gamma + w) + (a\alpha^2 + b\beta^2 + c\gamma^2 + 2u\alpha + 2v\beta + 2w\gamma + d) = 0.$$

This equation must be homogeneous, being the equation of the cone with its vertex at the origin.

Thus $\quad ax + u = 0,\ b\beta + v = 0,\ c\gamma + w = 0 \quad$ ...(1)

and $\quad a\alpha^2 + b\beta^2 + c\gamma^2 + 2u\alpha + 2v\beta + 2w\gamma + d = 0.$

$$\Rightarrow \quad \alpha(a\alpha + u) + \beta(b\beta + v) + \gamma(c\gamma + w) + u\alpha + v\beta + w\gamma + d = 0.$$

$$\Rightarrow \quad u\alpha + v\beta + w\gamma + d = 0, \text{ using (1)}$$

Hence $\dfrac{u^2}{a} + \dfrac{v^2}{b} + \dfrac{w^2}{c} = d$

(∴ by (1) $\alpha = -u/a,\ \beta = -v/b,\ \gamma = -w/c$)

Example 5:

Find the equation of the cone which passes through the three coordinate axes and the two lines

$$\frac{x}{1} = \frac{y}{-2} = \frac{z}{3},\ \frac{x}{3} = \frac{y}{-1} = \frac{z}{1}$$

Solution:

The equation of the cone through the axes is

$$fyz + gzx + hxy = 0. \quad ...(1)$$

Since (1) contains the given lines, therefore

$$f(-2)(3) + g(3)(1) + h(1)(-2) = 0 \Rightarrow 6f - 3g + 2h = 0, \quad ...(2)$$

and $f(-1)(1) + g(1)(3) + h(3)(-1) = 0 \Rightarrow f - 3g + 3h = 0.$ ...(3)

Solving (2) and (3) for f and g, we obtain

$$\frac{f}{-9-(-6)} = \frac{g}{2-18} = \frac{h}{-18+3} \quad (= k, \text{ say})$$

Substituting $f = -3k$, $g = -16k$, $h = -15k$ in (1), we get

$3yz + 16zx + 15xy = 0$, which is the required cone.

Example 6:

Find the equations to the lines in which the plane

$2x + y - z = 0$ cuts the cone $4x^2 - y^2 + 3z^2 = 0$.

Solution:

Let $$\frac{x}{l} = \frac{y}{m} = \frac{z}{n} \quad ...(1)$$

be one of the two lines in which the given place cuts the given cone so that we have

$$4l^2 - m^2 + 3n^2 - 0, \quad ...(2)$$

and $$2l + m + n \quad ...(3)$$

[$\therefore$ line is $\perp$ to the normal to the plane]

From (2) and (3), $4l^2 - m^2 + 3(2l + m)^2 = 0$

$$\Rightarrow \quad 8l^2 + 6lm + m^2 = 0 \text{ or } \frac{8l^2}{m^2} = \frac{6l}{m} + 1 = 0.$$

$$\therefore \quad \frac{l}{m} = \frac{-6 \pm \sqrt{36-32}}{16} = -\frac{1}{4} \text{ or } -\frac{1}{2}.$$

From (3), $\dfrac{2l}{m} + 1 = \dfrac{n}{m}$.

Now $$\frac{l}{m} = -\frac{1}{4} \Rightarrow \frac{n}{m} = \frac{1}{2}$$

and so $$\frac{l}{-1} = \frac{n}{4} = \frac{n}{2},$$

and $\quad \frac{l}{m} = -\frac{1}{2} \Rightarrow \frac{n}{m} = 0$

and so $\quad \frac{l}{-1} = \frac{m}{2} = \frac{n}{0}$.

Hence, by (1), the two required lines are

$$\frac{x}{-1} = \frac{y}{4} = \frac{z}{2} \text{ and } \frac{x}{-1} = \frac{y}{2} = \frac{z}{0}.$$

Example 7:

Find the angle between the lines of intersection of $2x + y - z = 0$ and $4x^2 - y^2 + 3z^2 = 0$.

Solution:

The two lines of intersection are

$$\frac{x}{-1} = \frac{y}{4} = \frac{z}{2} \text{ and } \frac{x}{-1} = \frac{y}{2} = \frac{z}{0}.$$

The angle between these two lines is

$$\cos\theta = \frac{(-1)(-1) + (4)(2) + (2)(0)}{\sqrt{(-1)^2 + 4^2 + 2^2}\sqrt{(-1)^2 + 2^2 + 0^2}}$$

$$= \frac{9}{\sqrt{21}\sqrt{5}} = \frac{9}{\sqrt{105}} \text{ Hence } \theta = \cos^{-1}\left(\frac{9}{\sqrt{105}}\right) \text{ is the required angle.}$$

Example 8:

Find the equations of the lines in which the pla ıe $x + 7y - 5z = 0$ cuts the cone $3yz + 14x - 30xy = 0$.

Solution:

$$\text{Let } \frac{x}{l} = \frac{y}{m} = \frac{z}{n} \qquad ...(1)$$

be one of the two lines in which the given plane cuts the given cone so that we have

$$l + 7m - 5m = 0 \text{ or } l = 5n - 7m, \qquad ...(2)$$

and $$3nm + 14nl - 30lm = 0. \qquad ...(3)$$

From (2) and (3), we obtain

$$3nm + 14n(5n - 7m) - 30m(5n - 7m) = 0$$

$$\Rightarrow \qquad 6m^2 - 7mn + 2n^2 = 0$$

$$\Rightarrow \qquad \frac{6m^2}{n^2} + \frac{7m}{n} + 2 = 0.$$

$$\therefore \qquad \frac{m}{n} = \frac{7 \pm \sqrt{49-48}}{12} = \frac{2}{3}, \frac{1}{2}.$$

From (2), $\frac{l}{n} = 5 - 7\frac{m}{n}$

Now $\quad \frac{m}{n} = \frac{2}{3} \Rightarrow \frac{l}{n} = \frac{1}{3}$ and so $\frac{l}{1} = \frac{m}{2} = \frac{n}{3}$,

and $\quad \frac{m}{n} = \frac{1}{2} \Rightarrow \frac{l}{n} = \frac{3}{2}$ and so $\frac{l}{n} = \frac{m}{1} = \frac{n}{2}$.

Hence, by (1), the required lines are

$$\frac{x}{1} = \frac{y}{2} = \frac{z}{3} \quad \text{and} \quad \frac{x}{3} = \frac{y}{1} = \frac{z}{2}.$$

ANGLE BETWEEN THE LINES IN WHICH A PLANE CUTS A CONE

Let the plane be $\qquad ux + vy + wz = 0 \qquad$ ...(1)

and the cone be $f(x, y, z) \equiv ax^2 + by^2 + cz^2$

$$+ 2fyz + 2gzx + 2hxy = 0 \qquad \text{...(ii)}$$

The vertex of this cone is (0, 0, 0) which evidently lies on (i). Let l, m, n be the d.c.'s of the line through the vertex in which the plane (i) cuts the cone (ii), then its equations are given as $\frac{x}{l} = \frac{y}{m} = \frac{z}{n}$ ...(iii)

As (ii) is generator of the cone (ii), so its d.c.'s will satisfy (ii) and so we have $al^2 + bm^2 + cn^2 + 2fmn + 2gnl + 2hlm = 0$ ...(iv)

Also the line (iii) lies on plane (i), hence we have

$$ul + vm + wn = 0$$

Eliminating n between (iv) and (v), we obtain

$$al^2 + bm^2 + c\left(-\frac{ul + vm}{w}\right)^2 + 2hlm + 2(gl + fm)\left(-\frac{ul + vm}{w}\right) = 0$$

$$\Rightarrow \qquad l^2(aw^2 + cu^2 - 2guw) + 2lm(cuv - fuw - gvw + hw^2)$$

$$+ m^2(bw^2 + cv^2 - 2fvw) = 0$$

$$\Rightarrow (l^2/m^2)\ (aw^2 + cu^2 - 2guw) + 2\ (l/m)\ (cuv - fuw - gvw + hw^2)$$
$$+ (bw^2 + cv^2 - 2fvw) = 0 \qquad ...(v)$$

This equation being quadratic in (l/m) indicates theat the plane (i) cuts the cone (ii) in two lines and if their d.c.'s are l_1, m_1, n_1, then.

$$\frac{l_1 l_2}{m_1 m_2} = \text{product of roots of (vi)} = \frac{bw^2 + cv^2 - 2fvw}{aw^2 + cu^2 - 2guw}$$

$$\therefore \frac{l_1 l_2}{bw^2 + cv^2 - 2fvm} = \frac{m_1 m_2}{cu^2 + aw^2 - 2guw} = \frac{n_1 n_2}{av^2 + bu^2 - 2hvw} \qquad ...(vii)$$

$= \lambda$ (say), by symmetry third fraction is written

Also from (vi), $\dfrac{l_1}{m_1} + \dfrac{l_2}{m_2}$ = sum of roots of equation (vi) is

$$\Rightarrow \qquad \frac{l_1 m_2 + l_2 m_1}{m_1 m_2} = \frac{-2\ (cuv - fuw - gvw - hw^2)}{aw^2 + cu^2 - 2guw}$$

$$\Rightarrow \quad \frac{l_1 m_2 + l_2 m_1}{-2\ (cuv - fuw - gvw + hw^2)} = \frac{m_1 m_2}{cu^2 + aw^2 - 2guw} = \lambda, \text{ from (vi)}$$
$$...(viii)$$

$$\therefore\ (l_1 m_2 - l_2 m_1)^2 = (l_1 m_2 + m_1 l_2)^2 - 4 l_1 l_2\ m_1 m_2$$
$$= 4\lambda^2\ (cuv - fuw - gvw + hw^2)^2 - 4l^2\ (bw^2 + cv^2 - 2\ fvw)$$
$$\times\ (cu^2 + aw^2 - 2guw), \text{ from (vii) and (viii)}$$
$$= \quad 4w^2\lambda^2\ [-\ (Au^2 + Bv^2 + Cw^2 + 2Fvm + 2Gwu + 2Huv)]$$
$$= \quad 4w^2 l^2\ P^2,$$

where the capital letters A, B, C, etc. are the cofactors of the corresponding small letters a, b, c, etc. in the determinant

$$\begin{vmatrix} a & h & g \\ h & b & f \\ g & f & c \end{vmatrix} \text{ and } P^2 = \begin{vmatrix} a & h & g & u \\ h & b & f & v \\ g & f & c & w \\ u & v & w & 0 \end{vmatrix}$$

$$\therefore \qquad l_1 m_2 - l_2 m_1 = 2w\lambda P \qquad ...(i)$$

Again if θ be the angle between the lines of intersection then we have

$' + m_1 m_2 + n_1 n_2.$

$^{?} + cv^2 - 2fvw)$, from (vii)

$+ (c + a)\ v^2 + (a + b)\ w^2 - 2fvw - 2gwu - 2huv]$

$` + v^2 + w^2).$

CONDITION FOR THE CONE TO HAVE THREE MUTUALLY PERPENDICULAR GENERATORS

Let the equation of the cone be given as

$$f(x, y, z) \equiv ax^2 + by^2 + cz^2 + 2fyz + 2gzx + 2hxy = 0. \qquad ...(i)$$

Let $x/l = y/m = z/n$ be one of its generators, then its direction cosines will satisfy (i) and therefore we get

$$f(l, m, n) \equiv al^2 + bm^2 + cn^2 + 2fmn + 2gnl + 2hlm = 0. \qquad ...(ii)$$

Now the plane through the vertex (0, 0, 0) and perpendicular to this generator $x/l = y/n = z/n$ is $\quad lx + my + nz = 0. \qquad ...(iii)$

The plane (iii) cuts the cone (i) in two (or generators) which are at right angle to each other if

$$(a + b + c)(l^2 + m^2 + n^2) - f(l, m, n) = 0$$

$$\Rightarrow \quad (a + b + c)(l^2 + m^2 + n^2) = 0, \text{ from (ii)}$$

$$\Rightarrow \quad a + b + c = 0, \qquad \because \quad l^2 + m^2 + n^2 \neq 0.$$

This condition $a + b + c = 0$ is independent of l, m, n and therefore we conclude that the plane through vertex and perpendicular to any generator cuts the cone in two generators. Hence the cone has a set of three mutually perpendicular generators and as the condition $a + b + c = 0$ is independent of l, m, n so there can be an infinite number of set of three mutually perpendicular of the cone.

Example 1:

Show that the cone whose vertex is the origin and which passes through the curve of intersection of the sphere $x^2 + y^2 + z^2 = 3a^2$ and any plane at a distance a, from the origin has three mutually perpendicular generators.

Solution:

Any plane at a distance 'a' from the origin is

$$lx + my + nz = a. \qquad ...(1)$$

(Hence l, m, n are the direction cosines of the normal to the plane). The equation of the cone whose vertex is the origin and which passes through the intersection of the given sphere and the plane (1) is

$$x^2 + y^2 + z^2 = 3a^2 = 3(lx + my + nz)^2$$

$$\Rightarrow \quad (1 - 3l^2)x^2 + (1 - 3m^2)y^2 + (1 - 3m^2)z^2 - 6lmxy - 6mnyz$$

$$- 6nlxy = 0. \qquad ...(2)$$

Here $\quad a + b + c = (1 - 3l^2) + (1 - 3m^2) + (1 - 3n^2)$

$$= 3 - 3\,(l^2 + m^2 + n^2) = 3 - 3,\ 1 = 0.$$

Hence the cone (2) has three mutually perpendicular generators.

Example 2:

Find the locus of a point from which three mutually perpendicular lines can be drawn to intersect the central conic

$$ax^2 + by^2 = 1,\ z = 0.$$

Solution:

Let P(α, β, γ) be a point from which three mutually perpendicular lines can be drawn to intersect the given conic. The equation of the cone with vertex P(α, β, γ) and intersecting conic. $ax^2 + by^2 = 1$, $z = 0$ is

$$a(\alpha x - \gamma x)^2 + b\,(\beta z - \gamma y)^2 - (z - \gamma)^2 = 0.$$

It will have three mutually perpendicular lines if the sum of the coefficients of x^2, y^2 and z^2 is zero.

$\therefore$ $a\alpha^2 + a\gamma^2 + b\beta^2 + b\gamma^2 - 1 = 0$ or $a(\alpha^2 + \gamma^2) + b\,(\beta^2 + \gamma^2) = 1.$

Hence, the locus of P (α, β, γ) is $a(x^2 + z^2) + b\,(y^2 + z^2) = 1.$

Example 3:

Show that three mutually perpendicular tangent lines can be drawn to the sphere $x^2 + y^2 + z^2 = r^2$ from any point on the sphere $2(x^2 + y^2 + z^2) = 3r^2$.

Solution:

Let P (α, β, γ) be any point on the sphere

$$2(x^2 + y^2 + z^2) = 3r^2.$$

$$\therefore \quad 2(\alpha^2 + \beta^2 + \gamma^2) = 3r^2. \quad ...(1)$$

The equation of the enveloping cone drawn from the point P (α, β, γ) to the sphere $x^2 + y^2 + z^2 = r^2$ is $SS_1 = T^2$.

$$\Rightarrow \quad (x^2 + y^2 + z^2 - r^2)\,(\alpha^2 + \beta^2 + \gamma^2 - r^2) = (\alpha x + \beta y + \gamma z - r^2)^2.$$

This cone will have three mutually perpendicular tangent lines if

$$a + b + c$$

Now $\quad a + b + c = (\beta^2 + \gamma^2 - r^2) + (\alpha^2 + \gamma^2 - r^2) + (\alpha^2 + \beta^2 - \gamma^2)$

$$= 2(\alpha^2 + \beta^2 + \gamma^2) - 3r^2 = 0, \text{ by } (1).$$

Example 4:

Show that the cone whose vertex is at the origin and which passes through the curve of intersection of the sphere $x^2 + y^2 + z^2 = 3a^2$ and any plane distance 'a' from the origin has three mutually perpendicular generators.

Solution:

Any plane at a distance *a* from the origin is given as

$$lx + my + nz = a, \qquad \text{...(i)}$$

where l, m, n are the actual d.c.'s of its normal.

Also the given sphere is

$$x^2 + y^2 + z^2 = 3a^2 \qquad \text{...(ii)}$$

$\therefore$ The equation of the cone with vertex at (0, 0, 0) and the curve of intersection of (i) and (ii) as the guiding curve is given as

$$x^2 + y^2 + z^2 = 3\,(lx + my + nz)^2, \qquad \text{...(iii)}$$

making (ii) homogeneous with the help of (i).

If the cone given by (iii) has three mutually perpendicular generators then $\quad a + b + c = 0$

i.e., $\quad (3l^2 - 1) + (3m^2 - 1) + (3n^2 - 1) = 0$

$\Rightarrow \quad 2\,(l^2 + m^2 + n^2) - 3 = 0,$

which is true as $l^2 + m^2 + n^2 = 1$.

Example 5:

If $\dfrac{x}{1} = \dfrac{y}{2} = \dfrac{z}{3}$ *represents one of a set of three mutually perpendicular generators of cone* $5yz - 8zx - 3xy = 0$*, find the equations of the other two.*

Solution:

Since the sum of the coefficients of x^2, y^2 and z^2 in the given cone is zero, so the cone has three mutually perpendicular generators.

The equation of the plane through the origin and perpendicular to the given line $\dfrac{x}{1} = \dfrac{y}{2} = \dfrac{z}{3}$ is

$$1(x - 0) + 2\,(y - 0) + 3\,(z - 0) = 0 \text{ or } x + 2y + 3z = 0. \qquad \text{...(1)}$$

Let $\dfrac{x}{l} = \dfrac{y}{m} = \dfrac{z}{n}$ be one of the two lines in which the plane (1) cuts the cone $5yz - 8zx - 3xy = 0$.

$\therefore \quad l + 2m + 3n = 0,\ 5mn - 8nl - 3lm = 0$

From these equations, we obtain

$$5mn + 8n\,(2m + 3n) + 3m\,(2m + 3n) = 0$$

$$\Rightarrow \qquad m^2 + 5mn + 4n^2 = 0 \text{ or } \frac{m^2}{n^2} + 5\frac{m}{n} + 4 = 0$$

$$\Rightarrow \qquad \left(\frac{m}{n}+4\right)\left(\frac{m}{n}+1\right) = 0. \qquad \therefore \frac{m}{n} = -1 \text{ or } -4.$$

Now $\quad l + 2m + 3n = 0 \Rightarrow \dfrac{l}{n} = -3 - 2\dfrac{m}{n}.$

If $\quad \dfrac{m}{n} = -1$, then $\dfrac{l}{n} = -1$ and so $\dfrac{l}{n} = \dfrac{m}{1} = \dfrac{n}{-1}.$

If $\quad \dfrac{m}{n} = -4$, then $\dfrac{l}{n} = 5$ and so $\dfrac{l}{5} = \dfrac{m}{1} = \dfrac{n}{1}.$

Hence the other two perpendicular generators are

$$\frac{x}{1} = \frac{y}{1} = \frac{z}{-1}; \ \frac{x}{5} = \frac{y}{-4} = \frac{z}{1}.$$

Example 6:

Prove that the plane $lx + my + n = 0$ cuts the cone

$$(b - c)\, x^2 + (c - a)\, y^2 + (a - b)\, z^2 + 2fyz + 2gzx + 2hxy = 0$$

in perpendicular lines if

$$(b - c)\, l^2 + (c - a)\, m^2 + (a - b)\, n^2 + 2fm + 2gnl + 2hlm = 0.$$

Solution:

The sum of the coefficients of x^2, y^2 and z^2 in the equation of the given cone

$$= (b - c) + (c - a) + (a - b) = 0.$$

Thus the given cone has three mutually perpendicular generators. Thus given plane will cut the given cone is perpendicular lines if the normal to the plane through the vertex O viz.

$$\frac{x}{l} = \frac{y}{m} = \frac{z}{n} \text{ is a generator of the given cone.}$$

Hence $(b - c)\, l^2 + (c - a)\, m^2 + (a - b)\, n^2 + 2fmn + 2gnl + 2hlm = 0.$

Example 7:

Show that the two straight lines represented by the equations $ax + by + cz = 0$ and $yz + zx + xy = 0$ will be perpendicular if $\dfrac{1}{a} + \dfrac{1}{b} + \dfrac{1}{c} = 0.$

Solution:

The sum of the coefficients of x^2, y^2, z^2 in the equation of the given cone $yz + zx + xy = 0$ is zero. Thus the given plane will cut the given come in perpendicular lines if the normal to the plane through the vertex O viz,

$$\frac{x}{a} = \frac{y}{b} = \frac{z}{c} \text{ is a generator of the given cone.}$$

Hence $bc + ca + ab = 0$ or $\frac{1}{a} + \frac{1}{b} + \frac{1}{c} = 0$.

TANGENT LINES AND TANGENT PLANE

Let the equation of the cone by given as

$$f(x, y, z) \equiv ax^2 + by^2 + cz^2 + 2fyz + 2gzx + 2hxy = 0 \quad \text{...(i)}$$

and that of a line through any point P (α, β, γ) be

$$\frac{x-\alpha}{l} = \frac{y-\beta}{m} = \frac{z-\gamma}{n}. \quad \text{...(ii)}$$

Any point on (ii) is $(\alpha + lr, \beta + mr, \gamma + nr)$. The point in which the line (ii) meets the cone (i) are determined as

$$a(\alpha + lr)^2 + b(\beta + mr)^2 + c(\gamma + nr)^2 + 2f(\beta + mr)(\gamma + nr) + 2g(\gamma + nr)(\alpha + lr) + 2h(\alpha + lr)(\beta + mr) = 0$$

$$\Rightarrow \quad r^2 [al^2 + bm^2 + cn^2 + 2fmn + 2gnl + 2hlm] + 2r[(a\alpha + h\beta + g\gamma) l + (h\alpha + b\beta + f\gamma) m + (g\alpha + f\beta + c\gamma) n] + f(\alpha, \beta, \gamma) = 0 \quad \text{...(iii)}$$

This being a quadratic in r, the line (ii) in general meets the cone (i) in two points. If P (α, β, γ) lies on (i), then $f(\alpha, \beta, \gamma) = 0$ and consequently from (ii) one value of r is zero. Further if

$$l(a\alpha + h\beta + g\gamma) + m(h\alpha + b\beta + f\gamma) + n(g\alpha + f\beta + c\gamma) = 0 \quad \text{....(iv)}$$

then the other root of (iii) is also zero *i.e.*, the points of intersection coincide at P.

∴ In this case the line is a tangent to the cone at P. Hence (iv) is the condition for the line (ii) to be a tangent line at P (α, β, γ) to the cone (i).

Again the locus of all such tangent lines at P *i.e.*, the tangent plane to the cone at P is obtained by eliminating l, m, n between (iv) and (ii). Thus we have

$$(x - \alpha)(a\alpha + h\beta + g\gamma) + (y - \beta)(h\alpha + b\beta + f\gamma) + (z - \gamma) + (g\alpha + f\beta + c\gamma) = 0$$

$\Rightarrow$ $x(a\alpha + h\beta + g\gamma) + y(h\alpha + b\beta + f\gamma) + z(g\alpha + f\beta + c\gamma) = 0,$

since $f(\alpha, \beta, \gamma) = 0$

$\Rightarrow$ $a\alpha x + b\beta y + c\gamma z + f(\gamma y + \beta z) + g(\alpha z + \gamma x) + h(\beta x + \alpha y) = 0,$

which is the equation of the tangent plane to the cone at P (α, β, γ).

Corollary : This plane passes through the vertex O (0, 0, 0) of the cone and thus the generator OP lies on this cone. Also we can show that the equation of the tangent plane at (ka, kb, kg) is the same as that of this plane and hence this plane is the tangent plane to the surface at every point of this generator OP and is called the generator of contact.

All tangent planes of the cone pass through the vertex and it is called a singular point of the surface.

CONDITION OF TANGENCY

To find the condition that the plane $ux + vy + wz = 0$ ...(i)

be a tangent plane to the cone

$$f(x, y, z) \equiv ax^2 + by^2 + cz^2 + 2fyz + 2gzx + 2hxy = 0 \quad ...(ii)$$

Let the given plane be a tangent plane to the cone at the point P (α, β, γ), say. The equation of the tangent plane to (ii) at P (α, β, γ) is given by

$$a\alpha x + b\beta y + c\gamma z + f(y\gamma + z\beta) + g(z\alpha + x\gamma) + h(x\beta + y\alpha) = 0$$

$$\Rightarrow (a\alpha + h\beta + g\gamma)x + (h\alpha + b\beta + f\gamma)y + (g\alpha + f\beta + c\gamma)z = 0 \quad ...(iii)$$

If (i) touches (ii) at P (α, β, γ), then (i) and (iii) represent the same plane and hence comparing the coefficients in (i) and (iii) we get

$$\frac{a\alpha + h\beta + g\gamma}{u} = \frac{h\alpha + b\beta + f\gamma}{v} = \frac{g\alpha + f\beta + c\gamma}{w} = k \text{ (say)}$$

$$\therefore a\alpha + h\beta + g\gamma - uk = 0 \quad ...(iv)$$

$$h\alpha + b\beta + f\gamma - vk = 0 \quad ...(v)$$

$$g\alpha + f\beta + c\gamma - wk = 0 \quad ...(vi)$$

Also P (α, β, γ) lies on (i), so we have $u\alpha + v\beta + w\gamma = 0$ ...(vii)

Eliminating $\alpha, \beta, \gamma, -k$ between (iv), (v), (vi) and (vii) we obtain

$$\begin{vmatrix} a & h & g & u \\ h & b & f & v \\ g & f & c & w \\ u & v & w & 0 \end{vmatrix} = 0 \quad ...(viii)$$

$\Rightarrow$ $Au^2 + By^2 + Cw^2 + 2Fvw + 2Gwu + 2Huv = 0,$ ...(ix) on expanding the determinant and where capital letters denote the cofactors of the corresponding small letters in the determinant

$$D = \begin{vmatrix} a & h & g \\ h & b & f \\ g & f & g \end{vmatrix}$$

i.e., $A = bc - f^2$, $B = ca - g^2$, $C = ab - h^2$, $F = gh - af$, $G = hf - bg$

and $\quad H = fg - ch.$

Hence (viii) or (ix) is the condition that the plane (i) touches the cone (ii).

RECIPROCAL CONE

Definition : *The locus of the lines through the vertex, at right angles to the tangent planes of the given cone is called the reciprocal cone of the given cone.*

Let $\quad ux + vy + wz = 0 \qquad$ ...(i)

be a tangent plane to the cone

$$f(x, y, z) \equiv ax^2 + by^2 + cz^2 + 2fyz + 2gzx + 2hxy = 0 \quad \text{...(ii)}$$

then we have

$$Au^2 + Bv^2 + Cw^2 + 2Fvw + 2Gwu + 2Huv = 0 \qquad \text{...(iii)}$$

Now the equation of the normal to the plane (i) passing through the vertex (0, 0, 0) of the cone (ii) are given by

$$\frac{x}{u} = \frac{y}{v} = \frac{z}{w} \qquad \text{...(iv)}$$

∴ The equation of the locus of (iv) is obtained by eliminating u, v and w between (iii) and (iv) and is given as

$$Ax^2 + By^2 + Cz^2 + 2Fyz + 2Gzx + 2Hxy = 0, \qquad \text{...(v)}$$

which is again a cone with its vertex at the origin.

The cone whose equation is given by (v) is know as the reciprocal cone of the given by (ii).

Working Rule : To write down the equation of the reciprocal cone of the cone $ax^2 + by^2 + cz^2 + 2fyz + 2gzx + 2hxy = 0$, we should simply replace the small letters a, b, c etc. by the capital letters A, B, C etc.; where capital letters denote the cofactors of the corresponding small letters in the determinant

$$D = \begin{vmatrix} a & h & g \\ h & b & f \\ g & f & c \end{vmatrix}$$

To find the reciprocal cone of $Ax^2 + By^2 + Cz^2 + 2Fyz + 2Gzx + 2Hxy = 0$, *where A, B, C etc. are the cofactors of the corresponding small letters in the determinant*

$$D = \begin{vmatrix} a & h & g \\ h & b & f \\ g & f & c \end{vmatrix}$$

Let A', B', C' etc. be the cofactors of A, B, C etc. in the determinant

$$\begin{vmatrix} A & H & G \\ H & B & F \\ G & F & C \end{vmatrix}$$

Then $A' = Bc - F^2 = (ca - g^2)(ab - h^2) - (gh - af)^2$

$= a^2bc - abg^2 - cah^2 + g^2h^2 - g^2h^2 - a^2f^2 + 2fgah$

$= a(abc + 2fgh - af^2 - bg^2 - ch^2) = aD$

Similarly we can prove that $B' = bD$, $C' = cD$, $F' = fD$, $G' = gD$ and $H' = hD$.

Now by the working rule given above the equation of the reciprocal cone of $Ax^2 + By^2 + Cz^2 + 2Fyz + 2Gzx + 2Hxy = 0$ is

$$A'x^2 + B'y^2 + C'z^2 + 2F'yz + 2G'zx + 2H'xy = 0, \qquad ...(i)$$

where A', B', C' are etc. are the cofactors of A, B, C etc. in the determinant

$$\begin{vmatrix} A & H & G \\ H & B & F \\ G & F & C \end{vmatrix}$$

and from above we know $A' = aD$, $B' = bD$ etc.

∴ From (i), the required reciprocal cone is given as

$(aD)x^2 + (bD)y^2 + (cD)z^2 + 2(fD)yz + 2(gD)xz + 2(hD)xy = 0$

$\Rightarrow \qquad ax^2 + by^2 + cz^2 + 2fyz + 2gzx + 2hxy = 0, D \neq 0 \qquad ...(ii)$

Hence we observe that the cones

$$ax^2 + by^2 + cz^2 + 2fyz + 2gzx + 2hxy = 0$$

and $\qquad Ax^2 + By^2 + Cz^2 + 2Fyz + 2Gzx + 2Hxy = 0$ are such that each is the locus of the normals drawn through the vertex (the origin) to the tangent planes to the other and due to this property they are known as reciprocal cones.

Corollary : The condition for the cone $ax^2 + by^2 + cz^2 + 2fyz + 2gzx + 2hxy = 0$ to have three mutually perpendicular tangent planes is the same as that of the reciprocal cone $Ax^2 + By^2 + Cz^2 + 2Fyz + 2Gzx + 2Hxy = 0$ to have three mutually perpendicular generators and that is given by

$$A + B + C = 0$$

i.e., $(bc - f^2) + (ca - g^2) + (ab - h^2) = 0$

i.e., $bc + ca + ab = f^2 + g^2 + h^2.$

Example 1:

Prove that the equation $\sqrt{(fx)} \pm \sqrt{(gy)} \pm \sqrt{(hz)} = 0$ *represents a cone which touches the co-ordinate planes.*

And that the equation of the reciprocal cone is $fyz + gzx + hxy = 0$

Solution:

The given equation can be written as $\sqrt{(fx)} \pm \sqrt{(gy)} \pm \sqrt{(hz)}$

$\Rightarrow$ $fx + by \pm 2\sqrt{(fgxy)} = hz$, squaring both sides

$\Rightarrow$ $(fx + gy - hz)^2 = \left[\pm 2\sqrt{(fgxy)}\right]^2 = 4\ fgxy$

$\Rightarrow$ $f^2x^2 + g^2y^2 + h^2z^2 - 2ghyz - 2hfzx - 2fgxy = 0.$ ...(i)

The equation is a homogeneous equation of second degree hence it represents a quadratic cone.

The co-ordinate plane $x = 0$ meets (i) where

$$g^2y^2 + h^2z^2 - 2ghyz = 0 \quad \text{or } (gy - hz)^2 = 0$$

which being a perfect square it follows that the plane $x = 0$ touches it. Similarly we can show that $y = 0$, $z = 0$ also touch the cone (i).

Again for the cone (i), we have

$$'a' = f^2,\ 'b' = g^2,\ 'c' = h^2,\ 'f' = -gh,\ 'g' = -hf,\ 'h' = -fg$$

$\therefore$ $A = bc - f^2 = g^2h^2 - (-gh)^2 = 0$, Similarly $B = 0 = C$

$$F = gh - af = (-hf) - (fg) - f^2(-gh) = 2f^2 gh.$$

Similarly, we have $G = 2g^2hf$, $H = 2h^2fg$.

$\therefore$ The required equation of the cone reciprocal to (i) is

$$Ax^2 + By^2 + Cz^2 + 2Fyz + 2Gzx + 2Hxy = 0$$

$\Rightarrow$ $2f^2ghz + 2g^2hfzx + 2h^2fgxy = 0$

$\Rightarrow$ $fyz + gzx + hxy = 0.$ **Hence proved.**

ENVELOPING CONE

Definition : *The locus of tangent lines drawn from a given point to a given surface is known as the enveloping cone or the tangent cone of that surface with the given point as its vertex.*

(a) To find the equation of the enveloping cone of the sphere $x^2 + y^2 + z^2 = a^2$ with vertex at the point (α, β, γ).

The equations of a line through (α, β, γ) are given as

$$\frac{x-\alpha}{l} = \frac{y-\beta}{m} = \frac{z-\gamma}{n} = r \text{ (say)} \qquad \text{...(i)}$$

Any point on its line is given by $(\alpha + lr, \beta + mr, \gamma + nr)$. ...(ii)

If the line (i) meets the given sphere at a distance r from point (α, β, γ) then the point given by (ii) must lie on the given sphere and so we have

$$(\alpha + lr)^2 + (\beta + mr)^2 + (\gamma + nr)^2 = a^2$$

$$\Rightarrow \quad r^2(l^2 + m^2 + n^2) + 2r(l\alpha + m\beta + n\gamma) + (\alpha^2 + \beta^2 + \gamma^2 - a^2) = 0 \text{ ...(iii)}$$

If the line (i) is a tangent to the given sphere, then the line (i) should meet the sphere in two coincident points, the condition for the same is that the roots of (iii) are equal, *i.e.*, "$B^2 = 4AC$" Thus, we have

i.e., $\quad 4(l\alpha + m\beta + n\gamma)^2 = 4(l^2 + m^2 + n^2)(\alpha^2 + \beta^2 + \gamma^2 - a^2)$ ...(iv)

∴ The locus of the line (i) which is tangent to the given sphere is obtained by eliminating l, m, n between (i) and (iv) and is given by

$$[\alpha(x - \alpha) + \beta(y - \beta) + \gamma(z - \gamma)]^2$$
$$= [(x - \alpha)^2 + (y - \beta)^2 + (z - \gamma)^2](\alpha^2 + \beta^2 + \gamma^2) \qquad \text{...(v)}$$

If $S \equiv x^2 + y^2 + z^2 - a^2$, $S_1 = \alpha^2 + \beta^2 + \gamma^2 - a^2$ and

$T = \alpha x + \beta y + \gamma z - a^2$, then the equation (v) can be written as

$$(T - S_1)^2 = (S + Sr - 2T)S_1 \qquad \textbf{(Note)}$$

$$\Rightarrow \quad T^2 - 2TS_1 + S_1^2 = SS_1 + S_1^2 - 2T_1$$

$$\Rightarrow \quad SS_1 = T^2$$

$$\Rightarrow \quad (x^2 + y^2 + z^2 - a^2)(\alpha^2 + \beta^2 + \gamma^2 - a^2) = (\alpha x + \beta y + \gamma z - a^2)^2$$

(b) To find the equation of the enveloping cone for the coincoid $ax^2 + by^2 + cz^2 = 1$ from the point (α, β, γ).

The equation of a line through the point (α, β, γ) is given as

$$\frac{x-\alpha}{l} = \frac{y-\beta}{m} = \frac{z-\gamma}{n} = r \text{ (say)} \qquad \text{...(i)}$$

Any point on this line is $(\alpha + lr, \beta + mr, \gamma + nr)$ ...(ii)

If the line (i) meets the given conicoid $ax^2 + by^2 + cz^2 = 1$ at a distance r from point (α, β, γ), then the point given by (ii) must lie on the given conicoid and so we get $a(\alpha + lr)^2 + b(\beta + mr)^2 + c(\gamma + nr)^2 = 1$

$$\Rightarrow \quad r^2(al^2 + bm^2 + cn^2) + 2r(al\alpha + bm\beta + cn\gamma) + (a\alpha^2 + b\beta^2 + c\gamma^2 - 1) = 0 \quad \text{...(iii)}$$

If the line (i) is a tangent to the given conicoid, then the line (i) should meet the conicoid in two coincident points, the condition for the same is that the roots of (iii) are equal,

i.e., Hence $B^2 = 4AC$

i.e., $4(al\alpha + bm\beta + cn\gamma)^2 = 4(al^2 + bm^2 + cn^2)(a\alpha^2 + b\beta^2 + c\gamma^2 - 1)$ ...(iv)

$\therefore$ The locus of the line (i) which is tangent to the given conicoid is obtained by eliminating l, m, n between (i) and (iv) and is

$$[a(x - \alpha)\alpha + b(y - \beta)\beta + c(z - \gamma)\gamma]^2$$
$$= [a(x - \alpha)^2 + b(y - \beta)^2 + c(z - \gamma)^2](a\alpha^2 + b\beta^2 + c\gamma^2 - 1) \quad \text{...(v)}$$

If $S \equiv ax^2 + by^2 + cz^2 - 1$; $S_1 = a\alpha^2 + b\beta^2 + c\gamma^2 - 1$ and

$T = a\alpha x + b\beta g + c\gamma z - 1$, then the equation (v) can be written as

$$(T - S_1)^2 = (S + S_1 - 2T)S_1 \quad \textbf{(Note)}$$

$$\Rightarrow \quad T^2 - 2TS_1 + S_1^2 + SS_1 + S_1^2 - 2TS_1 \text{ or } SS_1 = T^2 \quad \text{...(ii)}$$

$$\Rightarrow \quad (ax^2 + by^2 + cz^2 - 1)(a\alpha^2 + b\beta^2 + c\gamma^2 - 1)$$
$$= (a\alpha x + b\beta y + c\gamma z - 1)^2.$$

Remark:

If we expand the above determinant, we obtain

$$Al^2 + Bm^2 + Cn^2 + 2Fmn + 2Gnl + 2Hlm = 0,$$

where $A = bc - f^2$, $B = ca - g^2$,

$C = ab - h^2$,

$F = gh - af$, $G = hf - bg$,

$H = fg - ch$.

Notice that A, B, C, F, G, H are the cofactors of a, b, c, f, g, h respectively in the determinant

$$D = \begin{vmatrix} a & h & g \\ h & b & f \\ g & f & c \end{vmatrix}.$$

Hence the plane $lx + my + nz = 0$ *is a tangent plane to the cone*

$$ax^2 + by^2 + cz^2 + 2fyz + 2gzx + 2hxy = 0 \text{ if}$$

$$Al^2 + Bm^2 + Cn^2 + 2Fmn + 2Gnl + 2Hlm = 0.$$

RECIPROCAL CONES

Definition : *Two cones are called reciprocal cones if each is the locus of the normals drawn through the vertex to the tangent planes of the other.*

Theorem:

The reciprocal cone of the cone

$$ax^2 + by^2 + cz^2 + 2fyz + 2gzx - 2hxy = 0 \text{ is}$$

$$Ax^2 + By^2 + Cz^2 + 2Fyz + 2Gzx + 2Hxy = 0,$$

where A, B, C, F, G, H are the cofactors of a, b, c, f, g, h respectively in the determinant Δ.

Proof:

Let $lx + my + nz = 0$ be a tangent plane to the cone

$$ax^2 + by^2 + cz^2 + 2fyz + 2gzx + 2hxy = 0. \qquad ...(1)$$

$$\therefore \quad Al^2 + Bm^2 + Cn^2 + 2Fmn + 2Gnl + 2Hlm = 0. \qquad ...(2)$$

Any normal through the vertex (0, 0, 0) to the plane

$$lx + my + nz = 0 \text{ is } x/l = y/m = z/n. \qquad ...(3)$$

Eliminating l, m, n between (2) and (3), we obtain

$$Ax^2 + By^2 + Cz^2 + 2Fyz + 2Gzx + 2Hxy = 0. \qquad ...(4)$$

This is the required reciprocal cone of the given cone (1).

Remark: It can be similarly shown that the reciprocal cone of the cone (4) is the cone (1).

Example 1:

Show that the cone

$$ax^2 + by^2 + cz^2 + 2fyz + 2gzx + 2hxy = 0$$

has three mutually perpendicular tangent planes if

$$bc + ca + ab = f^2 + g^2 + h^2$$

Solution:

The given cone has three mutually perpendicular tangent planes if its reciprocal cone

$$Ax^2 + By^2 + Cz^2 + 2Fyz + 2Gzx + 2Hxy = 0$$

has three mutually perpendicular generators.

This requires $A + B + C \Rightarrow (bc - f^2) + (ca - g^2) + (ab - h^2) = 0.$

Hence $\quad bc + ca + ab = f^2 + g^2 + h^2.$

Example 2:

Prove that the equation $\sqrt{fx} \pm \sqrt{gy} \pm \sqrt{hz} = 0$ *represents a cone which touches the coordinate planes, and that the equation of the reciprocal cone is* $fyz + gzx + hxy = 0$.

Solution:

The cone which touches the coordinate planes is

$$f^2x^2 + g^2y^2 + h^2z^2 - 2ghyz - 2hfzx - 2fgxy = 0 \qquad ...(1)$$

$$\Rightarrow \qquad \sqrt{fx} \pm \sqrt{gy} \pm \sqrt{hz} = 0.$$

Now we shall find the reciprocal cone of (1).

We have $\Delta = \begin{vmatrix} f^2 & -fg & -hf \\ -fg & g^2 & -gh \\ -hf & -gh & h^2 \end{vmatrix}$

$\Rightarrow \qquad A = B = C = 0,$

$F = 2f^2gh,\ G = 2g^2hf,$

and $\quad H = 2h^2fg.$

Hence the reciprocal cone of (1) is

$$Ax^2 + By^2 + Cz^2 + 2Fyz + 2Gzx + 2Hxy = 0$$

$\Rightarrow \qquad 4f^2ghyz + 4g^2hfzx + 4h^2fgxy = 0,$

$\Rightarrow \qquad fyz + gzx + hxy = 0,$ (cancel out 4 fgh).

Example 3:

Show that the general equation of a conic which touches the three coordinate plane is

$$\sqrt{fx} \pm \sqrt{gy} \pm \sqrt{hz} = 0.$$

Solution:

The general equation of cone through the three coordinate axes is

$$fyz + gzx + hxy = 0.$$

The general equation of a cone which touches the three coordinate planes is

$$f^2x^2 + g^2y^2 + h^2z^2 - 2ghyz - 2hfzx - 2fgxy = 0.$$

(Take a = f, b = g, c = h).

$$\Rightarrow \quad (fx + gy - hz)^2 = 4fgxy$$

$$\Rightarrow \quad (fx + gy - hz) = \pm 2\sqrt{fgxy}$$

$$\Rightarrow \quad fx + gy \pm 2\sqrt{fgxy} = hz$$

$$\Rightarrow \quad \left(\sqrt{fx} \pm \sqrt{gy}\right)^2 = hz$$

$$\Rightarrow \quad \sqrt{fx} \pm \sqrt{gy} = \pm\sqrt{hz}$$

Hence $\sqrt{fx} \pm \sqrt{gy} \pm \sqrt{hz} = 0$.

Example 4:

Show that the locus of the lines of intersection of tangent planes to the cone $ax^2 + by^2 + cz^2 = 0$, which touch along perpendicular generator is the cone

$$a^2(b + c)x^2 + b^2(c + a)y^2 + c^2(a^2 + b^2)z^2 = 0.$$

Solution:

Let the tangent planes along two perpendicular generators of the cone intersect in the line

$$\frac{x}{l} = \frac{y}{m} = \frac{z}{n} (= r). \qquad ...(i)$$

Any point on this line is P(lr, mr, nr). The equation of the pair of tangent planes which pass through the given line is $SS_1 = T^2$ *i.e.*, $(ax^2 + by^2 + cz^2)(al^2r^2 + bm^2r^2 + cn^2r^2) = (alrx + bmry + cnrz)^2$.

Take $\alpha = lr$, $\beta = mr$, $\gamma = nr$

i.e., $(ax^2 + by^2 + cz^2)(al^2 + bm^2 + cn^2) = (ux + vy + wz)^2$,

where $u = al$, $v = bm$, $w = cn$. ...(ii)

The equation of the plane which passes through the two generators is

$$ux + vy + wz = 0. \qquad ...(iii)$$

The plane (iii) cuts the cone $ax^2 + by^2 + cz^2 = 0$ in perpendicular generators if

$$(b + c)u^2 + (c + a)v^2 + (a + b)w^2 = 0$$

$$\Rightarrow \qquad a^2(b^2 + c)l^2 + b^2(c + a)m^2 + c^2(a + b)n^2 = 0, \text{ [(using (2)] } \quad ...(iv)$$

Eliminating l, m, n between (1) and (4), we obtain

$$a^2(b + c)x^2 + b^2(c + a)y^2 + c^2(a + b)z^2 = 0.$$

Example 5:

Show that the planes which cut the cone $ax^2 + by^2 + cz^2 = 0$ in perpendicular generators touch the cone

$$\frac{x^2}{b+c} + \frac{y^2}{c+a} + \frac{z^2}{a+b} = 0.$$

Solution:

Any plane $lx + my + nz = 0$ cuts the cone $ax^2 + by^2 + cz^2 = 0$ in perpendicular generators if

$$(b + c)l^2 + (c + a)m^2 + (a + b)n^2 = 0. \qquad ...(1)$$

Now the plane $lx + my + nz = 0$ touches the cone

$$\frac{x^2}{b+c} + \frac{y^2}{c+a} + \frac{z^2}{a+b} = 0, \text{ if}$$

$$\begin{vmatrix} \frac{1}{b+c} & 0 & 0 & l \\ 0 & \frac{1}{c+a} & 0 & m \\ 0 & 0 & \frac{1}{a+b} & n \\ l & m & n & 0 \end{vmatrix} e = 0.$$

$$\text{L.H.S.} = \frac{1}{(b+c)} \begin{vmatrix} \frac{1}{c+a} & 0 & m \\ 0 & \frac{1}{a+b} & n \\ m & n & 0 \end{vmatrix} - l \begin{vmatrix} 0 & \frac{1}{c+a} & 0 \\ 0 & 0 & \frac{1}{a+b} \\ l & m & n \end{vmatrix}$$

$$= \frac{1}{(b+c)} \left[\frac{-n^2}{c+a} - \frac{m^2}{a+b} \right] - \frac{l^2}{(c+a)(a+b)}$$

$$= -\frac{1}{(a+b)(b+c)(c+a)} [(b + c)l^2 + (c + a)m^2 + (a + b)n^2]$$

$= 0$, using (1). **Hence Proved.**

Example 6:

Show that the plane $lx + my + nz = 0$ cuts the cone $ax^2 + by^2 + cz^2 = 0$ in perpendicular generators if

$$(b + c)\, l^2 + (c + a)m^2 + (a + b)n^2 = 0.$$

Solution:

Let $\frac{x}{\alpha} = \frac{y}{\beta} = \frac{z}{\gamma}$ be one of the two lines in which the given plane cuts the given cone, Then

$$l\alpha + m\beta + n\gamma = 0 \text{ and } a\alpha^2 + b\beta^2 + c\gamma^2 = 0.$$

From these equations, we get

$$a\alpha^2 + b\beta^2 + \frac{c}{n^2}(l\alpha + m\beta)^2 = 0$$

$$\Rightarrow \quad \alpha^2(an^2 + cl^2) + 2clm\alpha\beta + (bn^2 + cm^2)\beta^2 = 0.$$

On dividing the above equation by β^2, the equation becomes quadratic in $\frac{\alpha}{\beta}$. If the two roots are $\frac{\alpha_1}{\beta_1}$ and $\frac{\alpha_2}{\beta_2}$, then

$$\frac{\alpha_1\alpha_2}{\beta_1\beta_2} = \frac{bn^2 + cm^2}{an^2 + cl^2}$$

$$\Rightarrow \quad \frac{\alpha_1\alpha_2}{bn^2 + cm^2} = \frac{\beta_1\beta_2}{cl^2 + an^2} = \frac{\gamma_1\gamma_2}{am^2 + bl^2} \quad \text{(by symmetry)}$$

The two lines will be perpendicular if $\alpha_1\alpha_2 + \beta_1\beta_2 + \gamma_1\gamma_2 = 0$,

$$\Rightarrow \quad (bn^2 + cm^2) + (cl^2 + an^2) + (am^2 + bl^2) = 0$$

Hence $(b + c)\, l^2 + (c + a)m^2 + (a + b)n^2 = 0$.

Example 7:

Show that $x/l=y/m=z/n$ is the line of intersection of the tangent planes to the cone

$$ax^2 + by^2 + cz^2 + 2fyz + 2gzx + 2hxy = 0$$

along the lines in which it is cut by the plane

$$x(al + hm + gn) + y\,(hl + bm + fn) + z\,(gl + fm + cn) = 0.$$

Solution:

The tangent plane at any point (α, β, γ) of the given cone is

$$x(a\alpha + h\beta + g\gamma) + y\,(h\alpha + b\beta + f\gamma) + z\,(g\alpha + f\beta + c\gamma) = 0.$$

It will contain the given line $x/l = y/m = z/n$ if

$l(a\alpha + h\beta + g\gamma) + m(h\alpha + b\beta + f\gamma) + n(g\alpha + f\beta + c\gamma) = 0.$

$\Rightarrow \quad \alpha(al + hm + gn) + \beta(hl + bm + fn) + \gamma(gl + fm + cn) = 0.$

Hence the point (a, b, g) lies on the plane

$$x(al + hm + gn) + y(hl + bm + fn) + z(gl + fm + cn) = 0.$$

Example 8:

Show that the locus of the line of intersection of perpendicular tangent planes to the cone $ax^2 + by^2 + cz^2 = 0$ is the cone

$$a(b + c)x^2 + b(c + a)y^2 + c(a + b)z^2 = 0.$$

Solution:

Let the *tangent planes intersect in the line*

$$\frac{x}{l} = \frac{y}{m} = \frac{z}{n}. \quad ...(i)$$

The equation of the pair of tangent planes which pass through the line (1) is $SS_1 = T^2$.

$$\therefore \quad (ax^2 + by^2 + cz^2)(al^2 + bm^2 + cn^2) = (alx + bmy + cnz)^2.$$

These two planes are perpendicular if the sum of the coefficients of x^2, y^2 and z^2 is zero.

Thus $(a^2l^2 + abm^2 + acn^2 - a^2l^2) + (abl^2 + b^2m^2 + bcn^2 - b^2m^2)$

$$+ (cal^2 + cbm^2 + c^2n^2 - c^2n^2) = 0$$

$$\Rightarrow \quad (ab + ac)l^2 + (bc + ba)m^2 + (ca + cb)n^2 = 0$$

$$\Rightarrow \quad a(b + c)l^2 + b(c + a)m^2 + c(a + b)n^2 = 0. \quad ...(ii)$$

Eliminating l, m, n between (1) and (2), we obtain

$$a(b + c)x^2 + b(c + a)y^2 + c(a + b)z^2 = 0.$$

Example 9:

The sections of the enveloping cone of the surface $x^2/a^2 + y^2/b^2 + z^2/c^2 = 1$ whose vertex is $P(x_1, y_1, z_1)$ by the plane $z = 0$ is (i) rectangular hyperbola, (ii) a parabola and (iii) a circle. Find the locus of the vertex P.

Solution:

For the given surface we have

$$S = x^2/a^2 + y^2/b^2 + z^2/c^2 - 1;$$

$$S_1 = x_1^2/a^2 + y_1^2/b^2 + z_1^2/c^2 - 1.$$

and $\quad T\ (xx_1/a^2) + (yy_1/b^2) + (zz_1/c^2) - 1$

$\therefore$ The enveloping cone is given by $SS_1 = T_2$

i.e., $\left(\frac{x^2}{a^2}+\frac{y^2}{b^2}+\frac{z^2}{c^2}-1\right)\left(\frac{x_1^2}{a^2}+\frac{y_1^2}{b^2}+\frac{z_1^2}{c^2}-1\right)=\left(\frac{xx_1}{a^2}+\frac{yy_1}{b^2}+\frac{zz_1}{c^2}-1\right)^2$

Its section by the plane z = 0 is given by

$$z = 0,\ \left(\frac{x^2}{a^2}+\frac{y^2}{b^2}-1\right)\left(\frac{x_1^2}{a^2}+\frac{y_1^2}{b^2}+\frac{z_1^2}{c^2}-1\right)=\left(\frac{xx_1}{a^2}+\frac{yy_1}{b^2}-1\right)^2 \quad ...(A)$$

(i) If the equations (A) represent a rectangular hyperbola then the sum of the coefficients of x^2 and y^2 should be zero.

i.e., $$\frac{1}{a^2}\left(\frac{y_1^2}{b^2}+\frac{z_1^2}{c^2}-1\right)+\frac{1}{b^2}\left(\frac{x_1^2}{a^2}+\frac{z_1^2}{c^2}-1\right)=0 \quad \textbf{(Note)}$$

$$\Rightarrow \frac{x_1^2+y_1^2}{a^2b^2}+\frac{1}{c^2}\left(\frac{1}{a^2}+\frac{1}{b^2}\right)z_1^2=\frac{1}{a^2}+\frac{1}{b^2}$$

$$\Rightarrow \frac{x_1^2+y_1^2}{a^2+b^2}+\frac{z_1^2}{c^2}=1, \text{ dividing each term by } a^2+b^2.$$

$\therefore$ The required locus of P (x_1, y_1, z_1) is $\frac{x^2+y^2}{a^2+b^2}+\frac{z^2}{c^2}=1.$

(ii) If the equations (A) represent a parabola, then we should have "$h^2 = ab$".

Here 'a' = coeff. of $x^2 = \frac{1}{a^2}\left(\frac{y_1^2}{b^2}+\frac{z_1^2}{c^2}-1\right)$

'b' = coeff. of $y^2 = \frac{1}{b^2}\left(\frac{x_1^2}{a^2}+\frac{z_1^2}{c^2}-1\right)$

and $\quad$ 'h' = coeff. $2xy = x_1y_1/a^2b^2$.

$\therefore$ If the equations (A) represent parabola, then $h^2 = ab$ hence

i.e., $$\frac{x_1^2y_1^2}{a^4b^4}=\frac{1}{a^2b^2}\left(\frac{y_1^2}{b^2}+\frac{z_1^2}{c^2}-1\right)\left(\frac{x_1^2}{a^2}+\frac{z_1^2}{c^2}-1\right)$$

$$\Rightarrow \frac{x_1^2}{a^2}\left(\frac{z_1^2}{c^2}-1\right)+\frac{y_1^2}{b^2}\left(\frac{z_1^2}{c^2}-1\right)+\left(\frac{z_1^2}{c^2}-1\right)^2=0. \quad \textbf{(Note)}$$

$$\Rightarrow \left(\frac{z_1^2}{c^2}-1\right)\left[\frac{x_1^2}{a^2}+\frac{y_1^2}{b^2}+\frac{z_1^2}{c^2}-1\right]=0$$

$\Rightarrow$ $(z_1^2/c^2)-1=0$, since $x_1^2/a^2+y_1^2/b^2+z_1^2/c^2-1\neq 0$ as

P (x_1, y_1, z_1) does not lie on the given starface

$\Rightarrow$ $z_1^2=c^2$ or $z_1=\pm c$ **Ans.**

$\therefore$ The locus of P (x_1, y_1, z_1) is $z = \pm c$

(iii) If (A) represents a circle then the coefficients of x^2 and y^2 should be equal and coefficient of xy should be zero.

i.e.,
$$\frac{1}{a^2}\left(\frac{y_1^2}{b^2}+\frac{z_1^2}{c^2}-1\right)=\frac{1}{b^2}\left(\frac{x_1^2}{a^2}+\frac{z_1^2}{c^2}-1\right) \quad ...(B)$$

and $x_1y_1/a^2b^2 = 0$ ...(C)

From (C) either $x_1 = 0$ or $y_1 = 0$

If $x_1 = 0$, then from (B) we have $\frac{1}{a^2}\left(\frac{y_1^2}{b^2}+\frac{z_1^2}{c^2}-1\right)=\frac{1}{b^2}\left(\frac{z_1^2}{c^2}-1\right)$

$\therefore$ The locus of P (x_1, y_1, z_1) is $x = 0$, $\frac{y^2}{b^2-a^2}+\frac{z^2}{c^2}=1$ **Ans.**

If $y_1 = 0$, then from (B) we have $\frac{1}{a^2}\left(\frac{z_1^2}{c^2}-1\right)=\frac{1}{b^2}\left(\frac{x_1^2}{a^2}+\frac{z_1^2}{c^2}-1\right)$

$\therefore$ The locus of P (x_1, y_1, z_1) is $y = 0$, $\frac{x^2}{a^2-b^2}+\frac{z^2}{c^2}=1$. **Ans.**

Example 10:

Find the condition that the plane $lx + my + nz = 0$ may touch the cone $ax^2 + by^2 + cz^2 = 0$.

Solution:

The given plane $lx + my + nz = 0$ touches the given cone

$ax^2 + by^2 + cz^2 = 0$ if

$Al^2 + Bm^2 + Cn^2 + 2Fmn + 2Gnl + 2Hlm = 0.$ ...(1)

We have
$$D=\begin{vmatrix} a & 0 & 0\\ 0 & b & 0\\ 0 & 0 & c\end{vmatrix}.$$

Here A = cofactor of a in the determinant $\Delta = bc$.

Similarly $B = ca$ and $C = ab$.

Clearly $F = G = H = 0$. Substituting in (1), we obtain

$$bcl^2 + cam^2 + abn^2 = 0.$$

Hence $l^2/a + m^2/b + n^2/c = 0$ is the required condition.

Example 11:

Show that the general equation of a cone which touches the three coordinate planes is

$$a^2x^2 + b^2y^2 + c^2z^2 - 2bcyz - 2cazx - 2abxy = 0.$$

Solution:

The general equation of a cone through the three coordinate axes is $ayz + bzx + cxy = 0$.

Its reciprocal cone is the cone which touches the three coordinate planes.

$$a^2x^2 + b^2y^2 + c^2z^2 - 2bcyz - 2cazx - 2abxy = 0$$

is the required cone.

Example 12:

Prove that the cones $ayz + bzx+cxy =0$ *and* $(ax)^{1/2} +(by)^{1/2}+(cz)^{1/2} =0$ *are reciprocal.*

Solution:

The reciprocal cone of $ayz + bzx + cxy = 0$ is

$$Ax^2 + By^2 + Cz^2 + 2Fyz + 2Gzx + 2Hxy = 0. \qquad ...(1)$$

We have $a = b = c = 0$, $f = a/2$, $g = b/2$ and $h = c/2$.

$$\therefore \qquad \Delta = \begin{vmatrix} 0 & c/2 & b/2 \\ a/2 & 0 & a/2 \\ b/2 & a/2 & 0 \end{vmatrix}$$

$$\Rightarrow \qquad A = -a^2/4,\ B = -b^2/4,$$

$$C = -c^24,\ F = bc/4,$$

$$G = ca/4,\ H = ab/4.$$

Substituting these values in (1), we obtain

$$a^2x^2 + b^2y^2 + c^2z^2 - 2bcyz - 2cazx - 2abxy = 0 \qquad ...(2)$$

$$\Rightarrow \qquad (ax + by - cz)^2 = 4abxy$$

$$\Rightarrow \qquad ax + by - cz = \pm 2\sqrt{abxy}$$

$$\Rightarrow \qquad ax + by + 2\sqrt{abxy} = cz \text{ (taking – ive sign)}$$

$$\Rightarrow \qquad \left(\sqrt{ax}+\sqrt{by}\right)^2 = cz \Rightarrow \sqrt{ax} + \sqrt{by} = \pm \sqrt{cz}$$

Hence $\sqrt{ax} + \sqrt{by} + \sqrt{cz} = 0$ (taking – ve sign).

Example 13:

Prove that tangent planes to the cone

$$x^2 - y^2 + 2z^2 - 3yz + 4zx - 5xy = 0$$

are perpendicular to the generators of the cone

$$17x^2 + 8y^2 + 29z^2 + 28yz - 46zx - 16xy = 0.$$

Solution:

In other words, we have to show that the given cones are reciprocal. The reciprocal cone of

$$x^2 - y^2 + 2z^2 - 3yz + 4zx - 5xy = 0 \qquad ...(1)$$

is $\qquad Ax^2 + By^2 + Cz^2 + 2Fyz + 2Gzx + 2Hxy = 0. \qquad ...(2)$

From (1), a = 1, b = – 1, c = 2, f = – 3/2, g = 2, h = – 5/2.

$$\text{Now} \qquad \Delta = \begin{vmatrix} 1 & -5/2 & 2 \\ -5/2 & -1 & -3/2 \\ 2 & -3/2 & 2 \end{vmatrix}.$$

We have

$$A = \begin{vmatrix} -1 & -3/2 \\ -3/2 & 2 \end{vmatrix} = -\frac{17}{4}, \; B = \begin{vmatrix} 1 & 2 \\ 2 & 2 \end{vmatrix} = -2,$$

$$C = \begin{vmatrix} 1 & -5/2 \\ -5/2 & -1 \end{vmatrix} = -\frac{29}{4}, \; F = -\begin{vmatrix} 1 & -5/2 \\ 2 & -3/2 \end{vmatrix} = \frac{-7}{2},$$

$$G = \begin{vmatrix} -5/2 & -1 \\ 2 & -3/2 \end{vmatrix} = \frac{23}{4}, \; H = -\begin{vmatrix} -5/2 & -3/2 \\ 2 & 2 \end{vmatrix} = 2.$$

Substituting these values in (2), we obtain

$$17x^2 + 8y^2 + 29z^2 + 28yz - 46zx - 16xy = 0.$$

This is the reciprocal cone of the given cone.

Example 14:

Prove that the cones $ax^2 + by^2 + cz^2 = 0$ and $x^2/a + y^2/b + z^2/c = 0$ are reciprocal.

Solution:

The reciprocal cone of the cone

$$ax^2 + by^2 + cz^2 = 0 \text{ is}$$

$$Ax^2 + By^2 + Cz^2 + 2Fyz + 2Gzx + 2Hxy = 0. \qquad ...(1)$$

We have $\Delta = \begin{vmatrix} a & 0 & 0 \\ 0 & b & 0 \\ 0 & 0 & c \end{vmatrix}$.

$\therefore \qquad A = bc,\ B = ca,\ C = ab,\ F = G = H = 0.$

Substituting these values in (1), we obtain

$$bcx^2 + cay^2 + abz^2 = 0$$

$\Rightarrow \qquad x^2/a + y^2/b + z^2/c = 0,$

which is the reciprocal cone of $ax^2 + by^2 + cz^2 = 0$.

Example 15:

Prove that the locus of points from which three mutually perpendicular planes can be drawn to touch the ellipse

$$\frac{x^2}{a^2} + \frac{y^2}{b^2} = 1,\ z = 0$$

is the sphere $x^2 + y^2 + z^2 = a^2 + b^2$.

Solution:

First we obtain the equation of the cone with vertex (α, β, γ) and the given guiding curve.

Shift the origin to the vertex by means of the transformations

$$x = X + \alpha,\ y = Y + \beta \text{ and } z = Z + \gamma.$$

We have $\dfrac{(X+\alpha)^2}{a^2} + \dfrac{(Y+\beta)^2}{b^2} = 1$ and $Z + \gamma = 0 \Rightarrow -\dfrac{Z}{\gamma} = 1$.

Thus the equation of the cone with its vertex at the origin is

$$\frac{X^2}{a^2} + \frac{Y^2}{b^2} + 2\left(\frac{\alpha X}{a^2} + \frac{\beta Y}{b^2}\right)\left(-\frac{Z}{\gamma}\right) + \left(\frac{\alpha^2}{a^2} + \frac{\beta^2}{b^2} - 1\right)\left(-\frac{Z}{\gamma}\right)^2 = 0. \qquad ...(1)$$

The cone $ax^2 + by^2 + cz^2 + 2fyz + 2gzx + 2hxy = 0 \qquad ...(2)$

has three mutually perpendicular tangent planes if

$$bc + ca + ab = f^2 + g^2 + h^2 \qquad ...(3)$$

Comparing (1) and (2), we get

$$a = \frac{1}{a^2},\ b = \frac{1}{b^2},\ c = \frac{1}{\gamma^2}\left(\frac{\alpha^2}{a^2} + \frac{\beta^2}{b^2} - 1\right),$$

$$f = \frac{\beta}{b^2\gamma},\ g = -\frac{\alpha}{a^2\gamma},\ h = 0.$$

Substituting these values in (3), we obtain

$$\frac{1}{b^2}\left\{\frac{1}{\gamma^2}\left(\frac{\alpha^2}{a^2} + \frac{\beta^2}{b^2} - 1\right)\right\} + \frac{1}{a^2}\left\{\frac{1}{\gamma^2}\left(\frac{\alpha^2}{a^2} + \frac{\beta^2}{b^2} - 1\right)\right\}$$

$$+ \frac{1}{a^2 b^2} = \left(-\frac{\beta}{b^2\gamma}\right)^2 + \left(-\frac{\alpha}{a^2\gamma}\right)^2 + 0$$

$$\Rightarrow \qquad \frac{1}{\gamma^2}\left(\frac{\alpha^2}{a^2} + \frac{\beta^2}{b^2} - 1\right)\left(\frac{1}{a^2} + \frac{1}{b^2}\right) + \frac{1}{a^2b^2} = \frac{1}{\gamma^2}\left(\frac{\alpha^2}{a^4} + \frac{\beta^2}{b^4}\right)$$

$$\Rightarrow \qquad \frac{1}{\gamma^2}\left(\frac{\alpha^2}{a^2b^2} + \frac{\beta^2}{a^2b^2} - \frac{1}{a^2} - \frac{1}{b^2}\right)\frac{1}{a^2b^2} = 0$$

$$\Rightarrow \qquad \alpha^2 + \beta^2 - b^2 - a^2 + \gamma^2 = 0 \text{ or } \alpha^2 + \beta^2 + \gamma^2 = a^2 + b^2.$$

Hence the locus of (α, β, γ) is $x^2 + y^2 + z^2 = a^2 + b^2$.

RIGHT CIRCULAR CONE

Definition : *The surface generated by a line passing through a fixed point (called* **vertex***) and making a constant angle with a fixed line through the vertex is known as* the ***right circle cone***. *The fixed line is called the* **axis *of the cone*** *and the constant angle is called the* ***semi-vertical angle*** *of the cone.*

To Prove That the Section of a Right Cone by a Plane Perpendicular to its Axis is a Circle

Let V be the vertex and VO the axis of the right circular cone. Let θ be its semi-vertical angle. Let any plane at right angles to the axis meet the axis in O. Let Q be any point on this section. Join OQ. Now OQ is a line on the plane section of the cone which is at right angles to the axis VO and so OQ is also at right angles to VO. Also $\angle OVQ = \theta$, so OQ = VO tan θ. But θ is constant and hence OQ is also of constant length for all positions of Q on the section of the cone through O at right angles to the axis VO.

Hence the locus of Q is a circle with O as the centre and radius OQ.

For this fact, the cone is called right circular cone.

Equation of the Right Cone

Let the coordinates of the vertex V be (α, β, γ) and equations of the axis VO be given as $\frac{x-\alpha}{l} = \frac{y-\beta}{m} = \frac{z-\gamma}{n}$.

Let the semi-vertical angle of the cone be θ.

Let P (x, y, z) be any point on the cone, then the generator VP makes on angle θ with axis VO. The direction ratios of the line VP are $x - \alpha$, $y - \beta$, $z - \gamma$

and those of the axis VO from (i) are l, m, n. Hence

$$\therefore \quad \cos\theta = \frac{l(x-\alpha) + m(y-\beta) + n(z-\gamma)}{\sqrt{(l^2+m^2+n^2)}\sqrt{\{(x-\alpha^2)^2+(y-\beta)^2+(z-\gamma)^2\}}}$$

Hence, the required equation of the right circular cone is

given by $[l(x - \alpha) + m(y - \beta) + n(z - \gamma)]^2$

$$= \cos^2\theta\,(l^2 + m^2 + n^2)\,[(x - \alpha)^2 + (y - \beta)^2 + (z - \gamma)^2] \qquad \text{...(ii)}$$

Corollary : If the vertex is at the origin, then the above equation (ii) of the right circular cone reduces to

$$(lx + my + nz)^2 = \cos^2\theta\,(l^2 + m^2 + n^2)(x^2 + y^2 + z^2)$$

$$= (1 - \sin^2\theta)(l^2 + m^2 + n^2)(x^2 + y^2 + z^2)$$

$$\Rightarrow \quad (l^2 + m^2 + n^2)(x^2 + y^2 + z^2)\sin^2\theta$$

$$= (l^2 + m^2 + n^2)(x^2 + y^2 + z^2) - (lx + my + nz)^2$$

$$= (mz - ny)^2 + (nx - lz)^2 + (ly - mx)^2, \qquad \text{...(iii)}$$

Further if z-axis be the axis of the cone, then we have $l = 0 = m$ and then the equation (iii) reduces to $n^2(x^2 + y^2 + z^2)\sin^2\theta = n^2y^2 + n^2x^2$

$$\Rightarrow \quad z^2\sin^2\theta = (x^2 + y^2)(1 - \sin^2\theta) \quad \Rightarrow \quad x^2 + y^2 = z^2\tan^2\theta \qquad \text{...(iv)}$$

Example:

Find the equation of the cone formed by rotating the $2x + 3y = 6$, $z = 0$ about the y-axis.

Solution:

The axis of the cone is y-axis, whose d.c.'s are 0, 1, 0 and equation of any generator is given as $2x + 3y = 6$, $z = 0$

i.e., $$\frac{x}{3} = \frac{y-2}{-2} = \frac{z}{0}.$$

The point of intersection of y-axis (*i.e.,* x = 0, z = 0) and the line (i) (0, 2, 0), so the co-ordinates of the vertex of the cone is V (0, 2, 0).

Let θ be the semi-vertical angle of the cone, then θ is the angle between y-axis and the line (i), so we have

$$\cos\theta = \frac{0.3 + 1.(-2) + 0.0}{\sqrt{(0+1+0)}\sqrt{\{3^2 + (-2)^2 + (0)^2\}}} = \frac{-2}{\sqrt{(13)}}$$

Also let P (x, y, z) be any point on the cone, then VP is a generator of the cone and its direction ratios are x – 0, y – 2, z – 0, or x, y – 2, z..

Also θ is the angle between y-axis and VP, so we get

$$\cos\theta = \frac{0.x + 1.(y-2) + 0.z}{\sqrt{(0+1+0)}\sqrt{\{x^2 + (y-2)^2 + z^2\}}} \quad ...(iii)$$

Equating the value of cos θ from (ii) and (iii) we get equation of the cone as

$$\frac{-2}{\sqrt{(13)}} = \frac{y-2}{\sqrt{\{x^2 + (y-2)^2 + z^2\}}}$$

⇒ $4\{x^2 + (y-2)^2 + z^2\} = 13(y-2)^2$, squaring both sides.

⇒ $4x^2 - 9(x-2)^2 - 4z^2 = 0.$ **Ans.**

STANDARD EQUATION OF A CONE

The equation $ax^2 + by^2 + cz^2 = 0$ is called the standard equation of a cone. It is particular case of the general equation of the cone whose vertex is origin.

We can prove as particular cases of 6.01 and 6.11 or independently by similar methods that

(i) The tangent planes to the cone at (x_1, y_1, z_1) is $aax_1 + byy_1 + czz_1 = 0$

(ii) The condition for the plane ux + vy + wz = 0 to touch this cone is

$$bcu^2 + cav^2 + abw^2 = 0.$$

(iii) the polar plane of (x_1, y_1, z_1) is $axx_1 + byy_1 + czz_1 = 0$.

(iv) the equation of the pair of tangent planes passing through the line OP, where O is (0, 0, 0) and P is (x_1, y_1, z_1) is $SS_1 = T^2$

i.e., $$(ax^2 + by^2 + cz^2)\left(ax_1^2 + by_1^2 + cz_1^2\right) = (axx_1 + byy_1 + czz_1)^2.$$

Example 1:

Prove that the equation of a right circular cone with vertex (2, 1, – 3), axis parallel to y-axis and semi-vertical angle 45° is $x^2 + z^2 - 4x + 2y + 6z + 12 = 0$.

Solution:

We have

$$\alpha = 2,\ \beta = 1,\ \gamma = -3,\ l = 0,$$

$$m = 1,\ n = 0,\ \cos 45^o = \frac{1}{\sqrt{2}}$$

The required cone is

$$[0\,(x-2) + 1\,(y-1) + 0\,(z+3)]^2 = 0\,(0+1+0)$$

$$[(x-2)^2 + (y-1)^2 + (z+3)^2] \times \frac{1}{2}$$

$$\Rightarrow \quad 2\,(y^2 - 2y + 1) = (x^2 + y^2 + z^2 - 4x - 2y + 6z + 14).$$

Hence $x^2 - y^2 + z^2 - 4x + 2y + 6z + 12 = 0$.

Example 2:

Find the equation of the right circular cone with vertex (2, 3, 1), axis parallel to the line $-x = y/2 = z$ and one of its generators having direction cosines proportional to (1, – 1, 1).

Solution:

The axis of the given cone is

$$\frac{x-2}{-1} = \frac{y-3}{2} = \frac{z-1}{1}$$

and the direction ratios of a generator are 1, – 1, 1. The semi-vertical angle θ of the cone is

$$\cos\theta = \frac{-1.1 + 2.-1 + 1.1}{\sqrt{1+4+1}\,\sqrt{1+1+1}} \qquad \Rightarrow \cos^2\theta = \frac{2}{9}.$$

We have $l = -1$, $m = 2$, $n = 1$, $\alpha = 2$, $\beta = 3$ and $\gamma = 1$.

The required cone is

$$[-1\,(x-2) + 2\,(y-3) + 1\,(z-1)]^2 = (1+4+1)\,[(x-2)^2$$

$$+ (y-3)^2 + (z-1)^2 \,.\, \frac{2}{9}$$

$$\Rightarrow \quad 3\,(-x + 2y + z - 5)^2 + 4\,(x^2 + y^2 + z^2 - 4x - 6y - 2z + 41).$$

Hence $x^2 - 8y^2 + z^2 + 12xy - 12yz + 6zx - 46x + 36y + 22z - 19 = 0$ is the required equation of the right circular cone.

Example 3:

Find the equation of the right circular cone generated by straight lines drawn from the origin to cut the circle through the three points (1, 2, 2), (2, 1, – 2) and (2, – 2, 1).

Solution:

Let l, m, n be the direction cosines of the axis. The direction ratios of the given lines are (1, 2, 2), (2, 1, – 2) and (2, – 2, 1).

$$\therefore \cos\theta = \frac{l+2m+2n}{\sqrt{l^2+m^2+n^2}\,\sqrt{1+4+4}} = \frac{1}{3}(l + 2m + 2n). \qquad ...(1)$$

Similarly

$$\cos\theta = \frac{1}{3}(2l + m - 2n) \text{ and } \cos\theta = \frac{1}{3}(2l - 2m + n).$$

Thus

$$\text{(i) } \frac{1}{3}(l + 2m + 2n) = \text{(ii) } \frac{1}{3}(2l + m - 2n) = \text{(iii) } \frac{1}{3}(2l - 2m + n)$$

Now (i) = (ii) $\Rightarrow l - m - 4n = 0$,

and (ii) = (iii) $\Rightarrow 0l + m - n = 0$.

Solving these, $\frac{l}{5} = \frac{m}{1} = \frac{n}{1} = \frac{1}{\sqrt{27}}$.

$$\therefore l = \frac{5}{\sqrt{27}},\ m = \frac{1}{\sqrt{27}} \text{ and } n = \frac{1}{\sqrt{27}}, \qquad ...(2)$$

From (1) and (2), $\cos\theta = \frac{3}{\sqrt{27}}$.

Hence the required cone with its vertex at the origin is

$$\frac{1}{27}(5x + y + z)^2 = 1.\,(x^2 + y^2 + z^2)\,.\,\frac{9}{27}$$

$$\Rightarrow \quad 8x^2 - 4y^2 - 4z^2 + 5xy + yz + 5zx = 0.$$

Example 4:

If α is the semi-vertical angle of a circular cone which passes through the lines Oy, Oz, x = y = z; show that

$\cos \alpha = (9 - 4\sqrt{3})^{1\,2}$.

Solution:

Let l, m, n be the direction cosines of the axis. The direction ratios of the given lines are (0, 1, 0), (0, 0, 1) and (1, 1, 1) respectively.

$$\therefore \qquad \cos \alpha = \frac{0.l + 1.m + 0.n}{\sqrt{l^2 + m^2 + n^2}\ 0 + 1 + 0} = m.$$

Similarly $\cos \alpha = n$ and $\cos \alpha = \dfrac{l + m + n}{\sqrt{3}}$.

Thus we have $m = n = \dfrac{1}{\sqrt{3}}(l + m + n)$. These equations give

$$l + (1 - \sqrt{3})\, m + n = 0,$$

$$0.1 + 1.m - 1.n = 0.$$

Solving these,

$$\frac{l}{\sqrt{3} - 2} = \frac{m}{1} = \frac{n}{1} = \frac{1}{\sqrt{\left(\sqrt{3} - 2\right)^2 + 1^2 + 1^2}}.$$

Hence

$$\cos \alpha = m = \frac{1}{\sqrt{\left(\sqrt{3} - 2\right)^2 + 1 + 1}} = \frac{1}{\sqrt{9 - 4\sqrt{3}}} = \left(9 - 4\sqrt{3}\right)^{-1/2}$$

Example 5:

Find the equation of the right circular cone which passes through (1, 1, 2) and has its vertex at the origin and axis the line $\dfrac{x}{2} = -\dfrac{y}{4} = \dfrac{z}{3}$.

Solution:

The given point is P (1, 1, 2) and the direction ratios of the given axis are 2, – 4, 3. The direction ratios of the generator OP are 1 – 0, 1–0, 2 – 0 *i.e.*, 1, 1, 2.

$$\therefore \qquad \cos \theta = \frac{1.2 - 1.4 + 2.3}{\sqrt{1 + 1 + 4}\ \sqrt{4 + 16 + 9}} = \frac{4}{\sqrt{6}\ \sqrt{29}}$$

The equation of the required cone with vertex at the origin is

$$(lx + my + nz)^2 = (l^2 + m^2 + n^2)(x^2 + y^2 + z^2)\cos^2 \theta$$

$$\Rightarrow \quad (2x - 4y + 3z)^2 = (4 + 16 + 9)(x^2 + y^2 + z^2)\frac{16}{6 \times 29}$$

$$\Rightarrow \quad 3(4x^2 + 16y^2 + 9z^2 - 16xy - 24yz + 12zx) = 8(x^2 + y^2 + z^2)$$

Hence $\quad 4x^2 + 40y^2 + 19z^2 - 48xy - 72yz + 36zx = 0$

is the required equation of the right circular cone.

Example 6:

Lines are drawn through the origin with direction cosines proportional to (1, 2, 2), (2, 3, 6), (3, 4, 12). Show that the axis of the right circular cone through them has direction cosines $-1/\sqrt{3}$, $1/\sqrt{3}$, $1/\sqrt{3}$ and that the semi-vertical angle of the cone is $\cos^{-1}(1/\sqrt{3})$. Obtain the equation of the cone also.

Solution:

Let l, m, n be the direction cosines of the axis so that $l^2 + m^2 + n^2 = 1$. It is given that the direction ratios of a generator of the cone are 1, 2, 2.

$$\therefore \quad \cos\theta = \frac{1l + 2m + 2n}{\sqrt{l^2 + m^2 + n^2}.\sqrt{1+4+4}} = \frac{l + 2m + 2n}{3}. \quad ...(1)$$

Similarly

$$\cos\theta - \frac{2l + 3m + 6n}{7}, \cos\theta = \frac{3l + 4m + 12n}{13}.$$

Thus we have

$$\text{(i)} \ \frac{l + 2m + 2n}{3} = \text{(ii)} \ \frac{2l + 3m + 6n}{7} = \text{(iii)} \ \frac{3l + 4m + 12n}{13}.$$

Now (i) = (ii) $\Rightarrow \lambda + 5m - 4n = 0$,

and (ii) = (iii) $\Rightarrow 5l + 11m - 6n = 0$.

Solving these

$$\frac{l}{-30+44} = \frac{m}{-20+6} = \frac{n}{11-25}.$$

$$\Rightarrow \quad \frac{l}{-1} = \frac{m}{1} = \frac{n}{1} = \frac{1}{\sqrt{3}}.$$

Hence $l = -\frac{1}{\sqrt{3}}, m = \frac{1}{\sqrt{3}}, n = \frac{1}{\sqrt{3}}$

are the direction cosines of the axis of the cone.

Substituting these values in (1), we obtain

$$\cos\theta = \frac{1}{\sqrt{3}}. \text{ Hence } \theta = \cos^{-1}\left(\frac{1}{\sqrt{3}}\right)$$

is the required semi-vertical angle. The equation of the cone with its vertex at the origin is

$$\left(-\frac{1}{\sqrt{3}}x+\frac{1}{\sqrt{3}}y+\frac{1}{\sqrt{3}}z\right)^2 = 1.\ (x^2+y^2+z^2)\ .\ \frac{1}{3}.$$

Hence $xy - yz + zx = 0$ is the required cone.

Example 7:

Find the equations of the circular cones which contain the three coordinate axes as generators.

Solution:

Consider an axis of the given cone which is equal inclined to the positive side of the three axes and making an angle with each of them. So $\cos\theta = \frac{1}{\sqrt{3}}$ and the equations of the axis are

$$\frac{x}{1} = \frac{y}{1} = \frac{z}{1}. \qquad ...(i)$$

The equation of the cone whose axis is (i) is

$$(1.x + 1.y + 1.z)^2 = (1 + 1 + 1)\ (x^2 + y^2 + z^2).\ \frac{1}{3}.$$

$$\Rightarrow \qquad xy + yz + zx = 0.$$

Other possible equations for axis of the cone are

$$\frac{x}{1} = \frac{y}{-1} = \frac{z}{-1};\ \frac{x}{1} = \frac{y}{1} = \frac{z}{1};\ \frac{x}{1} = \frac{y}{1} = \frac{z}{-1}.$$

The corresponding cones are

$$(x - y - z)^2 = x^2 + y^2 + z^2,\ (x - y + z)^2 = x^2 + y^2 + z^2,$$
$$(x + y - z)^2 = x^2 + y^2 + z^2 \qquad \text{(respectively)}$$

$$\Rightarrow \qquad xy - yz + zx = 0;\ xy + yz - zx = 0;$$
$$xy - yz - zx = 0.$$

Hence $xy \pm yz \pm zx = 0$ are the required equations.

Example 8:

The axis of a right cone, vertex O, makes equal angles with the coordinate axes, and the cone passes through the line drawn from O with direction cosines proportional to (1, – 2, 2). Find the equation to the cone.

Solution:

We are given that l = m = n. The direction ratios of a generator of the cone are 1, – 2, 2. Now

$$\cos\theta = \frac{l.1 - l.2 + l.2}{\sqrt{1+4+4}\,\sqrt{l^2+l^2+l^2}} = \frac{1}{3\sqrt{3}}$$

The required cone with vertex at the origin is

$$(lx + ly + lz)^2 = (l^2 + l^2 + l^2)(x^2 + y^2 + z^2) \cdot \frac{1}{27}$$

$\Rightarrow$ $\quad 9(x + y + z)^2 = x^2 + y^2 + z^2.$

Hence $\quad 4(x^2 + y^2 + z^2) + 9(xy + yz + zx) = 0$ is the required cone.

SOME SOLVED EXAMPLES

Example 1:

The vertex of cone is (a, b, c) and the yz-plane cuts it in the curve F (y, z) = 0, x = 0, show that the xz-plane cuts it in the curve.

$$y = 0,\ F\left[\frac{bx}{x-a}, \frac{cx-az}{x-a}\right] = 0.$$

Solution:

Any line through (a, b, c) is given as $\dfrac{x-a}{l} = \dfrac{y-b}{m} = \dfrac{z-c}{n}$

Its meets x = 0 in the point $\left[0,\ b - \dfrac{am}{l},\ c - \dfrac{an}{l}\right]$ and if it lies on the given curve F (y, z) = 0, then we have $F\left[b - \dfrac{am}{l},\ c - \dfrac{an}{l}\right]$ ...(ii)

Eliminating l, m, n between (i) and (ii), we have

$$f\left[b - a\left(\frac{y-b}{x-a}\right),\ c - a\left(\frac{z-c}{x-a}\right)\right] = 0$$

or $$F\left[\frac{bx-ay}{x-a}, \frac{cx-az}{x-a}\right] = 0.$$

It meets zx-plane *i.e.*, y = 0 in the curve then we have

$$F\left[\frac{bx}{x-a}, \frac{cx-az}{x-a}\right] = 0, \; y = 0$$ **Hence proved.**

Example 2:

Find the equation to the cone whose vertex is the origin and base the circle $x = a$, $y^2 + z^2 = b^2$ and show that the section of the cone by a plane parallel to the plane XOY is a hyperbola.

Solution:

One line through (0, 0, 0) is $\frac{x}{l} = \frac{y}{m} = \frac{z}{n}$.

It meet the plane x = a at $\left(a, \frac{am}{l}, \frac{an}{l}\right)$ and if it lies on $y^2 + z^2 = b^2$, x = a

we have $$\frac{b^2m^2}{l^2} + \frac{a^2n^2}{l^2} = b^2$$

Eliminating l, m, n between (i) and (ii) we get

$$a^2\left(\frac{y^2}{x^2} + \frac{z^2}{x^2}\right) = b^2 \text{ or } a^2(y^2 + z^2) = b^2x^2, \qquad \text{...(iii)}$$

which is the required equation of the cone.

Again consider a plane z = c which is parallel to the plane XOY. The section of the cone (iii) by this plane is $a^2(y^2 + c^2) = b^2x^2$

$\Rightarrow$ $b^2x^2 - a^2y^2 = a^2c^2$, which is evidently a hyperbola.

Example 3:

Two cones with a common vertex pass through the curves $z^2 = 4ax$, $y = 0$ and $z^2 = 4by$, $x = 0$. The plane $z = 0$ meets them in two conics which intersect in four concyclic points. Show that the vertex lies on the surface $z(x/a + y/b) = 4(x^2 + y^2)$.

Solution:

Let (α, β, γ) be the common vertex of the cones, then their equations are $(y\gamma - z\beta)^2 = 4a(\alpha y - \beta x)(y - \beta)$

and $$(x\gamma - \alpha z)^2 = 4a(x\beta - \alpha y)(x - \alpha).$$

The plane z = 0 *i.e.*, xy-plane meets these cones in conics given by

$$S \equiv y^2\gamma^2 - 4a(\alpha y - \beta x)(y - \beta) = 0, \; z = 0$$

and $$S' \equiv x^2\gamma^2 - 4b(x\beta - \alpha y)(x - \alpha) = 0, \; z = 0.$$

The equations of any curve through the points of intersection of these conics S = 0 and S' = 0 in the xy-plane is $S = \lambda S' = 0$, $z = 0$,

$$\Rightarrow \quad [y^2\gamma^2 - 4a\,(\alpha y - \beta x)\,(y - \beta)] + \lambda\,[x^2y^2 - 4ab\,(x\beta - \alpha y)\,(x - \alpha)] = 0,\ z = 0.$$

If this curve is a circle, then the coefficients of x^2 and y^2 must be equal and that of xy should vanish which gives

$$\lambda\,(\gamma^2 - 4b\beta) = (\gamma^2 - 4a\alpha) \qquad \text{...(i)}$$

Eliminating λ between (i) and (ii), we get

$$[-(4\alpha\beta)/(4b\alpha)]\,(\gamma^2 - 4b\beta) = (\gamma^2 - 4a\alpha)$$

$$\Rightarrow \quad a\beta\,(y^2 - b\beta) + b\alpha\,(y^2 - 4a\alpha) = 0$$

$$\Rightarrow \quad x^2\,(\alpha_1 a + \beta_1 b) = 4\,(\alpha^2 + \beta^2)$$

The locus of the common vertex is

$$x^2\left(\frac{x}{a}+\frac{y}{b}\right) = 4\,(x^2 + y^2).$$

Example 4:

Prove that line $x = pz + q$, $y = rz + s$ intersects the conic $z = 0$, $ax^2 + by^2 = 1$ if $aq^2 + bs^2 = 1$.

Hence show that the coordinates of any point on a line which intersects the conic and passes through the point (α, β, γ) satisfy the equation

$$a\,(\gamma x - \alpha z)^2 + b\,(\gamma y - \beta z)^2 - (z - \gamma)^2.$$

Solution:

Equation of the given line $x = pz + q$, $y = rz + s$ ...(i)

meets the plane $z = 0$ at $x = q$, $y = s$ *i.e.*, in the point (q, s, 0).

If this point (q, s, 0) lies on the given conic

$$ax^2 + by^2 = 1,\ z = 0 \qquad \text{...(ii)}$$

then we have $\quad aq^2 + bs^2 + 1.$ ...(iii)

which is the required condition. **Hence proved.**

Again if the line (i) passes through (α, β, γ),

then $\quad \alpha = p\gamma + q,\ \beta = r\gamma + s$ ...(iv)

From (i) and (iv) we have

$$x\gamma - \alpha z = \gamma\,(pz + q) - z\,(p\gamma + q) = q\,(y - z)$$

and $\quad y\gamma - \beta z = \gamma\,(rz + s) - z\,(r\gamma + s) = s\,(y - z)$

These give $q = \dfrac{\gamma x - \alpha z}{\gamma - z}, s = \dfrac{y\gamma - \beta z}{\gamma - z}$

Substituting these values in (iii), we obtain the required locus as

$$a\left(\frac{\gamma x - \alpha z}{\gamma - z}\right)^2 + b\left(\frac{\gamma y - \beta z}{\gamma - z}\right)^2 = 1$$

$\Rightarrow$ $a(\gamma x - \alpha z)^2 + b(\gamma y - \beta z)^2 = (z - \gamma)^2$. **Hence proved.**

Example 5:

Find the equation of a cone whose vertex is the point (α, β, γ) and whose generating lines pass through the conic i.e., whose base curve is $x^2/a^2 + y^2/b^2 = 1$, $z = 0$.

Solution:

Any line through (α, β, γ) is given by $\dfrac{x-\alpha}{l} = \dfrac{y-\beta}{m} = \dfrac{z-\gamma}{n}$ ...(i)

It meets the plane $z = 0$ at $\left(\alpha - \dfrac{l\gamma}{n}, \beta - \dfrac{m\gamma}{n}, 0\right)$ and if it lies on the given conic we have $\dfrac{1}{a^2}\left(\alpha - \dfrac{l\gamma}{n}\right)^2 + \dfrac{1}{b^2}\left(\beta - \dfrac{m\gamma}{n}\right)^2 = 1.$ ...(ii)

Eliminating l, m, n between (i) and (ii) we get the equation of the required cone as

$$\frac{1}{a^2}\left[\alpha - \left(\frac{x-\alpha}{z-\gamma}\right)\gamma\right]^2 + \frac{1}{b^2}\left[\beta - \left(\frac{y-\beta}{z-\gamma}\right)\gamma\right]^2 = 1$$

$\Rightarrow$ $b^2(\alpha z - \gamma x)^2 + a^2(\beta z - \gamma y)^2 = a^2b^2(z - \gamma)^2$ **Ans.**

Example 6:

Find the equation of the cone whose vertex is (α, β, γ) and base

$ax^2 + by^2 = 1$, $z = 0$.

Solution:

Any line through (α, β, γ) is given by $\dfrac{x-\alpha}{l} = \dfrac{y-\beta}{m} = \dfrac{z-\gamma}{n}$ ...(i)

It meets the plane $z = 0$ at $\left(\alpha - \dfrac{l\gamma}{n}.\beta - \dfrac{m\gamma}{n}, 0\right)$ and if lies on the given conic we have $a\left(\alpha-\frac{l\gamma}{n}\right)^2+b\left(\beta-\frac{m\gamma}{n}\right)^2=1$

Eliminating l, m, n between (i) and (ii) we have

$$a\left[\alpha-\left(\frac{x-\alpha}{z-\gamma}\right)\gamma\right]^2+b\left[\beta-\left(\frac{y-\beta}{z-\gamma}\right)\gamma\right]^2=1$$

$\Rightarrow$ $$a\,(\alpha z-\gamma x)^2+b\,(\beta z-\gamma y)^2=(z-\gamma)^2$$ **Ans.**

Example 7:

Find the equation of a cone whose vertex is (α, β, γ) and base $y^2 = 4ax$, $z = 0$. (Avadh 95; Bundelkhand 95, 91; Lucknow 91, 90;

Solution:

Any line through (α, β, γ) is $\frac{x-\alpha}{l}=\frac{y-\beta}{m}=\frac{z-\gamma}{n}$ ...(i)

It meets the plane $z = 0$ at $\left(\alpha-\frac{l\gamma}{n}, \beta-\frac{m\gamma}{n}, 0\right)$ and if it lies on

$y^2 = 4ax$, $z = 0$, then $\left(\beta-\frac{m\gamma}{n}\right)^2=4a\left(\alpha-\frac{l\gamma}{n}\right)$. ...(ii)

Eliminating l, m, n between (i) and (ii) by putting

$\frac{x-\alpha}{z-\gamma}$ and $\frac{y-\beta}{z-\gamma}$ for $\frac{l}{n}$ and $\frac{m}{n}$ respectively

We have the equation of cone

$$\left[\beta-\left(\frac{y-\beta}{z-\gamma}\right)\gamma\right]^2=4a\left[\alpha-\left(\frac{x-\alpha}{z-\gamma}\right)\gamma\right]$$

$\Rightarrow$ $(\beta z - y\gamma)^2 = 4a\,(\alpha z - x\gamma)\,(z - \gamma)$ **Ans.**

Example 8:

Prove that a line which passes through (α, β, γ) and intersects the parabola $z^2 = 4ax$, $y = 0$ lies on the cone

$$(\beta z-\gamma y)^2=4b\,(\beta-y)\,(\beta x-\alpha y).$$

Solution:

Any line through (α, β, γ) is $\frac{x-\alpha}{l}=\frac{y-\beta}{m}=\frac{z-\gamma}{n}$ ...(i)

It meets the plane y = 0 at $\left(\alpha - \frac{l\beta}{m}, 0, \gamma - \frac{n\beta}{m}\right)$ and if it lies on

$z^2 = 4ax$, y = 0, then we have $\left(\gamma - \frac{n\beta}{m}\right)^2 = 4a\left(\alpha - \frac{l\beta}{m}\right)$ ...(ii)

Eliminating l, m, n between (i) and (ii) putting,

$$\left[\gamma - \left(\frac{z-\gamma}{y-\beta}\right)\beta\right]^2 = 4a\left[\alpha - \left(\frac{x-\alpha}{y-\beta}\right)\beta\right].$$

$\Rightarrow$ $(\gamma y - \beta z)^2 = 4a\,(\alpha y - \beta x)\,(y - \beta).$ **Hence proved.**

which is the equation of cone.

Example 9:

The section of a cone with vertex at P and guiding curve $(x^2/a^2) + (y^2/b^2) = 1$, $z = 0$ by the plane $x = 0$ is a rectangular hyperbola. Show that the locus of P is $(x^2/a^2) + \{(y^2 + z^2)/b^2\} = 1$.

Solution:

Let the vertex P of the cone by (α, β, γ).

Any line through P (α, β, γ) is $\frac{x-\alpha}{l} = \frac{y-\beta}{m} = \frac{z-\gamma}{n}$ ...(i)

This line meets the plane z = 0 is $\left(\alpha - \frac{l\gamma}{n}, \beta - \frac{m\gamma}{n}, 0\right)$ and if this point lies on the given curve $(x^2/a^2) + (y^2/b^2) = 1$, we have

$$\frac{1}{a^2}\left(\alpha - \frac{l\gamma}{n}\right)^2 + \frac{1}{b^2}\left(\beta - \frac{m\gamma}{n}\right)^2 = 1. \quad \text{...(ii)}$$

Eliminating l, m, n between (i) and (ii), the equation of the cone is

given as $\frac{1}{a^2}\left[\alpha - \left(\frac{x-\alpha}{z-\gamma}\right)\gamma\right]^2 + \frac{1}{b^2}\left[\beta - \frac{y-\beta}{z-\gamma}\gamma\right]^2 = 1$

$\Rightarrow$ $b^2\,(\alpha z - x\gamma)^2 + a^2\,(\beta z - y\gamma)^2 = a^2b^2\,(z - \gamma)^2$ ...(iii)

The section of this cone by the plane x = 0 gives the conic on yz-plane as

$b^2a^2z^2 + a^2\,(\beta z - y\gamma)^2 = a^2b^2\,(z - \gamma)^2$

$\Rightarrow$ $a^2\gamma^2y^2 + (b^2\alpha^2 + a^2\beta^2 - a^2b^2)\,z^2 - 2a^2\beta\gamma yz + 2a^2b^2\gamma z - a^2b^2\gamma^2 = 0.$

If it represents a rectangular hyperbola on the yz-plane, then the sum of the coefficients of y^2 and must be zero *i.e.,* $a^2\gamma^2 + (b^2\alpha^2 + a^2\beta^2 - a^2b^2) = 0$ **(Note)**

$\Rightarrow$ $(\alpha^2/a^2) + [(\beta^2 + \gamma^2)/b^2] = 1.$

$\therefore$ The locus of P (α, β, γ) is $(x^2/a^2) + [(y^2 + z^2)/b^2] = 1$. **Hence proved.**

Example 10:

A cone has as base the circle $x^2 + y^2 + 2ax + 2by = 0$, $z = 0$ and passes through the fixed point (0, 0, c). If the section of the cone by zx-plane is a rectangular hyperbola, prove that the vertex lies on a fixed circle.

Solution:

Let (α, β, γ) be the vertex of the cone. Any line through (α, β, γ) is given as

$$\frac{x-\alpha}{l} = \frac{y-\beta}{m} = \frac{z-\gamma}{n} \quad \text{...(i)}$$

It meets the plane z = 0 in $\left(\alpha - \frac{l\gamma}{n}, \beta - \frac{m\gamma}{n}, 0\right)$ and if this point lies on the given, conic we have

$$\left(\alpha - \frac{l\gamma}{n}\right)^2 + \left(\beta - \frac{m\gamma}{n}\right)^2 + 2a\left(\alpha - \frac{l\gamma}{n}\right) + 2a\left(\beta - \frac{m\gamma}{n}\right) = 0 \quad \text{...(ii)}$$

Eliminating l, m, n between (i) and (ii), the equation of the cone is

$$\left[\alpha - \left(\frac{x-\alpha}{z-\gamma}\right)\gamma\right]^2 + \left[\beta - \left(\frac{y-\beta}{z-\gamma}\right)\gamma\right]^2 + 2a\left[\alpha - \left(\frac{x-\alpha}{z \quad \gamma}\right)\gamma\right]$$
$$+2b\left[\beta - \left(\frac{y-\beta}{z-\gamma}\right)\gamma\right] = 0$$

$$\Rightarrow \quad (\alpha x - x\gamma)^2 + (\beta z - \gamma z)^2 + 2a(\alpha z - xy)(z - \gamma) + 2b(\beta z - y\gamma)(z - \gamma) = 0.$$

If this cone passes through (0, 0, c), then

$$(\alpha c)^2 + (\beta c)^2 + 2a(\alpha c)(c - \gamma) + 2b(\beta c)(c - \gamma) = 0 \quad \text{...(iii)}$$

Again the section of the cone by zx-plane *i.e.,* y = 0 is

$$(\alpha z - \gamma x)^2 + (\beta z)^2 + 2a(\alpha z - x\gamma)(z - \gamma) + 2b(\beta z)(z - \gamma) = 0$$

and if this section is a rectangular hyperbola in the zx-plane, then the sum of the coefficients of x^2 and z^2 should be zero

i.e., $\gamma^2 + (\alpha^2 + \beta^2 + 2a\alpha + 2b\beta) = 0$ ...(iv)

$\therefore$ The locus of (α, β, γ) from (iii) and (iv) is

$$c(x^2 + y^2) + 2ax(c - z) + 2by(c - z) = 0 \qquad ...(v)$$

and $$x^2 + y^2 + z^2 + 2ax + 2by = 0 \qquad ...(vi)$$

Multiplying (vi) by c and subtracting (v) from the result so obtained, we get

$$cz^2 + 2azx + 2byz = 0$$

$$\Rightarrow \quad 2ax + 2by + cz = 0, \qquad ...(vii)$$

which is the equation of a plane. Therefore the required locus of the vertex is given by (vi) [or (v)] and (vii) which taken together represent a circle.

Example 11:

Find the equation of the cone whose vertex is (1, 2, 3) and guiding curve is the circle $x^2 + y^2 + z^2 = 4$, $x + y + z = 1$.

Solution:

Any generator through (1, 2, 3) is

$$\frac{x-1}{l} = \frac{y-2}{m} = \frac{z-3}{n} = \frac{x+y+z-6}{l+m+n} \qquad ...(i)$$

If it meets the plane $x + y + z = 1$, then from (i) we have

$$\frac{x-1}{l} = \frac{y-2}{m} = \frac{z-3}{n} = \frac{1-6}{l+m+n} = \frac{-5}{l+m+n}$$

which gives $x = 1 - 1\,[5l/(l + m + n)]$, $y = 2 - [5m/(l + m + n)]$ and $z = 3 - [5n/(l + m + n)]$, *i.e.*, the generator (i) meets the plane

$$x + y + z = 1$$

in the point $\left[\frac{m+n-4l}{l+m+n}, \frac{2l-3m+2n}{l+m+n}, \frac{3l+3m-2n}{l+m+n}\right]$

If this point lies on $x^2 + y^2 + z^2 = 4$ we get

$$(m + n - 4l)^2 + (2l - 3m + 2n)^2 + (3l + 3m - 2n)^2 = 4(l + m + n)^2 \qquad ...(ii)$$

Eiiminating l, m, n between (i) and (ii) we get the required equation as

$$[(y - 2) + (z - 3) - 4(x - 1)]^2 + [2(x - 1) - 3(y - 2) + 2(z - 3)]^2$$
$$+ [3(x - 1) + 3(y - 2) - 2(z - 3)]^2 = 4[(x - 1) + (y - 2) + (z - 3)]^2$$

$$\Rightarrow \quad (y + z - 4x - 1)^2 + (2x - 3y + 2z - 2)^2 + (3x + 3y - 2z - 3)^2 = 4(x + y + z - 6)^2$$

$$\Rightarrow \quad 5x^2 + 3y^2 + z^2 - 6yz - 4zx - 2xy + 6x + 8y + 10z - 26 = 0.$$ **Ans.**

Example 12:

Prove that the equation

$$ax^2 + by^2 + cz^2 + 2ux + 2vy + 2wz + d = 0$$

represents a cone if $u^2/a + v^2/b + w^2/c = d$.

Solution:

Let $F(x, y, z, t) = ax^2 + by^2 + cz^2 + 2uxt + 2vyt + 2wzt + dt^2 = 0$

$\therefore \quad \dfrac{\partial F}{\partial x} = 0$ for t = 1 gives $2ax + 2u = 0$ or $x = -u/a$ ...(i)

Similarly $\dfrac{\partial F}{\partial y} = 0$ for t = 1 gives $y - - v/b$; ...(ii)

$\dfrac{\partial F}{\partial z} = 0$ for t = 1 gives $z = -w/c$ ...(iii)

and $\dfrac{\partial F}{\partial t} = 0$ for t = 1 gives $ux + vy + wz + d = -0$ ...(iv)

Substituting the values x, y, z from (i), (ii), (iii) in (iv) we get the required condition as $u(-u/a) + v(-v/b) + w(-w/c) + d = 0$

$\Rightarrow \quad (u^2/a) + (v^2/b) + (w^2/c) = d.$ **Hence proved.**

Example 13:

Find the equation of the cone with vertex at (2a, b, c) and passing through the curve $x^2 + y^2 = 4a^2$ *and* $z = 0$. *Find b and c if the cone also passes through the curve* $y^2 = 4a(z + a)$, $x = 0$. *Also show that the cone is cut by the plane* $y = 0$ *in two straight lines and the angle* θ *between them is given by* $\tan\theta = 2$.

Solution:

Any line through (2a, b, c) is $\dfrac{x-2a}{l} = \dfrac{y-b}{m} = \dfrac{z-c}{n}$ ...(i)

It meets z = 0 in the point $\left[2a - \dfrac{lc}{n}, b - \dfrac{mc}{n}, 0\right]$ and if it lies on the curve

$x^2 + y^2 = 4a^2$, $z = 0$, then we get $\left(2a - \dfrac{lc}{n}\right)\left(b - \dfrac{mc}{n}\right)^2 = 4a^2$, ...(ii)

Eliminating l, m, n between (i) and (ii) we get the required cone as

$$\left[2a - \left(\frac{x-2a}{z-c}\right)c\right]^2 + \left[b - \left(\frac{y-b}{z-c}\right)c\right]^2 = 4a^2$$

$\Rightarrow \qquad (2az - cx)^2 + (bz - yc)^2 = 4a^2 (z - c)^2 \qquad ...(iii)$

If this cone passes through $y^2 = 4a (z + a)$, $x = 0$, then putting $x = 0$ in (iii), we get $(2az)^2 + (bz - yc)^2 = 4a^2 (z - c)^2$

$\Rightarrow \qquad b^2z^2 + c^2y^2 - 2bcyz - 4a^2c^2 + 8a^2zc = 0 \qquad ...(iv)$

If it is the same as $y^2 = 4a (z + a)$, then comparing this with (iv) we have $b^2 = 0$ *i.e.*, $b = 0$ which reduces (iv) to

$$c^2y^2 = 4a^2c^2 - 8a^2zc \quad \text{or} \quad y^2 = -(8a^2/c)\left(z - \frac{1}{2}c\right).$$

Comparing this with $y^2 = 4a (z + a)$, we get $-(8a^2/c) = 4a$ and $-\frac{1}{2}c = a$ which gives $c = -2a$. Hence $b = 0$, $c = -2a$.

Substituting these values of b and c in (iii), the equation of cone intersecting the given conics reduces to $(2ax + 2az)^2 + 4a^2y^2 = 4a^2 (z + 2a)^2$

$\Rightarrow \qquad x^2 + y^2 + 2zx - 4az - 4a^2 = 0. \qquad ...(v)$

The plane $y = 0$ cuts cone (v) in $x^2 + 2zx - 4az - 4a^2 = 0$, $y = 0$

$\Rightarrow \qquad (x^2 - 4a^2) + 2z (x - 2a) = 0, \ y = 0$

$\Rightarrow \qquad (x - 2a) [x + 2a + 2z] = 0, \ y = 0$

$\Rightarrow$ $x - 2a = 0$, $y = 0$ and $x + 2a + 2z = 0$, $y = 0$ which give the required lines. These lines lie in the plane $y = 0$ and their combined equation is given by

$$y = 0, \ x^2 + 2zx - 4ax - 4a^2 = 0.$$

$\therefore$ If θ be the angle between these lines, then

$$\tan\theta = \frac{"2\sqrt{(h^2 - ab)}"}{a + b} = \frac{2\sqrt{(1^2 - 0)}}{1 + 0} = 2.$$ **Hence proved.**

Example 14:

Show that the equation $2y^2 - 8yz - 4zx - 8xy + 6x - 4y - 2z + 5 = 0$ represents a cone whose vertex is $(-7/6, 1/3, 5/6)$.

Solution:

Making the given equation homogeneous, we get

$$F(x, y, z, t) \equiv 2y^2 - 8yz - 4zx - 8xy + 6xt - 4yt - 2zt + 5t^2 = 0$$

$\therefore \qquad \frac{\partial F}{\partial x} = 0$ gives $-4z - 8y + 6t = 0 \Rightarrow 2z + 4y - 3t = 0$

$\frac{\partial F}{\partial y} = 0$ gives $4y - 8z - 8x - 4t = 0 \Rightarrow y - 2z - 2x - t = 0$

$$\frac{\partial F}{\partial z} = 0 \text{ gives } - 8y - 4x - 2t = 0 \Rightarrow 4y + 2x + t = 0$$

$$\frac{\partial F}{\partial t} = 0 \text{ gives } 6x - 4y - 2z + 10t = 0 \Rightarrow 3x - 2y - z + 5t = 0$$

Putting $t = 1$, we get $2z + 4y - 3 = 0$ ...(i)

$y - 2z - 2x - 1 = 0$ ...(ii)

$4y + 2x + 1 = 0$ ...(iii)

and $3x - 2y - z + 5 = 0$ ...(iv)

Adding (i) and (v), we get $9y - 3 = 0 \Rightarrow y = 1/3$

$\therefore$ From (v) $2z = 5y = 5\ (1/3) \Rightarrow z = 5/6$

And from (iii), $2x = -4 - 1 = -4\ (1/3) - 1 = -7/3 \Rightarrow x = -7/6$

These values viz. $x = -7/6$, $y = 1/3$, $z = 5/6$ satisfy (iv) and so the given equation represents a cone its vertex is $(-7/6, 1/3, 5/6)$. **Hence proved.**

Example 15:

Find the equations to the planes in which the plane $2x + y - z = 0$ cuts the cone $4x^2 - y^2 + 3z^2 = 0$.

Find also the angle between the lines of section.

Solution:

Let the plane $2x + y - z = 0$ ...(i)

cut the cone $4x^2 - y^2 + 3z^2 = 0$ ...(ii)

in a line $x/l = y/m = z/n$ ...(iii)

Then $2l + m - n = 0$ and $4l^2 - m^2 + 3n^2 = 0$ ...(iv)

Eliminating n between these relations, we get

$$4l^2 - m^2 + (2l + m)^2 = 0 \qquad \textbf{(Note)}$$

$$\Rightarrow \quad 16l^2 + 12lm + 2m^2 = 0$$

$$\Rightarrow \quad 8\ (l/m)^2 + 6\ (l/m) + 1 = 0$$

$$\Rightarrow \quad l/m = (1/16)\ [-6 \pm \sqrt{\{36-32\}}\ ] = -\frac{1}{4}, -\frac{1}{2}$$

$$\Rightarrow \quad l = -\frac{1}{4}m \quad \text{and} \quad l = -\frac{1}{2}m \qquad \text{...(v)}$$

From (iv), $n = 2l + m = 2\left(-\frac{1}{4}m\right) + m$, when $l = -\frac{1}{4}m$

$\Rightarrow \quad n = \frac{1}{2} m. \qquad \therefore\ l : m : n = -\frac{1}{4}\ m : m : \frac{1}{2} m$

i.e., $\qquad l : m : n = -1 : 4 : 2$ **(Note)**

Again when $l = -(1/2)$ m, then from (iv), we get

$$n = 2l + m = 2\left(-\frac{1}{2} m\right) + m = 0.m$$

$$\therefore \quad l : m : n = -(1/2)\ m : m : 0\ .\ m = -1 : 2 : 0$$

Hence the required equations of the lines are

$$\frac{x}{-1} = \frac{y}{4} = \frac{z}{2}$$

and $$\frac{x}{-1} = \frac{y}{2} = \frac{z}{0}.$$ **Ans.**

Also if θ be the required angle between these lines, then

$$\cos\theta = \frac{(-1)(-1) + (4)(2) + (2)(0)}{\sqrt{\left[(-1)^2 + 4^2 + 2^2\right]}\sqrt{\left[(-1)^2 + 2^2 + 0^2\right]}} = \frac{9}{\sqrt{(21)}\sqrt{(5)}}$$ **Ans.**

Example 16:

Find the condition that the plane $ux + vy + wz = 0$ may meet the cone $f(x, y, z) \equiv ax^2 + by^2 + cz^2 + 2fyz + 2gzx + 2hxy =$ in two mutually perpendicular lines.

Solution :

Let $x/l = y/m = z/n$ be one of the lines in which the plane ux + vy + wz = meets the cone f (x, y, z) = 0, then we have

$$ul + vm + wn = 0 \qquad ...(i)$$

and $$f(l, m, n) \equiv al^2 + bm^2 + cn^2 + 2fmn + 2gnl + 2hlm = 0 \qquad ...(ii)$$

Eliminating n between (i) and (ii) we get

$$al^2 + bm^2 + c\left(-\frac{ul + vm}{w}\right)^2 + 2fm\left(-\frac{ul + vm}{w}\right) + 2g\left(-\frac{ul + vm}{w}\right)l + 2hlm = 0$$

$$\Rightarrow \quad aw^2l^2 + bw^2m^2 + c(ul + vm)^2 - 2fwm(ul + vm) - 2gwl(ul + vm) + 2hw^2lm = 0$$

$$\Rightarrow \quad (aw^2 + cu^2 - 2guw)\ l^2 + 2(cuv - fuw - gvw + hw^2)\ lm + (bw^2 + cv^2 - 2fvw)\ m^2 = 0$$

$\Rightarrow \quad (aw^2 + cu^2 - 2guw)\,(l/m^2) + 2\,(cuv - fuw - gvw + hw^2)\,(l/m)$
$+ (bw^2 + cv^2 - 2fvw) = 0,$

which is a quadratic equation in l/m and if its roots are l_1/m_1 and l_2/m_2, then

we get $\dfrac{l_1}{m_1}\cdot\dfrac{l_2}{m_2}$ = product of the roots = $\dfrac{bw^2 + cv^2 - 2fvw}{aw^2 + cu^2 - 2guw}$

$\Rightarrow \quad \dfrac{l_1 l_2}{bw^2 + cv^2 - 2fvw} = \dfrac{m_1 m_2}{cu^2 + aw^2 - 2gwu} = \dfrac{n_1 n_2}{av^2 + bu^2 - 2luv}$, by symmetry, **(Note)**

If the lines of intersection of the given plane and cone are mutually perpendicular, then $l_1 l_2 + m_1 m_2 + n_1 n_2 = 0$

i.e., $(bw^2 + cv^2 - 2fvw) + (cu^2 + aw^2 - 2gwu) + (av^2 + bu^2 - 2huv) = 0$

$\Rightarrow \quad a\,(v^2 + w^2) + b\,(w^2 + u^2) + c\,(u^2 + v^2) = 2\,(fvw + gwu + huv),$

which is the required condition.

Example 17:

Prove that the condition that the plane $ux + vy + wz = 0$ may cut the cone $ax^2 + by^2 + cz^2 = 0$ in perpendicular generators if

$$(b + c)\,u^2\,(c + a)\,v^2 + (a + b)\,w^2 = 0$$

Solution :

Let $x/l = y/m = z/n$ be one of the lines in which the plane $ux + vy + wz = 0$ meets the cone $ax^2 + by^2 + cz^2 = 0$, then we have

$$ul + vm + wn = 0 \qquad \text{...(i)}$$

and $$al^2 + bm^2 + cn^2 = 0 \qquad \text{...(ii)}$$

Eliminating n between (i) and (ii), we get

$$al^2 + bm^2 + c\,[-(ul + vm)/w]^2 = 0$$

$\Rightarrow \quad (aw^2 + cu^2)\,l^2 + 2cuvlm + (bw^2 + cv^2)\,m^2 = 0$

$\Rightarrow \quad (aw^2 + cu^2)\,(l^2/m^2) + 2cuv(l/m)\,(bw^2 + cv^2) = 0$

If its roots are l_1/m_1 and $_2/m_2$, then we have

$$\frac{l_1}{m_1}\cdot\frac{l_2}{m_2} = \text{product of the roots} = \frac{bw^2 + cv^2}{aw^2 + cu^2}$$

$\Rightarrow \quad \dfrac{l_1 l_2}{bw^2 + cv^2} = \dfrac{m_1 m_2}{cu^2 + aw^2} = \dfrac{n_1 n_2}{av^2 + bu^2}$, by symmetry.

If the lines are perpendicular then $l_1 l_2 + m_1 m_2 + n_1 n_2 = 0$

i.e., $(bw^2 + cv^2) + (cu^2 + aw^2) + (av^2 + bu^2) = 0$

$\Rightarrow$ $(b + c)\, u^2 + (c + a)\, v^2\, (a + b)\, w^2 = 0$ **Hence proved.**

Example 18:

Prove that the plane $ax + by + cz = 0$ cuts the cone $yz + zx + xy = 0$ in perpendicular lines if $1/a + 1/b + 1/c = 0$.

Solution:

Let the plane $ax + by + cz = 0$ cut the cone

$$yz + zx + xy = 0 \text{ in a line } x/l = y/m = z/n.$$

Then $mn + nl + xy = 0$ and $al + bm + cn = 0$

Eliminating n between these relations, we get

$$(m + l)\,[-(al + bm)/c] + lm + 0$$

$$\Rightarrow \quad al^2 + (a + b - c)\, lm + bm^2 = 0$$

$$\Rightarrow \quad a(l/m)^2 + (a + b - c)\,(l/m) + b = 0$$

$$\frac{l_1}{m_1} \cdot \frac{l_2}{m_2} = \text{product of the roots} = \frac{b}{a}$$

$$\Rightarrow \quad \frac{l_1 l_2}{1/a} = \frac{m_1 m_2}{1/b} = \frac{n_1 n_2}{1/c}, \text{ by symmetry.}$$

If these are at right angles, then $l_1 l_2 + m_1 m_2 + n_1 n_2 = 0$

$\Rightarrow$ $(1/a) + (1/b) + (1/c) = 0.$ **Hence proved.**

Example 19:

Find the angle between the lines of section of the plane $6x - y - 2z = 0$ and the cone $108x^2 - 7y^2 - 20z^2 = 0$.

Solution:

Let $x/l = y/m = z/n$ be one of the lines in which the given plane cuts the given cone, then we have

$$6l - m - 2n = 0 \qquad \text{...(i)}$$

and $$108l^2 - 7m^2 - 20n^2 = 0 \qquad \text{...(ii)}$$

Eliminating n between (i) and (ii) we get

$$108l^2 - 7m^2 - 20\left[\frac{1}{2}(6l - m)^2\right] = 0$$

$$\Rightarrow \quad 6l^2 - 5lm + m^2 = 0 \Rightarrow (3l - m)\,(2l - m) = 0$$

When $3l - m = 0$, from (i) we have $3l - 2n = 0$

$$\therefore \qquad 3l = m = 2n \Rightarrow \frac{l}{(1/3)} = \frac{m}{1} = \frac{n}{(1/2)}$$

When $\quad 2l - m = 0$, rom (i) we have $2l - n = 0$

$$\therefore \qquad 2l = m = n \quad \Rightarrow \quad \frac{l}{(1/2)} = \frac{m}{1} = \frac{n}{2}$$

$$\cos\theta = \frac{l_1 l_2 + m_1 m_2 + n_1 n_2}{\sqrt{(l_1^2 + m_1^2 + n_1^2)}\sqrt{(l_2^2 + m_2^2 + n_2^2)}}$$

$$= \frac{\frac{1}{3}.\frac{1}{2} + 1.1 + \frac{1}{2}.1}{\sqrt{\left(\frac{1}{9} + 1 + \frac{1}{4}\right)}\sqrt{\left(\frac{1}{4} + 1 + 1\right)}} = \frac{\frac{1}{6} + 1 + \frac{1}{2}}{\sqrt{\left(\frac{49}{36}\right)}.\sqrt{\left(\frac{9}{4}\right)}} = \frac{(10/6)}{\frac{7}{6}.\frac{3}{2}} = \frac{20}{21}$$

$\Rightarrow \qquad \theta = \cos^{-1}(20/21)$ **Ans.**

Example 20:

The plaes $x/a + y/b + z/c = 1$ meets the coordinate axes in A, B, C. Prove that the equation of the cone generated by lines drawn from O to meet the circle ABC is

$$yz\left(\frac{b}{c} + \frac{c}{b}\right) + zx\left(\frac{c}{a} + \frac{a}{c}\right) + xy\left(\frac{a}{b} + \frac{b}{a}\right) = 0.$$

Or

Show that the equation of the cone whose vertex is the origin and whose base is the circle through the points (a, 0, 0), (0, b, 0), (0, 0, c) is

$$\Sigma a\,(b + c)\,yz = 0.$$

Or

The plane $x/a + y/b + z/c = 1$ cuts the coordinate axes in A, B, C. Prove that the lines passing through the origin are intersecting the circle ABC generate the following cone.

$$yz\left(\frac{b}{c} + \frac{c}{b}\right) + zx\left(\frac{c}{a} + \frac{a}{c}\right) + xy\left(\frac{a}{b} + \frac{b}{a}\right) = 0.$$

Solution:

The plane ABC is given by $x/a + y/b + z/c = 1$ ...(i)

It meets the axes A (a, 0, 0), B (0, b, 0), and C (0, 0, c). The equation of the sphere OABC is given as

$$x^2 + y^2 + z^2 - ax - by - cz = 0, \qquad ..;(ii)$$

The required cone is generated by the lines drawn from O to meet the circle ABC [given by (i) and (ii) together] and will be homogeneous. So making (ii) homogeneous with the help of (i), we get the required equation as

$$x^2 + y^2 + z^2 - (ax + by + cz)(x/a + y/b + z/c) = 0$$

$$\Rightarrow \qquad yz\left(\frac{b}{c}+\frac{c}{b}\right) + zx\left(\frac{c}{a}+\frac{a}{c}\right) + xy\left(\frac{a}{b}+\frac{b}{a}\right) = 0.$$ **Hence Proved.**

Example 21:

OP and OQ are two lines which remain perpendicular and move so that the plane OPQ passes through OZ. If OP describes the cone $f(y/x, z/x) = 0$, prove that OQ describes the cone

$$f\left\{\frac{y}{x}, \left(-\frac{x}{z} - \frac{y^2}{zx}\right)\right\} = 0$$

Solution:

Let the direction cosines of the lines OP and OQ passing through the origin be l_1, m_1, n_1 and l_2, m_2, n_2.

Since OP is perpendicular to OQ then we have

$$l_1 l_2 + m_1 m_2 + n_1 n_2 = 0$$

Also the equation of any plane through z-axis (*i.e.*, x = 0, y = 0) is given by

$$x + \lambda y = 0 \qquad ...(ii)$$

If (ii) represents the plane OPQ, then the line OP and OQ lie on the plane given by (ii) then we have

$$l_1 + \lambda m_1 = 0 \qquad ...(iii)$$

and $$l_2 + \lambda m_2 = 0 \qquad ...(iv)$$

Also it is given that OP describes the cone $f(y/x, z/x) = 0$ ...(v)

∴ The direction cosines l_1, m_1, n_1 of the generator OP must satisfy (v) and so we have

$$f(m_1/l_1, n_1/l_1) = 0 \qquad ...(vi)$$

Now, if we eliminate l_1, m_1, n_1 from (iii), (iv) and (vi) and obtain a relation between l_2, m_2, n_2, then we can find the locus of OQ whose d.c.'s are l_2, m_2, n_2.

From (iii) and (iv) we get $\frac{m_1}{l_1} = \frac{1}{\lambda} = \frac{m_2}{l_2}$ ...(vii)

Dividing each term of (i) by $l_1 n_2$ we get

$$\frac{l_2}{n_2} + \frac{m_1}{l_1} \cdot \frac{m_2}{n_2} + \frac{n_1}{l_1} = 0 \Rightarrow \frac{l_2}{n_2} + \frac{m_2}{l_2} \cdot \frac{m_2}{n_2} + \frac{n_1}{l_1} = 0, \text{ from} \quad \text{(vii)}$$

$$\Rightarrow \quad \frac{n_1}{l_1} = -\left(\frac{l_2}{n_2} + \frac{m_2^2}{l_2 n_2}\right) \quad \text{...(viii)}$$

Substituting the values of m_1/l_1 and n_1/l_1 in terms of l_2, m_2, n_2 in (vi) then we have

$$f\left[\frac{m_2}{l_2}, -\left(\frac{l_2}{n_2} + \frac{m_2^2}{l_2 n_2}\right)\right] = 0 \quad \text{...(ix)}$$

Also the equations of the line OQ whose d.c.'s ar l_2, m_2, n_2 and which passes through O (0, 0, 0) are $x/l_2 = y/m_2 = z/n_2$ ...(x)

∴ With the help of (x) the locus of OQ from (ix) is

given as $f\left[\frac{y}{x}, -\left(\frac{x}{z} + \frac{y^2}{zx}\right)\right] = 0$. **Hence proved.**

Example 22:

Find the equation to the cone whose vertex is the point (0, 0, 0) and whose base is the curve $x^2 + y^2 = 4$, $z = 2$.

Solution:

Let the equations of the given curve be written in the homogeneous form by introducing a new variable t as

$$x^2 + y^2 = 4r^2 \quad \text{...(i)}$$

and $\quad z = 2t$ ...(ii)

Substituting the value of t from (ii) in (i) we get the required equation as

$$x^2 + y^2 = z^2. \quad \textbf{Ans.}$$

Example 23:

Find the equation of the cone with vertex at (0, 0, 0) and passing through the circle given by $x^2 + y^2 + z^2 + x - 2y + 3z - 4 = 0$, $x - y + z = 2$.

Solution:

The equations of the circle are given as

$$x^2 + y^2 + z^2 + x - 2y + 3z - 4 = 0 \quad \text{...(i)}$$

and $x - y + z = 2.$ ...(ii)

Making (i) homogeneous with the help of (ii) we get the required equations as follows

$$(x^2 + y^2 + z^2) + (x - 2y + 3z) \left[\left[\frac{1}{2}(x-y+z)\right] - 4\left[\frac{1}{2}(x-y+z)\right]^2 = 0.$$

(Note)

$\Rightarrow$ $2(x^2 + y^2 + z^2) + (x - 2y + 3z)(x - y + z) - 2(x - y + z)^2 = 0$

$\Rightarrow$ $x^2 + 2y^2 + z^2 - xy - 3yz + 2zx = 0.$

Example 24:

Show that $2x^2 + 3y^2 - 5z^2 = 0$ represents a cone with its vertex at the origin. What relation must the direction cosines of the generators of this cone satisfy?

Solution:

As we know $2x^2 + 3y - 5z^2 = 0$ ...(i)

is a homogeneous equation of second degree in x, y and z, so it represents a cone with its vertex at origin.

Let l, m, n be the d. cosines of any generator of the cone given by (i), then l, m, n must satisfy (i). So the required relation is

$$2l^2 + 3m^2 - 5n^2 = 0.$$ **Ans.**

Example 25:

Find the equation of the cone with vertex at the origin and which passes through the curve $ax^2 + by^2 + cz^2 = 1$; $\alpha x^2 + \beta y^2 = 2z$.

Solution:

Let the equations of the curve be written in the homogeneous form as

$$ax^2 + by^2 + cz^2 = t^2 ...(i), \quad \alpha x^2 + \beta y^2 = 2zt$$ ...(ii)

Here a new variable t has been introduced to make them homogeneous.

(Note)

From (ii), $t = (\alpha x^2 + \beta y^2)/2z$. Substituting this value of t in (i) we get the required cone as

$$(ax^2 + by^2 + cz^2) 4z^2 = (\alpha x^2 + \beta y^2)^2.$$ **Ans.**

Example 26:

Show that a cone of second degree can be found to pass through any five concurrent lines.

Solution:

Let the point of concurrence of the five lines be the origin. Then their equations can be taken as follows

$$\frac{x}{lr} = \frac{y}{mr} = \frac{z}{nr}\ ;\ r = 1, 2, 3, 4, 5. \qquad ...(i)$$

Also we know that general second degree equation of a cone is given by

$$ax^2 + by^2 + cz^2 + 2fyz + 2gzx + 2hxy = 0$$

$$\Rightarrow \qquad x^2 + \frac{b}{a}y^2 + \frac{c}{a}z^2 + 2\left(\frac{f}{a}\right)yz + 2\left(\frac{g}{a}\right)zx + 2\left(\frac{h}{a}\right)xy = 0,$$

dividing each term by a we have

$\Rightarrow \quad x^2 + b_1y^2 + c_1z^2 + 2f_1yz + 2g_1zx + 2h_1xy + 0$, where $b_1 = b/a$ etc.

This equation of the cone contains five arbitrary constants b_1, c_1, f_1 g_1 and h_1 and so it can satisfy five independent conditions.

Now we also know that the d.c.'s of generators of a cone satisfy the equation of the cone, so any five lines passing through the origin given (i) are sufficient to evaluate the five arbitrary constants.

Hence a cone of second degree can be found to pass through any five concurrent lines. **Hence prroved.**

Example 27:

Find the equation of the cone which passes through three coordinates axes and the lines

$$x/1 = y/(-2) = z/3;\ x/3 = y/2 = z/(-1).$$

Solution:

If the cone passes through the coordinate axes, then its vertex must be the origin and so its equation must be of the form

$$ax^2 + by^2 + cz^2 + 2fyz + 2gzx + 2hxy = 0 \qquad ...(i)$$

Also the d.c.'s of the coordinate axes are (1, 0, 0), (0, 1, 0) and (0, 0, 1) which should satisfy (i).

So we get a = 0, b = 0, c = 0 and so from (i), the equation of the cone reduces to

$$fyz + gzx + hxy = 0 \qquad ...(ii)$$

Now if the given lines $x/1 + y/(-2) = z/3$ and $x/3 = y/2 = z/(-1)$ whose d.c.'s are 1, – 2, 3 and 3, 2, – 1 also be on (ii), then

We have f (– 2) (3) + g (3) (1) + h (1) (– 2) = 0

and $\quad f(2)(-1)+g(-1)(3)+h(3)(2)=0$

$\Rightarrow \quad 6f-3g+2h=0 \quad$ and $2f+3g-6h=0$

Solving these, we get $\dfrac{f}{18-6}=\dfrac{g}{4+36}=\dfrac{h}{18+6}$ or $\dfrac{f}{3}=\dfrac{g}{10}=\dfrac{h}{6}$

∴ From (ii), required equation is $3yz+10zx+6xy=0$. **Ans.**

Example 28:

Find the equation of the cone with vertex at the origin and which passes through the curve given by

$$(x^2/a^2)+(y^2/b^2)+(z^2/c^2)=1,\ (x^2/\alpha^2)+(y^2/\beta^2)=2z.$$

Solution:

Let the equations of the curve be written in the homogeneous form by introducing a new variable t as follows

$$\frac{x^2}{a^2}+\frac{y^2}{b^2}+\frac{z^2}{c^2}=t^2 \qquad ...(i)$$

and
$$\frac{x^2}{\alpha^2}+\frac{y^2}{\beta^2}=2zt \qquad ...(ii)$$

from (ii) we have $t=\dfrac{\beta^2x^2+\alpha^2y^2}{2\alpha^2\beta^2z}$. Substituting this value of t in (i), we obtain

$$\frac{x^2}{a^2}+\frac{y^2}{b^2}+\frac{z^2}{c^2}=\left(\frac{x^2\beta^2+\alpha^2y^2}{2\alpha^2\beta^2z}\right)^2$$

$$\Rightarrow \quad \frac{x^2}{a^2}+\frac{y^2}{b^2}+\frac{z^2}{c^2}=\frac{1}{4z^2}\left(\frac{x^2}{\alpha^2}+\frac{y^2}{\beta^2}\right)^2. \qquad \textbf{Ans.}$$

which is the required equation of the cone.

Example 29:

Find the equation of the cone which has vertex at the origin and passes through the curve given by

$$ax^2+by^2=2z,\ lx+my+nz=p.$$

Solution:

Making the equation of the given surface homogeneous with the help of the equation of the given plane, then we have

$$ax^2 + by^2 = 2z\,[(lx + my + nz)/p]$$

$\Rightarrow$ $$p\,(ax^2 + by^2) = 2z\,(lx + my + nz),$$

which is the required equation. **Ans.**

Example 30:

Find the equation of the cone whose vertex is the origin and base the circle $x = a$, $y^2 + z^2 = b^2$ and show that the section of the cone by a plane parallel to the plane XOY is a hyperbola.

Solution:

The equation of the cone is $y^2 + z^2 = b^2\,(x/a)^2$, making the equations of the circle homogeneous we have

i.e., $$a^2\,(y^2 + z^2) = b^2x^2 \Rightarrow b^2x^2 + a^2y^2 - a^2z^2 = 0.$$ **Ans.**

Its section by a plane $z = c$ parallel to the plane XOY is the conic $b^2x^2 - a^2y^2 - a^2c^2 = 0$, $z = c$, which is a hyperbola. **Ans.**

Example 31:

Find the equation to the cone whose vertex is at (0, 0, 0) and which passes through the curve

$$x^2 + y^2 + z^2 + x - 2y + 3z + 4,\ x^2 + y^2 + z^2 + 2x - 3y + 4z = 5.$$

Solution:

The curve is given by $x^2 + y^2 + z^2 + x - 2y + 3z = 4$. ...(i)

and $$x^2 + y^2 + z^2 + 2x - 3y + 4z = 5$$ ...(ii)

Substracting (i) from (ii) we get $x - y + z = 1$. ...(iii)

Now making (i) homogeneous with the help of (iii) we get the required equation as $x^2 + y^2 + z^2 + (x - 2y + 3z)\,(x - y + z) = 4\,(x - y + z)^2$

$\Rightarrow$ $$2x^2 + y^2 - 5xy - 3yz + 4zx = 0$$ **Ans.**

Example 32:

Find the equation of the cone whose vertex is (0, 0, 0) and which passes through the curve of intersection of the plane $lx + my + nz = p$ and the surface $ax^2 + by^2 + cz^2 = 1$.

Solution:

Making the equation of the given surface homogeneous with the help of the equation of the given plane, we have

$$ax^2 + by^2 + cz^2 = [(lx + my + nz)/p]^2$$ **(Note)**

$$\Rightarrow \quad p^2 (ax^2 + by^2 + cz^2) = (lx + mx + nz)^2,$$

which is the required equation. **Ans.**

Example 33:

Planes through OX, Oy include as angle α. Show that their line of intersection lies on the cone

$$z^2 (x^2 + y^2 + z^2) = x^2y^2 \tan^2 \alpha.$$

Solution:

The equation of any plane through OX *i.e.,* y = 0, z = 0 is

$$y + \lambda z = 0 \quad \text{or} \quad 0.x + 1.y + \lambda.z = 0 \qquad \text{...(i)}$$

Similarly the equation of the plane through OY is given as

$$z + \mu x = 0 \quad \text{or} \quad \mu.x + 0.y + 1.z = 0. \qquad \text{...(ii)}$$

Since a is the angle between these planes, so we have

$$\cos \alpha = \frac{0.\mu + 1.0 + \lambda.1}{\sqrt{(0^2 + 1^2 + \lambda^2)}.\sqrt{(\mu^2 + 0^2 + 1^2)}} = \frac{\lambda}{\sqrt{(1+1^2)}\sqrt{(1+\mu^2)}}$$

$$\Rightarrow \quad \sec^2 \alpha = \frac{(1+\lambda^2)(1+\lambda^2)}{\lambda^2} = \frac{1+\lambda^2+\mu^2+\lambda^2\mu^2}{\lambda^2}$$

$$\Rightarrow \quad \lambda^2 (1 + \tan^2 \alpha) = 1 + \lambda^2 + \mu^2 + \lambda^2 \mu^2$$

$$\Rightarrow \quad \lambda^2 \tan^2 \alpha = 1 + \mu^2 + \lambda^2 \mu^2 \qquad \text{...(iii)}$$

Now from (i) and (ii), we get $\lambda = - y/z$, $\mu = - z/x$,

Substituting these values of λ and μ in (iii), we get

$$(y^2/z^2) \tan^2 \alpha = 1 + (z^2/x^2) + (y^2/z^2)(z^2/x^2)$$

$$\Rightarrow \quad x^2y^2 \tan^2 \alpha = z^2 (x^2 + y^2 + z^2),$$

which being a homogeneous equation represents a cone with its vertex at (0, 0, 0). **Hence proved.**

Example 34:

Prove that the equation of the cone whose vertex is (0, 0, 0) and base the curve z = k, f (x, y) = 0 is f [xk/z), (yk/z)] = 0, where f (x, y) ° ax² + 2hxy + by² + 2gx + 2fy + c = 0.

Solution:

Making f (x, y) = 0, homogeneous with the help of z = k, we get the required equationof the cone as follows

$$ax^2 + 2hxy + by^2 + 2gx (z/k) + 2fy (z/k) + c (z^2/k^2) = 0$$

$\Rightarrow \quad ax^2k^2 + 2hxyk^2 + by^2k^2 + 2gxzk + 2fyzk + cz^2 = 0$

$\Rightarrow \quad a\,(xk/z)^2 + 2h\,(xk/z)\,(yk/z) + b\,(yk/z)^2 + 2g(xk/z) + 2f(yk/z) + c = 0$

$\Rightarrow \quad f\,[(xk/a), (yk/z)] = 0.$ **Hence proved.**

Example 35:

Show that the lines drawn through the point (α, β, γ) whose direction cosines satisfy $al^2 + bm^2 + cn^2 = 0$ generate the cone

$$a\,(x - \alpha)^2 + b\,(y - \beta)^2 + c\,(z - \gamma)^2 = 0$$

Solution:

Any line through (α, β, γ) is $\dfrac{x-\alpha}{l} = \dfrac{y-\beta}{m} = \dfrac{z-\gamma}{n}$ ...(i)

Its direction cosines satisfy $al^2 + bm^2 + cn^2 = 0$ ...(ii)

Eliminating l, m, n between (i) and (ii), we get the required equation as

$$a\,(x - \alpha)^2 + b\,(y - \beta)^2 + c\,(z - \gamma)^2 = 0.$$ **Hence proved.**

Example 36:

Find the equation of the cone with vetex at the origin and direction cosines of its generators satsifying the relation

$$3l^2 - 4m^2 + 5n^2 = 0.$$

Solution:

Any line through the origin is $\dfrac{x}{l} = \dfrac{y}{m} = \dfrac{z}{n}$. ...(i)

Its direction cosines satisfy $3l^2 - 4m^2 + 5n^2 = 0$ ...(ii)

Eliminating λ, m, n between (i) and (ii) we get the required equation as

$$3\,(x)^2 - 4\,(y)^2 + 5\,(z)^2 = 0 \quad \text{or} \quad 3x^2 - 4y^2 + 5z^2 = 0.$$ **Ans.**

Example 37:

Show that a cone of the second degree can be found to pass through any two sets of rectangular axes through the same origin.

Or

Show that a cone can be found to contain any two sets of three mutually perpendicular concurrent lines as generators.

Solution:

We can prove that the equation of a cone passing through one set of rectangular axes is $fyz + gzx + hxy = 0$. ...(i)

Let the direction ratios of another set of axes through the same origin be given as l_1, m_1, n_1; l_2, m_2, n_2; l_3, m_3, n_3.

The cone (i) passes through first two axes of this second set if

$$fm_1n_1 + gn_1l_1 + hl_1m_1 = 0 \quad \text{...(ii)}$$

and $$fm_2n_2 + gn_2l_2 + hl_2m_2 = 0 \quad \text{...(iii)}$$

Adding (ii) and (iii), we get $\Sigma\, f\,(m_1n_1 + m_2n_2) = 0$ ...(iv)

Also as these axes of second set are mutually perpendicular, so we have

$$m_1n_1 + m_2n_2 + m_3n_3 = 0 \text{ etc.}$$

Using these relations (iv) reduces to $f\,(m_3n_3) + g\,(n_3l_3) + h\,(l_3m_3) = 0$ which shows that the cone (i) passes through the third axis of the second set.

Hence proved.

Example 38:

Prove that the equation

$$4x^2 - y^2 + 2z^2 + 2xy - 3yz + 12x - 11y + 6z + 4 = 0,$$

represents a cone. Hence find its vertex.

Solution:

Making the given equation homogeneous, we get

$$F\,(x, y, z, t) \equiv 4x^2 - y^2 + 2z^2 + 2xy - 3yz + 12xt + 11yt + 6zt + 4t^2 = 0$$

$$\therefore \frac{\partial F}{\partial x} = 0 \text{ gives } 8x + 2y + 12t = 0 \Rightarrow 4x + y + 6t = 0,$$

$$\frac{\partial F}{\partial y} = 0 \text{ gives } -2y + 2x - 3z - 11t = 0 \Rightarrow 2x - 2y - 3z - 11t = 0$$

$$\frac{\partial F}{\partial z} = 0 \text{ gives } 4z - 3y + 6t = 0 \Rightarrow 3y - 4z - 6t = 0$$

and $$\frac{\partial F}{\partial t} = 0 \text{ gives } 12 - 11y + 6z + 8t = 0$$

Putting t = 1, these equations become

$$4x + y + 6 = 0 \quad \text{...(i)}$$

$$2x - 2y - 3z - 11 = 0 \quad \text{...(ii)}$$

$$3y - 4z - 6 = 0 \quad \text{...(iii)}$$

$$12x - 11y + 6z + 8 = 0 \quad \text{...(iv)}$$

From (ii) we get $4x - 4y - 6z - 22 = 0$

Subtracting (i) from it we get $5y + 6z + 28 = 0$

$\Rightarrow$ $$10y + 12z + 56 = 0 \qquad ...(v)$$

Multiplying (iii) by 3 we get $9y - 12z - 18 = 0$ ...(vi)

Adding (v) and (vi) we get $19y + 38 = 0$ or $y = -2$

$\therefore$ From (iii) we get $3(-2) - 4z - 6 = 0$ or $z = -3$

From (i) we get $4x + (-2) + 6 = 0$ or $x = -1$

These values viz. $x = -1$, $y = -2$, $z = -3$ satisfy (iv) and so the given equation represents a cone and its vertex is $(-1, -2, -3)$. **Ans.**

Example 39:

Find the equation of the cone generated by rotating the line $x/l + y/m = z/n$ about the line $x/a = y/b = z/c$ as axis.

Solution:

The axis of the cone is $x/a = y/b = z/c$ ...(i)

and that of any generator is $x/l = y/m = z/n$. ...(ii)

If θ be the semi-vertical angle of the cone, we have

$$\cos\theta = \frac{al + bm + cn}{\sqrt{(\Sigma a^2)}\sqrt{(\Sigma l^2)}} \qquad ...(iii)$$

Since both (i) and (ii) intersect at (0, 0, 0), so the vertex of the cone is 0 (0, 0, 0). Let P (x, y, z) be any point on the cone, then the direction ratios of the generator OP are $x - 0$, $y - 0$, $z - 0$ *i.e.,* x, y, z.

Also OP is inclined to (i) at an angle θ, therefore we get

$$\cos\theta = \frac{ax + by + cz}{\sqrt{(\Sigma a^2)}\sqrt{(\Sigma x^2)}} \qquad ...(iv)$$

Equating the values of $\cos\theta$ from (iii) and (iv) we get the required equation of the cone as $\dfrac{al + bm + cn}{\sqrt{(\Sigma l^2)}} = \dfrac{ax + by + cz}{\sqrt{(\Sigma x^2)}}$

$\Rightarrow$ $(al + bm + cn)^2 (x^2 + y^2 + z^2) = (ax + by + cz)^2 (l^2 + m^2 + n^2)$, squaring and cross multiplying. **Ans.**

Example 40:

Prove that the plane $z = 0$ cuts the enveloping cone of the sphere $x^2 + y^2 + z^2 = 11$ which has the vertex at (2, 4, 1) in a rectangular hyperbola.

Solution:

Here $S = x^2 + y^2 + z^2 - 11$, $S_1 = 2^2 + 4^2 + 1^2 - 11 = 10$

and $T = x(2) + y(4) + z(1) - 11 = 2x + 4y + z - 11$.

$\therefore$ The equation of the enveloping cone of the given sphere with vertex at (2, 4, 1) is $(x^2 + y^2 + z^2 - 11)(10) = (2x + 4y + z - 11)^2$.

...using $SS_1 = T^2$

The plane $z = 0$ meets this cone in the conic

$$10(x^2 + y^2 - 11) = (2x + 4y - 11)^2,\ z = 0$$

$$\Rightarrow \quad 6x^2 - 6y^2 - 16xy + 44x - 88y - 231 = 0,\ z = 0.$$

Since $\Delta \neq 0$ and sum of the coefficients of x^2 and y^2 in the first condition is zero, so these represent rectangular hyperbola. **Hence proved.**

Example 41:

Prove that the tangent lines from the origin of coordinates to the sphere $(x - a)^2 + (y - b)^2 + (z - c)^2 = k^2$ lie on the cone given by the equation $(a^2 + b^2 + c^2 - k^2)(x^2 + y^2 + z^2) = (ax + by + cz)^2$.

Solution:

Here we are to find the enveloping cone of the given sphere with vertex at (0, 0, 0).

Here $S \equiv (x - a)^2 (y - b)^2 + (z - c)^2 - k^2$

$$\Rightarrow \quad S \equiv x^2 + y^2 + z^2 - 2ax - 2by - 2cz + (a^2 + b^2 + c^2 - k^2).$$

$$S_1 = (a^2 + b^2 + c^2 - k^2)$$

and $T = x.0 + y.0 + z.0 - a(x + 0) - b(y + 0) - c(z + 0) + (a^2 + b^2 + c^2 - k^2)$

$$\Rightarrow \quad T = -ax - by - cz + (a^2 + b^2 + c^2 - k^2)$$

$\therefore$ The equation of the enveloping cone of the given sphere with vertex at (0, 0, 0) is "$SS_1 = T^2$"

$$\Rightarrow \quad [x^2 + y^2 + z^2 - 2ax - 2by - 2cz + (a^2 + b^2 + c^2 - k^2)](a^2 + b^2 + c^2 - k^2) = [-(ax + by + cz) + (a^2 + b^2 + c^2 - k^2)]^2$$

$$\Rightarrow \quad (x^2 + y^2 + z^2)(a^2 + b^2 + c^2 - k^2) - 2(ax + by + cz)(a^2 + b^2 + c^2 - k^2)^2 + (a^2 + b^2 + c^2 - k^2)^2 = (ax + by + cz)^2 + (a^2 + b^2 + c^2 - k^2)^2 - 2(ax + by + cz)(a^2 + b^2 + c^2 - k^2)$$

$$\Rightarrow \quad (x^2 + y^2 + z^2)(a^2 + b^2 + c^2 - k^2) = (ax + by + cz)^2.$$

Hence proved.

Example 42:

Find the equation of the right circular cone whose vertex is the origin and whose axis is the line $x = t$, $y = 2t$, $z = 3t$ and which has a vertical angle of 60°.

Or

Find the equation of the right circular cone, whose vertex is at the origin, whose axis is the line $x/1 = y/2 = z/3$ and whose semi-vertical angle is 60°.

Solution:

The vertex of the cone is V (0, 0, 0), its semi-vertical angle is 30° and its axis is given by

$$\frac{x}{1} = \frac{y}{2} = \frac{z}{3} = t \qquad ...(i)$$

Let P (x, y, z) be any point of the cone. Then the direction ratios of the generator VP are $x - 0$, $y - 0$, $z - 0$ *i.e.*, x, y, z.

Also are VP makes an angle of 30° with the line (i), so the required equation of the cone is

$$\cos 30^o = \frac{x.1 + y.2 + z.3}{\sqrt{(x^2 + y^2 + z^2)}.\sqrt{(1^2 + 2^2 + 3^2)}}$$

$$\Rightarrow \quad \frac{\sqrt{3}}{2} = \frac{x + 2y + 3z}{\sqrt{(x^2 + y^2 + z^2)}\ \sqrt{(14)}}$$

$$\Rightarrow \quad 42\ (x^2 + y^2 + z^2) = 4\ (x + 2y + 3z)^2$$

$$\Rightarrow \quad 38x^2 + 26y^2 + 6z^2 - 16xy - 48yz - 24zx = 0$$ **Ans.**

Example 43:

A right circular cone is passing through the point (1, 1, 1) and its vertex is at the point (1, 0, 1). The axis of the cone is equally inclined to the coordinate axes. Find the equation of the cone.

Solution:

The direction ratios of the generator passing through (1, 1, 1) and the vertex V (1, 0, 1) are $1 - 1$, $1 - 0$, $1 - 1$, *i.e.*, 0, 1, 0

Also the direction ratios of the axis of the cone which is equally inclined of the coordinate axes is 1, 1, 1.

$\therefore$ If θ be the semi-vertical angle of the cone, then we have

$$\cos\theta = \frac{0.1+1.1+0.1}{\sqrt{(0^2+1^2+0^2)}\sqrt{(1^2+1^2+1^2)}} = \frac{1}{\sqrt{3}} \quad \text{...(i)}$$

Also if P (x, y, z) be any point on the cone, then direction ratios of the generator VP are x – 1, y – 0, z – 1.

The generator is also inclined to the axis of the cone at an angle θ, so

$$\cos\theta = \frac{(x-1).1+(y-0).+(z-1).1}{\sqrt{\left[(x-1)^2+(y-0)^2+(z-1)^2\right]}\sqrt{(1^2+1^2+1^2)}}$$

$$\Rightarrow \quad \frac{1}{\sqrt{3}} = \frac{x+y+z-2}{\sqrt{\left[(x-1)^2+y^2+(z-1)^2\right]}.\sqrt{3}}, \text{ from (i)}$$

$$\Rightarrow \quad \sqrt{\left[(x-1)^2+y^2+(z-1)^2\right]} = x+y-z-2$$

$$\Rightarrow \quad (x-1)^2 + y^2\ (z-1)^2 = (x+y-z-2)^2, \text{ on squaring}$$

$$\Rightarrow \quad yz + zx - xy + x + 2y - 3z - 1 = 0, \text{ on simplifying.}$$ **Ans.**

Example 44:

Obtain the equation of the right circular cone with vertex at (1, – 2, – 1)., semi vertical angle 60º and the axis

$$\frac{1}{3}(x-1) = -\frac{1}{4}(y+2) = \frac{1}{5}(z+1).$$

Solution:

Do your self.

Ans. $7x^2 - 7y^2 - 25z^2 + 48xy + 80yz - 60zx + 22x + 4y + 17z + 78 = 0$

Example 45:

Find the equation to the right circular cone whose vertex is (2, – 3, 5), axis makes equal angles with the coordinate axes and semi-vertical angle is 30º.

Solution:

Let V (2, – 3, 5) be the vertex of the cone. Also the direction cosines of the axis of the cone are $1/\sqrt{3}$, $1/\sqrt{3}$, $1/\sqrt{3}$ as it makes equal angles with the coordinate axes.

Let P (x, y, z) be any point on the cone. Then the direction ratios of the generator VP are

$$x - 2,\ y + 3,\ z - 5.$$

Also VP is making an angle of 30° with axis of the cone, so we get the equation of the cone as

$$\cos 30^\circ = \frac{(x-2).(1/\sqrt{3})+(y+3)(1/\sqrt{3})+(z-5)(1/\sqrt{3})}{\sqrt{\left[(x-2)^2+(y+3)^2+(z-5)^2\right]}\sqrt{\left[(1/\sqrt{3})^2+(1/\sqrt{3})^2+(1/\sqrt{3})^2\right]}}$$

$$\Rightarrow \quad \frac{\sqrt{3}}{2} = \frac{(x-2)+(y+3)+(z-5)}{\sqrt{3}.\sqrt{\left[(x-2)^2+(y+3)^2+(z-5)^2\right]}}$$

$$\Rightarrow \quad 9\,[(x-2)^2 + (y+3)^2 + (z-5)^2] = 4\,[(x-2) + (y+3) + (z-5)]^2$$

$$\Rightarrow \quad 5\,(x^2 + y^2 + z^2) - 8\,(xy + yz + zx) - 4x + 86y - 58z + 278 = 0.$$

Example 46:

Find the equation to the right circular cone whose vertex is the origin, axis the axis of z and semi-vertical angle 30°.

Solution:

The vertex of the cone is V (0, 0, 0), semi-vertical angle is 30° and the axis is z-axis, whose d. cosines are 0, 0, 1.

∴ The equation of the axis of the cone is

$$\frac{x-0}{0} = \frac{y-0}{0} = \frac{z=0}{1} \quad i.e., \quad \frac{x}{0} = \frac{y}{0} = \frac{z}{1} \qquad \text{...(i)}$$

Let P (x, y, z) be any point on the cone. Then the direction ratios of the generator VP are x – 0, y – 0, z – 0, *i.e.*, x, y, z. Also VP makes an angle of 30° with the axis of the cone whose d.c.'s are 0, 0, 1, so the equation of the cone is

$$\cos 30^\circ = \frac{x.0 + y.0 + z.1}{\sqrt{(x^2+y^2+z^2)}.\sqrt{(0^2+0^2+1^2)}} = \frac{z}{\sqrt{(x^2+y^2+z^2)}}$$

$$\Rightarrow \quad \frac{\sqrt{3}}{2} = \frac{z}{\sqrt{(x^2+y^2+z^2)}}$$

$$\Rightarrow \quad 3\,(x^2 + y^2 + z^2) = (2z)^2$$

$$\Rightarrow \quad 3x^2 + 3y^2 - z^2 = 0$$

Ans.

Example 47:

Lines are drawn from O with direction cosines proportional to (1, 2, 2), (2, 3, 6), (3, 4, 12). Prove that the axis of the right circular cone through them has direction cosines $\left(1/\sqrt{3}, 1/\sqrt{3}, 1/\sqrt{3}\right)$ *and that the semi-vertical angle of the cone is* $\cos^{-1} 1/\sqrt{3}$.

Solution:

The vertex of the cone is O (0, 0, 0) and so let the equation of its axis be $\frac{x}{l} = \frac{y}{m} = \frac{z}{n}$, where l, m, n are direction cosines. ...(i)

Let θ be the semi-vertical angle of the cone. Then the given lines with direction ratios (1, 2, 2), (2, 3, 6), (3, 4, 12) or direction cosines (1.3, 1/3, 2/3), (2/7, 3/7, 6/7), (3/13, 4/13, 12/13) makes angle θ with (i)

∴ We have $\cos\theta = l\,(1/3) + m\,(2/3) + n\,(2/3)$ ...(ii)

$$= l\,(2/7) + m\,(3/7) + n\,(6/7)$$

$$= l\,(3/13) + m\,(4/13) + n\,(12/13).$$

From these we have

$$(1/3)\,l + (2/3)\,m + (2/3)\,n = (2/7)\,l + (3/7)\,m + (6/7)\,n$$

and $(1/3)\,l + m + (2/3)\,n = (3/13)\,l + (4/13)\,m + (12/13)\,n$

⇒ $[(1/3) - (2/7)]\,l + [(2/3) - (3/7)]\,m + [(2/3) - (6/7)]\,n = 0$

and $[(1/3) - (3/13)]\,l + [(2/3) - (4/13)]\,m + [(2/3) - (12/13)]\,n = 0$

⇒ $(1/21)\,l + (5/21)\,m - (4/21)\,n = 0$

and $(4/39)\,l + (14/39)\,m - (10/39)\,n = 0$

⇒ $1 + 5m - 4n = 0$ and $2l + 7m - 5n = 0$

Solving these simultaneously we get

$$\frac{l}{-25+28} = \frac{m}{-8+5} = \frac{n}{7-10} \Rightarrow \frac{l}{-1} = \frac{m}{1} = \frac{n}{1}$$

∴ Direction cosines of axis of the cone, from (i), are proportional to – 1, 1, 1 or the direction cosines are

$$\frac{-1.1.1}{\sqrt{\left[(-1)^2 + 1^2 + 1^2\right]}} \text{ i.e., } \frac{-1}{\sqrt{3}}, \frac{1}{\sqrt{3}}, \frac{1}{\sqrt{3}}$$ **Hence proved.**

Also from (ii) we have

$$\cos\theta = -\frac{1}{\sqrt{3}}(1/3) + \frac{1}{\sqrt{3}}(2/3) + \frac{1}{\sqrt{3}}(2/3)$$

$$\Rightarrow \quad \cos\theta = \frac{1}{3\sqrt{3}}(-1+2+2) = \frac{1}{\sqrt{3}} \Rightarrow \theta = \cos^{-1}\left(\frac{1}{\sqrt{3}}\right)$$

Hence proved.

Example 48:

The axis of a right circular cone, with origin O as vertex, makes equal angles with the co-ordinate axes and the cone passes through the line drawn from O with direction cosines proportional to 1, – 2, 3. Find is equation.

Solution:

The vertex of the cone is O (0, 0, 0) and since its axis makes equal angles with the co-ordinates axes, so the equations of its axis can be taken as

$$\frac{x}{1} = \frac{y}{1} = \frac{z}{1} \qquad (\because l = m = n) \qquad \textbf{(Note)}$$

Also the d.r.'s of its generators are given as 1, – 2, 2.

∴ If θ be the semi-vertical angle of the cone, then

$$\cos\theta = \frac{1.1 + 1.(-2) + 1.2}{\sqrt{(1+1+1)}\sqrt{\left\{1^2 + (-2)^2 + (2)^2\right\}}} = \frac{1}{3\sqrt{3}} \qquad \text{...(i)}$$

Also, if P (x, y, z) be any point on the cone, then OP is a generator and the d.r.'s of OP are x – 0, z – 0, z – 0, *i.e.*, x, y, z. Also OP makes an angle θ with the axis of the cone.

$$\therefore \qquad \cos\theta = \frac{x.1 + y.1 + z.1}{\sqrt{(1+1+1)}\sqrt{(x^2+y^2+z^2)}} \qquad \text{...(ii)}$$

Equating (i) and (ii), the required equation of the cone is

$$\frac{x+y+z}{\sqrt{3}\sqrt{(x^2+y^2+z^2)}} = \frac{1}{3\sqrt{3}}$$

$$\Rightarrow \qquad 9(x+y+z)^2 = x^2 + y^2 + z^2$$

$$\Rightarrow \qquad 4(x^2+y^2+z^2) + 9(yz + zx + xy) = 0. \qquad \textbf{Ans.}$$

Example 49:

Find the equation to the right circular cone whose vertex is O, axis OZ and semi-vertical angle α.

Solution:

Let P (x, y, z) be any point on the cone. From P draw PN perpendicular to z-axis.

Then $PN = OM = \sqrt{(x^2 + y^2)}$ and ON = z.

From the adjoining figure it is evident that

$$\tan \alpha = PN/ON$$

$$\Rightarrow \quad \tan \alpha = \sqrt{(x^2 + y^2)}/z$$

$$\Rightarrow \quad z \tan \alpha = \sqrt{(x^2 + y^2)}$$

$$\Rightarrow \quad x^2 + y^2 = z^2 \tan^2 \alpha.$$

Hence proved.

Example 50:

Find the equation of the right circular cone which pass through the point (1, 1, 2) and has its vertex at the origin, axis the line x/2 = y/(– 4) = z/3.

Solution:

The direction ratios of the generator passing through (1, 1, 2) and the vertex V (0, 0, 0) are 1 – 0, 1 – 0, 2 – 0 *i.e.,* 1, 1, 1

Also the direction ratios of the axis of the cone are 2, – 4, 3 (given).

∴ If θ be the semi-vertical angle of the cone, then we have

$$\cos \theta = \frac{1.2 + 1.(-4) + 2.3}{\sqrt{(1^2 + 1^2 + 2^2)}\sqrt{[2^2 + (-4)^2 + 3^2]}} = \frac{2}{\sqrt{(29)}}$$

Also, if P (x, y, z) be any point on the cone, then the direction ratios of generator VP are x – 0, y – 0, z – 0. *i.e.,* x, y, z. The generator is also inclined to the axis of the cone at an angle θ, so

$$\cos \theta = \frac{x.2 + y.(-4) + z.3}{\sqrt{(x^2 + y^2 + z^2)}\sqrt{[2^2 + (-4)^2 + 3^2]}} \qquad \text{...(ii)}$$

Equating (i) and (ii) we have the equation of the cone as

$$\frac{2x - 4y + 3z}{\sqrt{(x^2 + y^2 + z^2)}\sqrt{(29)}} = \frac{2}{\sqrt{(29)}}$$

$$(2x - 4y + 3z)^2 = 4(x^2 + y^2 + z^2)$$

$$12y^2 - 5x^2 - 16xy - 24yz + 12zx = 0.$$

Ans.

Example 51:

If a right circular cone has three mutually perpendicular generators, show that the semi-vertical angle is $tan^{1}\sqrt{2}$.

Solution:

Let vertex of the cone be origin and its axis be z-axis then if θ be the semi-vertical angle of the cone, its equation can be proved to be

$$x^2 + y^2 - z^2 \tan^2 \theta = 0.$$

If it has three mutually perpendicular generators, then

$$'a + b + c = 0'$$

i.e., $1 + 1 - \tan^2 \theta + 0 \Rightarrow \tan^2 \theta = 2 \Rightarrow \theta = \tan^{-1}\sqrt{2}$.

Example 52:

Show that the semi-vertical angle of the right circular cone which has three mutually perpendicular tangent planes is

$$tan^{-1}\left(1/\sqrt{2}\right) \text{ or } cot^{1}.$$

Solution:

Let the right circular cone be taken as

$$x^2 + y^2 - z^2 \tan^2\theta = 0. \qquad ...(i)$$

It has three mutually perpendicular tangent planes then its reciprocal cone

$$\frac{x^2}{1} + \frac{y^2}{1} - \frac{z^2}{\tan^2 \theta} = 0$$

must have three mutually perpendicular generators and the condition for the same is

$$1 + 1 - (1/\tan^2 \theta) = 0 \qquad ...(ii)$$

i.e., $\cot^2 \theta = 2 \Rightarrow \theta = \cos^{-1}\left(\sqrt{2}\right)$. **Hence proved.**

Also from (ii) we can get $\tan^2 \theta = 1/2$ or $\theta = \tan^{-1}\left(1/\sqrt{2}\right)$.

Example 53:

Prove that the equation of the cone through the coordinate axes and the lines through which the plane $ux + vy + wz = 0$ *cuts the cone* $ax^2 + by^2 + cz^2 + 2fyz + 2gzx + 2hxy = 0$ *is*

$$u\,(bw^2 + cv^2 - 2fwv)\,yz + v\,(cu^2 + aw^2 - 2guw)\,zx + w\,(av^2 + bu^2 - 2hvu)\,xy = 0.$$

Solution:

We know that the equation of the cone which passes through the coordinate axes is

$$Fyz + Gzx + Hxy = 0 \quad ...(i)$$

have been replaced by other three quantities F, G, H respectively.

Again if the plane ux + vy + wz = 0 cuts the cone $ax^2 + by^2 + cz^2 + 2fyz + 2gzx + 2hxy = 0$ in two lines whose d.c.'s are l_1, m_1, n_1 and l_2, m_2, n_2 then, we can prove that

$$\frac{l_1 l_2}{bw^2 + cv^2 - 2fvw} = \frac{m_1 m_2}{cu^2 + aw^2 - 2guw} = \frac{n_1 n_2}{av^2 + bu^2 - 2huv} \quad ...(ii)$$

and $\quad l_1 m_2 - l_2 m_1 = 2w\lambda P$ etc. which give

$$\frac{l_1 m_2 - l_2 m_1}{w} = \frac{m_1 n_2 - m_2 n_1}{u} = \frac{n_1 l_2 - n_2 l_1}{v} \quad ...(iii)$$

Also if these two lines with d.c.'s l_1, m_1, n_1, and l_2, m_2, n_2 lie on the cone (i), then their d.c.'s must satisfy the equation (i)

$$\therefore \quad F\, m_1 n_1 + G\, n_1 l_1 + H\, l_1 m_1 = 0 \quad ...(iv)$$

$$\text{and} \quad F\, m_2 n_2 + G\, n_2 l_2 + H\, l_2 m_2 = 0 \quad ...(v)$$

Solving (iv) and (v) we get

$$\frac{F}{l_1 l_2\ (m_1 n_2 - m_2 n_2)} = \frac{G}{m_1 m_2 (n_1 l_2 - n_2 l_1)} = \frac{H}{n_1 n_2\ (l_1 m_2 - l_2 m_1)}$$

$$\Rightarrow \frac{F}{(bw^2 + cv^2 - 2fvw)\ u} = \frac{G}{(cu^2 + aw^2 - 2gwu)\ v} = \frac{H}{(av^2 + bu^2 - 2huv)\ w}$$

Substituting these proportionate values of F, G and H in (i), the required equation is $u\ (bw^2 + cv^2 - 2fvw)\ yz + v\ (cu^2 + aw^2 - 2gwu)\ zx$

$+ w\ (av^2 + bu^2 - 2huv)\ xy = 0$ **Hence proved.**

Example 54:

Prove that the angle between the lines given by $x + y + z = 0$, $ayz + bzx + cxy = 0$ is $\pi/2$ if $a + b + c = 0$

and $\quad \pi/3$ if $1/a + 1/b + 1/c = 0$.

Solution:

Let the plane x + y + z = 0 cut the cone ayz + bzx + cxy = 0 in a line

$$x/l = y/m = z/n.$$

Then $l + m + n = 0$ and $amn + bnl + clm = 0$.

Eliminating n between these relations, we get

$$(am + bl)(-l - m) + clm = 0$$

$$\Rightarrow \quad bl^2 + (a + b - c)\, lm + am^2 = 0$$

$$\Rightarrow \quad b\,(l/m)^2 + (a + b - c)(l/m) + a = 0 \qquad ...(i)$$

If the roots of this equation are l_1/m_1 and l_2/m_2, then

$$\frac{l_1}{m_1} \cdot \frac{l_2}{m_2} = \text{product of the roots} = \frac{a}{b}$$

$$\Rightarrow \quad \frac{l_1 l_2}{a} = \frac{m_1 m_2}{b} = \frac{n_1 n_2}{c} \text{ (by symmetry)} = k \text{ (say)}, \qquad ...(ii)$$

If the angle between the lines is $\pi/2$, then we get

$$l_1 l_2 + m_1 m_2 + n_1 n_2 = 0$$

$$\Rightarrow \quad a + b + c = 0, \text{ from (ii).}$$

Again from (i) we get

$$\frac{l_1}{m_1} + \frac{l_2}{m_2} = \text{sum of the roots} = \frac{c - b - a}{b}$$

$$\Rightarrow \quad \frac{l_1 m_2 + l_2 m_1}{m_1 m_2} = \frac{c - b - a}{b}$$

$$\Rightarrow \quad \frac{l_2 m_2 + l_2 m_1}{c - b - a} = \frac{m_1 m_2}{b} = k \text{ from (ii).}$$

Now $(l_1 m_2 - l_2 m_1)^2 = (l_1 m_2 + l_2 m_1)^2 - 4 l_1 l_2 . m_1 m_2$

$= k^2 [c + b - a)^2 - 4ak.\, bk = k^2[(c - b - a)^2 - 4ab$

$= k^2 [a^2 + b^2 + c^2 - 2ab - 2bc - 2ca]$,

$$\text{Now } \tan\theta = \frac{\sqrt{\left[\Sigma\,(l_1 m_2 - l_2 m_1)^2\right]}}{l_1 l_2 + m_1 m_2 + n_1 n_2}$$

$$= \frac{\sqrt{\left[k^2\,\{3\,(a^2 + b^2 + c^2 - 2bc - 2ca - 2ab)\}\right]}}{k\,(a + b + c)} \qquad \textbf{(Note)}$$

If $\theta = \pi/3$, then

$$\tan^2(\pi/3) = 3\,(a^2 + b^2 + c^2 - 2bc - 2ca - 2ab)/(a + b + c)^2$$

$$\Rightarrow \quad 3(a + b + c)^2 = 3(a^2 + b^2 + c^2 - 2bc - 2ca - 2ab);$$

$\because \quad \tan(\pi/3) = \sqrt{3}$

$\Rightarrow \quad 4(bc + ca + ab) = 0 \Rightarrow 1/a + 1/b + 1/c = 0.$ **Hence proved.**

Example 55:

Show that the plane ax + by + cz = 0 cuts the cone yz + zx + xy = 0 in two lines inclined at an angle

$$\tan^{-1}\left[\frac{\sqrt{\{(a^2+b^2+c^2)(a^2+b^2+c^2-2bc-2ca-2ab)\}}}{bc+ca+ab}\right]$$

and by considering the value of this expression when a + b + c = 0, show that the cone is of revolution and that its axis is x = y = z and vertical angle

$\tan^{-1}(-2/\sqrt{2})$.

Solution:

Let $x/l = y/m = z/n$ be one of the lines in which the plane ax + by + cz = 0 cuts the cone yz + zx + xy = 0, then we have

$$al + bm + cn = 0 \qquad ...(i)$$

and $$mn + nl + lm = 0 \qquad ...(ii)$$

Eliminating n between (i) and (ii), we get

$$(m + l)[-(al + bm)/c] + lm = 0$$

$\Rightarrow \quad al^2 + (a + b - c)\,lm + bm^2 = 0$

$\Rightarrow \quad a\,(l/m)^2 + (a + b - c)\,(l/m) + b = 0$

If its roots are l_1/m_1 and l_2/m_2, then we have

$$\frac{l_1}{m_1}\cdot\frac{l_2}{m_2} = \text{product of the roots} = \frac{b}{a}$$

$$\Rightarrow \quad \frac{l_1 l_2}{1/a} = \frac{m_1 m_2}{1/b} = \frac{n_1 n_2}{1/c} \text{ (by symmetry)} = k \text{ (say).} \qquad ...(iii)$$

Also $$\frac{l_1}{m_1} + \frac{l_2}{m_2} = \text{sum of the roots} = \frac{c-b-a}{a}$$

$$\Rightarrow \quad \frac{l_1 m_2 + l_2 m_1}{m_1 m_2} = \frac{c-b-a}{a} \Rightarrow \frac{l_1 m_2 + l_2 m_1}{c-b-a} = \frac{m_1 m_2}{a}$$

$$\Rightarrow \quad \frac{l_1 m_2 + l_2 m_1}{(c-b-a)/(ab)} = \frac{m_1 m_2}{1/b} = k\text{, from (iii)} \qquad \textbf{(Note)}$$

$$\therefore \quad (l_1 m_2 - l_2 m_1) = (l_1 m_2 + l_2 m_1)^2 - 4 l_1 l_2 . m_1 m_2$$

$= k^2 [(c - b - a)/(ab)]^2 - 4 (k/a) (k/b)$

$= (k^2/a^2b^2) [(c - b - a)^2 - 4ab]$

$= (k^2/a^2b^2) [\lambda^2] = k^2\lambda^2/(a^2b^2),$

where $\quad \lambda^2 = a^2 + b^2 + c^2 - 2bc - 2ca - 2ab$ ...(iv)

$\therefore$ If θ be the angle between these two lines, then we have

$$\tan\theta = \frac{\sqrt{\left[(l_1 m_2 - l_2 m_1)^2\right]}}{l_1 l_2 + m_1 m_2 + n_1 n_2} = \frac{k\lambda\left[(1/a^2b^2) + (a/b^2c^2) + (1/c^2a^2)\right]}{k\left[1/a + 1/b + 1/c\right]}$$

$$\Rightarrow \quad \tan\theta = \frac{\lambda\sqrt{(c^2 + a^2 + b^2)}}{abc\,(1/a + 1/b + 1/c)}$$

$$\Rightarrow \quad \theta = \tan^{-1}\left[\frac{\lambda\sqrt{(c^2 + a^2 + b^2)}}{bc + ca + ab}\right], \qquad \text{...(v)}$$

where λ is given by (iv).

Again if $a + b + c = 0$, then $(a + b + c)^2 = 0$

$\Rightarrow \quad a^2 + b^2 + c^2 = -2(bc + ca + ab)$ **(Note)**

$\therefore$ From (iv), we get l \ –4 (bc + ca + ab).

Substituting these values of l and $(a + b + c)^2$ in (v), we get

$$\theta = \tan^{-1}\left[\frac{\sqrt{\{-4(bc + ca + ab)\}}\ \sqrt{\{-2(bc + ca + ab)\}}}{(bc + ca + ab)}\right]$$

$= \tan^{-1} [\sqrt{(8)}] = = \tan^{-1} [\square 2\sqrt{(2)}]$ ...(vi) **Hence proved.**

Again if $a + b + c = 0$, it means that the line $\frac{x}{1} = \frac{y}{1} = \frac{z}{1}$ lies on the plane $ax + by + cz = 0$ and we are to prove that this line is the axis of the cone *i.e.,* the angle between this line and any generator of the cone is half of angle θ given by (vi).

Now if $x/l = y/m = z/n$ is a generator of the cone, then $mn + nl + lm = 0$ and if α be the angle between this generator and the axis, then

$$\cos\alpha = \frac{1.l + 1.m + 1.n}{\sqrt{(1^2 + 1^2 + 1^2)}\ \sqrt{(l^2 + m^2 + n^2)}} = \frac{\Sigma l}{\sqrt{3}\ \sqrt{\left[(\Sigma l)^2 - 2\Sigma mn\right]}}$$

$= 1/\sqrt{3}, \qquad \because \Sigma mn = 0.$

$$\therefore \tan\alpha = \sqrt{2} \Rightarrow \quad \tan 2\alpha = \frac{2\tan\alpha}{1 - \tan^2\alpha} = \frac{2\sqrt{2}}{1 - 2} = -2\sqrt{2}.$$

$\therefore$ From (vi) $\tan 2\alpha = \tan \theta \Rightarrow \quad \alpha = (1/2)\,\theta$. **Hence proved.**

Example 46:

Prove that the equation of the planes through the origin perpendicular to the lines of section of the plane $lx + my + nz = 0$ and the cone $ax^2 + by^2 + cz^2 = 0$ is $x^2 (bn^2 + cm^2) + y^2 (cl^2 + an)^2 + z^2 (am^2 + 6l^2) - 2amnyz - 2bnlzx - 2clmxy = 0$.

Solution:

Let $x/\lambda = y/\mu = z/v$ be one of the lines in which the given plane and the given cone intersect. Then we have

$$l\lambda + m\mu + nn = 0 \text{ and } a\lambda^2 + b\mu^2 + cv^2 = 0$$

Eliminating v between these, we get

$$a\lambda^2 + b\mu^2 + c\,[(-\lambda l + \mu)/n]^2 = 0$$

$$\Rightarrow \quad (\lambda^2/\mu^2)\,(an^2 + cl^2) + 2lmc\,(\lambda/\mu) + (bn^2 + cm^2) = 0$$

$$\therefore \quad \frac{\lambda_1}{\mu_1}\cdot\frac{\lambda_2}{\mu_2} = \text{product of roots} = \frac{bn^2 + cm^2}{an^2 + cl^2}$$

$$\Rightarrow \quad \frac{l_1 l_2}{bn^2 + cm^2} = \frac{\mu_1\mu_2}{cl^2 + am^2} = \frac{v_1 v_2}{am^2 + bl^2} \text{ (by symmetry)} = K \text{ (say)...(i)}$$

$$\text{Also} \quad \frac{\lambda_1}{\mu_1} + \frac{\lambda_2}{\mu_2} = \text{sum of the roots} = \frac{-2lmc}{an^2 + cl^2}$$

$$\Rightarrow \quad \frac{\lambda_1\mu_2 + \lambda_2\mu_1}{\mu_1\mu_2} = \frac{-2lmc}{an^2 + cl^2} \Rightarrow \frac{\lambda_1\mu_2 + \lambda_2\mu_1}{-2lmc} = \frac{\mu_1\mu_2}{an^2 + cl^2}$$

$$\Rightarrow \quad \frac{\lambda_1\mu_2 + \lambda_2\mu_1}{-2lmc} = \frac{\mu_1\mu_2}{an^2 + cl^2} = k\text{, from (i)}$$

$$= \frac{\mu_1 v_2 + \mu_2 v_1}{-2mna} = \frac{v_1\lambda_2 + v_2\lambda_1}{-2nlb} \quad \text{...(ii)}$$

Now these lines are $x/\lambda_1 = y/\mu_1 = z/v_1$; $x/\lambda_2 = y/\mu_2 = z/v_2$

The planes through origin at right angles to these lines are

$$(\lambda_1 x + \mu_1 y + v_1 z)\,(\lambda_2 x + \mu_2 y + v_2 z) = 0$$

$$\Rightarrow \quad \lambda_1\lambda_2 x^2 + \mu_1\mu_2 y^2 + v_1 v_2 z^2 + (\lambda_1\mu_2 + \lambda_2\mu_1)\,xy + (\mu_1 v_2 + \mu_2 v_1)\,yz + (\lambda_1 v_2 + \lambda_2 v_1) = 1$$

$$\Rightarrow \quad (bn^2 + cm^2)\,x^2 + \ldots + \ldots + (-2mlc)\,xy + \ldots + \ldots = 0\text{, from (i) and (ii).}$$

Hence proved.

Example 57:

Find the equation of the cone generated by the straight lines drawn from the origin to cut the circle through the points A (1, 0, 0), B (0, 2, 0), C (2, 1, 1) and prove that the acute angle between two straight lines in which the plane x = 2y cuts the cone is $\cos^{-1}\sqrt{(5/14)}$.

Solution:

Let the equation of the sphere through A, B, C and origin be

$$x^2 + y^2 + z^2 + 2ux + 2vy + 2wz = 0. \quad \text{...(i)}$$

If it passes through A (1, 0, 0), then $1 + 2u = 0 \Rightarrow 2u = -1$

If it passes through B (0, 2, 0), then $4 + 4v = 0 \Rightarrow v = -1$

If it passes through C (2, 1, 1), then $6 + 4u + 2v + 8w = 0$

$$\Rightarrow \quad 6 + 4\left(-\frac{1}{2}\right) + 2(-1) + 8w = 0 \Rightarrow w = -1$$

∴ From (i) the equation of the sphere ABC is

$$x^2 + y^2 + z^2 - x - 2y - 2z = 0 \quad \text{...(ii)}$$

Also the equation of the plane through A (1, 0, 0) and B (0, 2, 0) can be taken as $\frac{x}{1} + \frac{y}{2} + \frac{z}{c} = 1$. **(Note)**

If it passes through C (2, 1, 1) then $2 + \frac{1}{2} + (1/c) = 1 \Rightarrow c = -\frac{2}{3}$.

∴ The plane ABC is given by $\frac{x}{1} + \frac{y}{2} - \frac{3z}{2} = 1$

$$\Rightarrow \quad 2x + y - 3z = 2 \quad \text{...(iii)}$$

∴ Making (ii) homogeneous with the help of (iii) we get the equation of the required cone as $x^2 + y^2 + z^2 - (x + 2y + 2z)\left[\frac{1}{2}(2x + y - 3z)\right] = 0$

$$\Rightarrow \quad 8z^2 + 4yz - zx - 5xy = 0. \quad \text{...(iv)}$$

If the plane x = 2y cuts this cone along a line $x/l = y/m = z/n$, then

$8n^2 + 4mn - nl - 5lm = 0$ and $l = 2m$ ...(v) **(Note)**

Eliminating l, we get $8n^2 + 4mn - n(2m) - 5(2m)m = 0$

$$\Rightarrow \quad 8n^2 + 2mn - 10m^2 = 0 \Rightarrow 2(n - m)(4n + 5m) = 0$$

When n = m, then from (v) we get $l/2 = m/1 = n/1$.

And when $4n + 5m = 0$, then from (v) get $l = 2m = -(8/5)n$

$$\Rightarrow \quad \frac{l}{2} = \frac{m}{1} = \frac{n}{-(5/4)}$$

$\therefore$ If θ be the required angle between these two lines, then

$$\cos\theta = \frac{\Sigma l_1 l_2}{\sqrt{(\Sigma l_1^2)}\sqrt{(\Sigma l_2^2)}} = \frac{2.2 + 1.1 + 1.(-5/4)}{\sqrt{(2^2+1^2+1^2)}\sqrt{\left[\left(2^2+1^2+(-5/4)^2\right)\right]}}$$

$$\Rightarrow \quad \cos\theta = \frac{(15/4)}{\sqrt{6}\sqrt{(105/16)}} = \sqrt{\left(\frac{5}{14}\right)} \Rightarrow \theta = \cos^{-1}\sqrt{\left(\frac{5}{14}\right)}. \qquad \textbf{Ans.}$$

Example 58:

Three points P, Q, R are taken on the ellipsoid $x^2/a^2 + y^2/b^2 + z^2/c^2 = 1$, so that line joining P, Q, R to the origin are mutually perpendicular. Prove that the plane PQR touches a fixed sphere.

Solution:

Let the equation of the plane PQR be

$$lx + my + nz = 1 \qquad \text{...(i)}$$

The equation of the cone with vertex at (0, 0, 0) and the curve of intersection of (i) and the given ellipsoid as the guiding curve is

$$\frac{x^2}{a^2} + \frac{y^2}{b^2} + \frac{z^2}{c^2} = (lx + my + nz)^2 \qquad \text{...(ii)}$$

making the equation of the ellipsoid homogeneous with the help of (i).

If the cone (i) has three mutually perpendiculars generators then

$$\left(l^2 - \frac{1}{a^2}\right) + \left(m^2 - \frac{1}{b^2}\right) + \left(n^2 - \frac{1}{c^2}\right) = 0 \quad \text{..."a + b + c = 0"}$$

$$\Rightarrow \quad l^2 + m^2 + n^2 = (1/a^2) + (1/b^2) + (1/c^2) = 1/\lambda^2, \text{ (say)} \qquad \text{...(iii)}$$

Now if the plane (i) touches the sphere $x^2 + y^2 + z^2 = \lambda^2$, then the length of the perpendicular from the centre (0, 0, 0) of the sphere to (i) must be equal to the ratios λ of the sphere.

i.e., $\dfrac{1}{\sqrt{(\lambda^2 + m^2 + n^2)}} = \lambda \Rightarrow l^2 + m^2 + n^2 = 1/\lambda^2$, which is true by virtue of (iii). Hence plane (i) touches the sphere $x^2 + y^2 + z^2 = \lambda^2$.

Hence proved.

Example 59:

Show that the locus of points from which three mutually perpendicular lines can be drawn to intersect a given circle $x^2 + y^2 = a^2$, $z = 0$ is a surface of revolution.

Solution:

Let P (α, β, γ) be the point whose locus is to be found.

Any line through P (α, β, γ) is $\frac{x-\alpha}{l} = \frac{y-\beta}{m} = \frac{z-\gamma}{n}$ ...(i)

and this meets the plane z = 0 in $\left(\alpha - \frac{l\gamma}{n}, \beta - \frac{m\gamma}{n}, 0\right)$ which lies on the circle if

$$\left(\alpha - \frac{l\gamma}{n}\right)^2 + \left(\beta - \frac{m\gamma}{n}\right)^2 = a^2 \quad \text{...(ii)}$$

Eliminating *l*, m, n between (i) and (ii) we get the locus of the line

$$\left[\alpha - \left(\frac{x-\alpha}{z-\gamma}\right)\gamma\right]^2 + \left[\beta - \left(\frac{y-\beta}{z-\gamma}\right)\gamma\right]^2 = a^2$$

$$\Rightarrow \quad (\alpha z - x\gamma)^2 + (\beta z - y\gamma)^2 = a^2 (z - \gamma)^2,$$

which is a cone with P (α, β, γ) as its vertex.

If (iii) has three mutually perpendicular generators, then

"a + b + c = 0" *i.e.*, $\gamma^2 + \gamma^2 + (\alpha^2 + \beta^2 - a^2) = 0$

$$\Rightarrow \quad \alpha^2 + \beta^2 + 2\gamma^2 = a^2$$

∴ Locus of P (α, β, γ) is $x^2 + y^2 + 2z^2 = a^2$, ...(iv)

which evidently represents a surface of revolution obtained by revolving the ellipse $x = 0$, $y^2 + 2z^2 = a^2 \Rightarrow y = 0$, $x^2 + 2z^2 = a^2$ about z-axis.

(**Note :** Later on you may read the chapter on conicoids and from there it would be clear that (iv) represents an ellipsoid, a surface of revolution).

Example 60:

Prove that $x^2 - y^2 + z^2 - 4x + 2y + 6z + 12 = 0$, represents a right circular cone whose vertex is the point (2, 1, – 3), whose axis is parallel to Oy and whose semi-vertical angle is 45°.

Solution:

The vertex of the cone is V (2, 1, – 3), its semi-vertical angle is 45° and its axis is parallel to Oy *i.e.*, the direction cosines of its axis are 0, 1, 0.

∴ The equation of the axis of the cone is $\dfrac{x-2}{0}=\dfrac{y-1}{1}=\dfrac{z+3}{0}$ ...(i)

Let P (x, y, z) be any point on the cone. Then the direction ratios of the generator VP are x – 2, y – 1, z + 3.

Also VP is making on angle of 45° with the axis whose d.c.'s are 0, 1, 0 so the equation of the cone is

$$\cos 45^\circ = \frac{(x-2).0+(y-1).1+(z+3).0}{\sqrt{(0^2+1^2+0^2)}\sqrt{\left[(x-2)^2+(y-1)^2+(z+3)^2\right]}}.$$

$$\Rightarrow \quad \frac{1}{\sqrt{2}}=\frac{(y-1)}{\sqrt{\left[(x-2)^2+(y-1)^2+(z+3)^2\right]}}$$

$\Rightarrow \quad (x-2)^2 + (y-1)^2 (z+3)^2 = 2(y-1)^2$, squaring and cross multiplying

$$\Rightarrow \quad (x-2)^2 - (y-1)^2 + (z+3)^2 = 0$$

$$x^2 - y^2 + z^2 - 4x + 2y + 6z + 12 = 0.$$

Example 61:

Find the locus of the point from which three mutually perpendicular tangent lines can be drawn to the surface

$$ax^2 + by^2 + 2cz = 0.$$

Solution:

Let P (x_1, y_1, z_1) be the point, then the three mutually perpendicular tangent lines drawn from it to the given surface are the three mutually perpendicular generators of the enveloping cone of the given surface with P (x_1, y_1, z_1) as its vertex.

The equation of enveloping cone is $SS_1 = T^2$.

i.e., $(ax^2 + by^2 + 2cz)(ax_1^2 + by_1^2 + 2cz_1) = [axx_1 + byy_1 + c(z + z_1)]^2$

If this cone has three mutually perpendicular generators, then sum of the coefficients of x^2, y^2 and z^2 is zero.

i.e., $a(by_1^2 + 2cz_1) + b(ax_1^2 + 2cz_1) + (-c^2) = 0$

∴ On generalising, we have the required locus of P (x_1, y_1, z_1) as

$$ab(x^2 + y^2) + 2c(a + b)z - c^2 = 0$$ **Ans.**

Example 62:

Find the enveloping cone of the sphere $x^2 + y^2 + z^2 + 2x - 2y - 2$ with the vertex at (1, 1, 1).

Solution:

Here $S = x^2 + y^2 + z^2 + 2x - 2y - 2.$

$\therefore$ $S_1 = 1^2 + 1^2 + 1^2 + 2.1 - 2.1 - 2 = 1$

and $T = x \,.\, 1 + y \,.\, 1 + z \,.\, 1 + (x + 1) - (y + 1) - 2$ **(Note)**

$= 2x + z - 2.$

$\therefore$ The required equation of the enveloping cone is SSi = T^2

$\Rightarrow$ $(x^2 + y^2 + z^2 + 2x - 2y - 2)\,(1) = (2x + z - 2)^2$

$\Rightarrow$ $x^2 + y^2 + z^2 + 2x - 2 = 4x^2 + z^2 + 4 + 4xz - 8x - 4z$

$\Rightarrow$ $3x^2 - y^2 + 4zx - 10x + 2y - 4z + 6 = 0.$ **Ans.**

Example 63:

Prove that the lines drawn from the origin so as to touch the sphere $x^2 + y^2 + z^2 + 2ux + 2vy + 2wz + d = 0$ lie on the cone

$$d\,(x^2 + y^2 + z^2) = (ux + vy + wz)^2$$

Solution:

Here we are to show that the given cone is the enveloping cone of the given sphere with vertex at (0, 0, 0).

$\therefore$ Here $S = x^2 + y^2 + z^2 + 2ux + 2vy + 2wz + d$; $S_1 = d$

and $T = x\,(0) + y\,(0) + z\,(0) + u\,(x + 0) + v\,(y + 0) + w\,(z + 0) + d$

$= ux + vy + wz + d$

$\therefore$ The enveloping cone of the given sphere with vertex at (0, 0, 0) is

$(x^2 + y^2 + z^2 + 2ux + 2vy + 2wz + d)\, d = (ux + vy + wz + d)^2$

$\Rightarrow$ $(x^2 + y^2 + z^2)\, d + (2ux + 2vy + 2wz)\, d + d^2 = (ux + vy + wz)^2 + d^2$

$+ 2d\,(ux + vy + wz)$ **(Note)**

$\Rightarrow$ $(x^2 + y^2 + z^2)\, d = (ux + vy + wz)^2.$ **Hence proved.**

Example 64:

Show that the three mutually perpendicular tangent lines can be drawn to the sphere $x^2 + y^2 + z^2 = r^2$ from any point on the sphere

$$x^2 + y^2 + z^2 = (3/2)\, r^2$$

Solution:

The given sphere are $x^2 + y^2 + z^2 = r^2$ ...(i)

and $x^2 + y^2 + z^2 = (3/2)\, r^2$ ...(ii)

Let P (x_1, y_1, z_1) be any point on the sphere (ii), then we have

$$x_1^2 + y_1^2 + z_1^2 = (3/2)\, r^2. \quad ...(iii)$$

Now the three mutually perpendicular tangent lines drawn from P (x_1, y_1, z_1) to the sphere (i) are the three mutually perpendicular generators of the enveloping cone of the sphere (i) with P (x_1, y_1, z_1) as the vertex.

The equation of this enveloping cone is "$SS_1 = T^2$"

i.e., $$(x^2 + y^2 + z^2 - r^2)\left(x_1^2 + y_1^2 + z_1^2 - r^2\right) = (xx_1 + yy_1 + zz_1 - r^2)^2$$

If this cone has three mutually perpendicular generators then the sum of the coefficients of x^2, y^2 and z^2 must be zero.

i.e., $$\left(y_1^2 + z_1^2 - r^2\right) + \left(x_1^2 + z_1^2 - r^2\right)\left(x_1^2 + y_1^2 - r^2\right) = 0$$

$$\Rightarrow \quad x_1^2 + y_1^2 + z_1^2 - (3/2)\, r^2 = 0,$$

which is true be virtue of (iii).

Example 65:

A line OP is such that the two planes through OP, each of which cuts the cone $ax^2 + by^2 + cz^2 = 0$ in perpendicular generators are perpendicular, prove that the locus of OP is a cone and find it.

Solution:

Let the equations of the line OP be $x/l = y/m = z/n$ ...(i)

The equation of any plane through (i) is $ux + vy + wz = 0$, ...(ii)

where $ul + vm + wn = 0$. ...(iii)

We know that if the plane (ii) cuts the cone $ax^2 + by^2 + cz^2 = 0$ in perpendicular generators then

$$(b + c)\, u^2 + (c + a)\, v^2 + (a + b)\, w^2 = 0. \quad ...(iv)$$

Now the direction ratios of the normal to the plane (ii) viz. u, v, w are given by the relations (ii) and (iv). Eliminating w between (iii) and (iv) we have $$(b + c)\, u^2 + (c + a)\, v^2 + (a + b)\, [-(ul + vm)/n]^2 = 0$$

$$\Rightarrow \quad (b + c)\, n^2u^2 + (c + a)\, n^2v^2 + (a - b)\, (ul + vm)^2 = 0$$

$$\Rightarrow \quad [(b + c)\, n^2 + (a + b)\, l^2]\, u^2 + 2\, (a + b)\, lmuv + [(c + a)\, n^2 + (a + b)\, m^2]\, v^2 = 0$$

$$\Rightarrow \quad [(b + c)\, n^2 + (a + b)\, l^2]\, (u/v)^2 + 2\, (a + b)\, lm\, (u/v) + [(c + a)\, n^2 + (a + b)\, m^2] = 0$$

which is a quadratic in u/v and if its roots are u_1/v_1, u_2/v_2, then we have

$$\frac{u_1}{v_1}\cdot\frac{u_2}{v_2} = \text{product of the roots.}$$

$$\Rightarrow \quad \frac{u_1u_2}{v_1v_2} = \frac{(c+a)\,n^2+(a+b)\,m^2}{(b+c)\,n^2+(a+b)\,l^2}$$

$$\Rightarrow \quad \frac{u_1u_2}{(a+b)\,m^2+(c+a)\,n^2} = \frac{v_1v_2}{(b+c)\,n^2+(a+b)\,l^2}$$

$$= \frac{w_1w_2}{(c+a)\,l^2+(b+c)\,m^2}, \text{ by symmetry} \quad ...(v)$$

Again if the two planes

$u_1x + v_1y + w_1z = 0$ and $u_2x + v_2y + w_2z = 0$ are perpendicular (given) then

$$u_1u_2 + v_1v_2 + w_1w_2 = 0$$

$$\Rightarrow \quad [(a + b)\,m^2 + (c + a)\,n^2] + [(b + c)\,n^2\,(a + b)\,l^2]$$
$$+ [(c + a)\,l^2 + (b + c)\,m^2] = 0, \text{ from (v)}$$

$$\Rightarrow \quad (2a + b + c)\,l^2 + (2b + c + a)\,m^2 + (2c + a + b)\,n^2 = 0. \quad ...(vi)$$

∴ The locus of OP *i.e.,* the line (i) is obtained by eliminating l, m, n between (i) and (vi) and is given by

$$(2a + b + c)\,x^2 + (2b + c + a)\,y^2 + (2c + a + b)\,z^2 = 0,$$

which evidently is a cone with vertex at origin, being homogeneous equation of second degree in x, y, z. **Ans.**

Example 66:

Find the reciprocal cone of the cone

$$(x^2/a^2) + (y^2/b^2) + (z^2/c^2) = 0.$$

Solution:

Let the cone reciprocal to $(x^2/a^2) + (y^2/b^2) + (z^2/c^2) = 0$ ...(i)

be $Ax^2 + By^2 + Cz^2 + 2Fyz + 2Gzx + 2Hxy = 0$ ...(ii)

where A = "$bc - f^2$" = $(1/b^2)\,(1/c^2) - (0)^2 = 1/(b^2c^2)$

Similarly $B = 1/(c^2a^2)$

and $C = 1/(a^2b^2)$

Also F = '$gh - af$' $= 0$.

Similarly $G = 0,\ H = 0$

$\therefore$ From (ii), the equation of cone reciprocal to (i) is

$$(1/b^2c^2)\,x^2 + (1/c^2a^2)\,y^2 + (1/a^2b^2)\,z^2 = 0$$

$\Rightarrow$ $$a^2x^2 + b^2y^2 + c^2z^2 = 0$$ **Ans.**

Example 67:

Find the condition that the plane $ux + vy + wz = 0$ may touch the cone $ax^2 + by^2 + cz^2 = 0$.

Solution:

The d.c.'s of the normal to the plane $ux + vy + wz = 0$ are u, v, w and if this plane touches the given cone then,

$$Au^2 + Bv^2 + Cw^2 + 2Fvw + 2Gwu + 2Huv = 0. \quad ...(1)$$

Here $A = bc - f^2 = bc$, since $f = 0$.

Similarly $B = ca$, $C = ab$,

$F = \text{"}gh - af\text{"} = 0$; $G = 0$, $H = 0$.

Example 68:

Prove that the perpendicular drawn from the origin to the tangent planes to the cone $ax^2 + by^2 + cz^2 = 0$ lies on the cone

$$x^2/a + y^2/b + z^2/c = 0.$$

Solution:

We know that the locus of the perpendicular (*i.e.*, normal) drawn from the origin, (*i.e.*, vertex) to the tangent planes to a given cone is called reciprocal cone. Hence we are to find the reciprocal cone to the cone $ax^2 + by^2 + cz^2 = 0$.

Example 69:

Prove that the perpendicular drawn from the origin to tangent planes to the cone $3x^2 + 4y^2 + 5z^2 + 2yz + 4zx + 6xy = 0$ lie on the cone

$$19x^2 + 11y^2 + 3z^2 + 6yz - 10zx - 26xy = 0.$$

Solution:

We are to find the reciprocal cone of the cone

$$3x^2 + 4y^2 + 5z^2 + 2yz + 4zx + 6xy + 0. \quad ...(i)$$

Here $a = 3$, $b = 4$, $c = 5$, $f = 1$, $g = 2$, $h = 3$

$\therefore$ $A = bc - f^2 = 4.5 - 1 = 19$, $B = ca - g^2 = 15 - 4 = 11$;

$C = ab - h^2 = 12 - 9 = 3$; $F = gh - af = 6 - 3 = 3$;

$$G = hf - bg = 3 - 8 = -5;\ H = fg - ch = 2 - 15 = -13.$$

$\therefore$ The equation of the reciprocal cone to (i) is

$$Ax^2 + By^2 + Cz^2 + 2Fyz + 2Gzx + 2Hxy = 0$$

$\Rightarrow$ $19x^2 + 11y^2 + 3z^2 + 6yz - 10zx - 26xy = 0$ **Hence proved.**

Example 70:

Find the equation of the right circular cone whose vertex is (0, 0, 0), axis Ox (i.e., x-axis) and semi-vertical angle is a.

Solution:

The vertex of the cone is O (0, 0, 0), its semi-vertical angle is α and direction cosines of its axis are 1, 0, 0.

Let P (x, y, z) be any point on the cone. Then the direction ratios of the generator OP are x – 0, y – 0, z – 0 *i.e.,* x, y, z.

Also OP is making an angle α with the axis of the cone, so we get the equation of the cone as

$$\cos\alpha = \frac{x.1 + y.0 + z.0}{\sqrt{(x^2+y^2+z^2)}\ \sqrt{(1+0+0)}} = \frac{x}{\sqrt{(x^2+y^2+z^2)}}$$

$\Rightarrow$ $(x^2 + y^2 + z^2)\cos^2\alpha = x^2$

$\Rightarrow$ $x^2 = (y^2 + z^2)\cot^2\alpha.$ **Ans.**

Example 71:

Find the equation of the right circular cone whose axis is x = y = z vertex is origin and whose semi-vertical angle is 45°.

Solution:

The vertex of the cone is V (0, 0, 0), its semi-vertical angle is 45° and its axis is given by $x = y = z.$ ...(i)

Let P (x, y, z) be any point on the cone. Then the direction ratios of the generator VP are x – 0, y – 0, z – 0 *i.e.,* x, y, z.

Also VP is making an angle of 45° with the line (i), so we get the equation of the cone as

$$\cos 45^\circ = \frac{x.1 + y.1 + z.1}{\sqrt{(x^2+y^2+z^2)}.\sqrt{(1^2+1^2+1^2)}}$$

$\Rightarrow$ $\dfrac{1}{\sqrt{2}} = \dfrac{x+y+z}{\sqrt{(x^2+y^2+z^2)}\ \sqrt{3}}$ $3(x^2 + y^2 + z^2) = 2(x + y + z)^2$

$\Rightarrow$ $x^2 + y^2 + z^2 - 4yz - 4zx - 4xy = 0.$ **Ans.**

Example 72:

Show that the locus of the line of intersection of perpendicular tangent planes to the cone $ax^2 + by^2 + cz^2 = 0$ is the cone

$$a(b + c)x^2 + b(c + a)y^2 + c(a + b)z^2 = 0.$$

Solution:

Let $ux + vy + wz = 0$ be any tangent plane to the cone

$$ax^2 + by^2 + cz^2 = 0.$$

Then the normal to this plane through the vertex of the cone viz $x/u = y/v = z/w$ will be a generator of the reciprocal cone

$$(x^2/a) + (y^2/b) + (z^2/c) = 0$$

$\therefore$ We have $(u^2/a) + (v^2/b) + (w^2/c) = 0$

$\Rightarrow$ $bcu^2 + cav^2 + abw^2 = 0.$ ...(i)

This equation being in quadratic u, v, w shows that there will be two tangent planes like $ux + vy + wz = 0$.

Let the line of the intersection of these two tangent planes be

$$x/l = y/m = z/n. \quad \text{...(ii)}$$

Since this line lies on the plane $ux + vy + wz = 0$

so we have $ul + vm + wn = 0$ ...(iii)

Now the direction of the normal to the plane viz. u, v, w are given by the relations (i) and (iii). Again if the two planes of the form $ux + vy + wz = 0$ be perpendicular, then we have $u_1u_2 + v_1v_2 + w_1w_2 = 0.$ ...(iv)

Now eliminating w between (i) and (iii) we get

$$bcu^2 + cav^2 + ab[-(ul + vm)/n]^2 = 0$$

$\Rightarrow$ $bcn^2u^2 + can^2v^2 + ab(ul + vm)^2 = 0$

$\Rightarrow$ $(bcn^2 + abl^2)u^2 + 2ablm\,uv + (can^2 + abm^2)v^2 = 0$

$\Rightarrow$ $(bcn^2 + abl^2)(u/v)^2 + 2ablm(u/v) + (can^2 + abm^2) = 0,$

which is a quadratic in u/v and if its roots are u_1/v_1 and u_2/v_2 then we get

$$\frac{u_1}{v_1}\cdot\frac{u_2}{v_2} = \frac{abm^2 + can^2}{ban^2 + abl^2}$$

$\Rightarrow$ $$\frac{u_1u_2}{abm^2 + can^2} = \frac{v_1v_2}{bcn^2 + abl^2} = \frac{w_1w_2}{cal^2 + bcm^2}, \text{ by symmetry}$$

$\therefore$ From (iv) we have

$$(abm^2 + can^2) + (bcn^2 + abl^2) + (cal^2 + bcm^2) = 0$$

$$\Rightarrow \quad a(b + c)\, l^2 + b(c + a)\, m^2 + c(a + b)\, n^2 = 0 \qquad \text{...(v)}$$

$\therefore$ The locus of the line of intersection (ii), obtained by eliminating l, m, n between (ii) and (v) is $a(b + c)x^2 + b(c + a)y^2 + c(a + b)z^2 = 0$, which evidently is a cone with vertex at origin, being homogeneous equation of second degree in x, y, z.

Example 73:

Show that the planes which cut $ax^2 + by^2 + cz^2 = 0$ perpendicular generators, touch the cone

$$S\,[x^2/(b + c)] = 0.$$

Solution:

We can prove that if the plane $ux + vy + wz = 0$ cuts the cone $ax^2 + by^2 + cz^2 = 0$ in perpendicular generators then

$$(b + c)u^2 + (c + a)v^2 + (a + b)w^2 = 0. \qquad \text{...(i)}$$

Again if the plane $ux + vy + wz = 0$ touches the cone

$$\frac{x^2}{(b+c)} + \frac{y^2}{(c+a)} + \frac{z^2}{(a+b)} = 0$$

then we must have $$\frac{u^2}{1/(b+c)} + \frac{v^2}{1/(c+a)} + \frac{w^2}{1/(a+b)} = 0 \qquad \textbf{(Note)}$$

$$\Rightarrow \quad (b + c)u^2 + (c + a)v^2 + (a + b)w^2 = 0.$$

which is the same as (i) **Hence proved.**

Example 74:

Prove that the cones $ax^2 + by^2 + cz^2 = 0$ and $x^2/a + y^2/b + z^2/c = 0$ are reciprocal to each other.

Solution:

Let the cone reciprocal to $ax^2 + by^2 + cz^2 = 0$ ...(i)

be $Ax^2 + By^2 + Cz^2 + 2Fyz + 2Gzx + 2Hxy = 0,$ ...(ii)

where $A = "bc - f^2" = bc$, here $f = 0$. Similarly $B = ca$ and $C = ab$.

Also $F = "gh - af" = 0$, $G = 0$, $H = 0$.

$\therefore$ From (ii) the equation of the cone reciprocal to (i) is

$$bcx^2 + cay + abz^2 = 0$$

$$\Rightarrow \quad x^2/a + y^2/b + z^2/c = 0.$$ **Hence proved.**

Example 75:

Show that the general equation to a cone which touches the coordinates planes is $a^2x^2 + b^2y^2 + c^2z^2 - 2bcyz - 2cazx - 2abxy = 0$.

Solution:

The cone which touches the co-ordinate planes is reciprocal to the cone which contains the three co-ordinate axes as these axes are normal to the co-ordinate planes.

Now the equation of the cone which contains the three coordinate axes is

$$fyz + gzx + hxy = 0. \qquad ...(i)$$

For this cone 'a' = 0; 'b' = 0; 'c' = 0; 'f' = $\frac{1}{2}$ f; 'g' = $\frac{1}{2}$ g and 'h' = $\frac{1}{2}$h.

$$\therefore \qquad A = \text{"bc} - f^2\text{"} = 0 - \frac{1}{4}f^2;\ B = -\frac{1}{4}g^2;\ C = -\frac{1}{4}h^2$$

and
$$F = gh - af = \frac{1}{4}gh;\ G = \frac{1}{4}hfl;\ \ H = \frac{1}{4}fg.$$

$\therefore$ The required equation of the cone reciprocal to (i) is

$$Ax^2 + By^2 + Cz^2 + 2Fyz + 2Gzx + 2Hxy = 0$$

$$\Rightarrow -\frac{1}{4}f^2x^2 - \frac{1}{4}g^2y^2 - \frac{1}{4}h^2z^2 + 2\left(\frac{1}{4}gh\right)yz + 2\left(\frac{1}{4}hf\right)zx + 2\left(\frac{1}{4}fg\right)xy = 0$$

$$\Rightarrow \qquad f^2x^2 + g^2y^2 + h^2z^2 - 2ghyz - 2hfzx - 2fgxy = 0.$$

which is of the form

$$a^2x^2 + b^2y^2 + c^2z^2 - 2bcyz - 2cazx - 2abxy = 0.$$ **Hence proved.**

Example 76:

Find the equation of the enveloping cone of the sphere given below whose vertex is at the origin

$$x^2 + y^2 + z^2 + 2ux + 2vy + 2wz + d = 0.$$

Solution:

Here $\quad S = x^2 + y^2 + z^2 + 2ux + 2vy + 2wz + d$

$\therefore \quad S_1 = 0 + 0 + 0 + 2u(0) + 2v(0) + 2w(0) + d = d$

and $\quad T = x.0 + y.0 + z.0 + u(x + 0) + v(y + 0) + w(z + 0) + d$

$\qquad = ux + vy + wz + d$

$\therefore$ Required equation of the enveloping cone is

$$SS_1 = T^2$$

$\Rightarrow \quad (x^2 + y^2 + z^2 + 2ux + 2vy + 2wz + d)(d) = (ux + vy + wz + d)^2$

Example 77:

If the plane $2x - y + cz = 0$ cuts the cone $yz + zx + xy = 0$ in perpendicular lines, find the value of c.

Solution:

Let the plane $2x - y + cz = 0$ cut the cone $yz + zx + xy = 0$ in a line

$$x/l = y/m = z/n.$$

Then $\quad 2l - m + cn = 0$

and $\quad mn - nl - lm = 0. \qquad$ ...(i)

Eliminating n between these equation we have

$$(2l + cn)\, n + nl + l\,(2l + cn) = 0$$

$\Rightarrow \quad 2l^2 + (c + 3)\, ln + cn^2 = 0 \Rightarrow 2\,(l/n)^2 + (c + 3)\,(l/n) + c = 0 \qquad$ ...(ii)

If the roots of this equation are (l_1/n_1) and (l_2/n_2), then

$$\frac{l_1}{n_1} \cdot \frac{l_2}{n_2} = \text{product of the roots} = \frac{c}{2} \Rightarrow \frac{l_1 l_2}{c} = \frac{n_1 n_2}{2} \qquad \text{...(iii)}$$

Eliminating l between the relations (i) we get

$2mn + n\,(m - cn) + m\,(m - cn) = 0 \qquad$ **(Note)**

$\Rightarrow \quad m^2 + (3 - c)\, mn - cn^2 = 0 \Rightarrow c\,(n/m)^2 + (c - 3)\,(n/m) - 1 = 0$

$\therefore$ If the roots of this equation are (n_1/m_1) and (n_2/m_2) then

$$\frac{n_1}{m_1} \cdot \frac{n_2}{m_2} = \text{product of the roots} = -\frac{1}{c}.$$

$$\Rightarrow \quad \frac{n_1 n_2}{1} = \frac{m_1 m_2}{-c}$$

$$\Rightarrow \quad \frac{n_1 n_2}{2} = \frac{m_1 m_2}{-2c} \qquad \text{...(iv)}$$

From (iii) and (iv) we get $\dfrac{l_1 l_2}{1} = \dfrac{m_1 m_2}{-2c} = \dfrac{n_1 n_2}{2} = k$ (say).

If the angle between the lines is a right angle, then we have

$l_1 l_2 + m_1 m_2 + n_1 n_2 = 0 \Rightarrow c + (-2c) + 2 = 0 \qquad \Rightarrow c = 2.$ **Ans.**

Example 78:

Find the angle between the lines given by $x + y + z = 0$ and

$$\frac{yz}{q-r}+\frac{zx}{r-p}+\frac{xy}{p-q}=0.$$

Solution:

The angle between the lines given by

$x + y + z = 0$ and $ayz + bzx + cxy = 0$,

where $a = 1/(q - r)$, $b = 1/(r - p)$, $c = 1/(p - q)$

Also here $1/a + 1/b + 1/c = 0$, which show that the angle between those lines is $\pi/3$. **Ans.**

Example 79:

Prove that the locus of the line of intersection of tangent planes to the cone $ax^2 + by^2 + cz^2 = 0$ which touch along perpendicular generators is the cone

$$a^2 (b + c) x^2 + b^2 (c + a) y^2 + c^2 (a + b) z^2 + = 0.$$

Solution:

We know that the tangent plane to a cone at any point touches it along the generator through that point. Let $x/l = y/m = z/n$ be the line of intersection of two tangent planes which touch the cone along perpendicular generators.

$\therefore$ The equations of the plane containing these two perpendicular generators is $ax . l + by . m + cz . n = 0$ ...(i)

Also the equation of the cone is $ax^2 + by^2 + cz^2 = 0$. ...(ii)

Now we can prove that the plane

$$ux + vy + wz = 0$$

cuts the cone $ax^2 + by^2 + cz^2 = 0$ in perpendicular generators if

$$(b + c) u^2 (c + a) v^2 + (a + b) w^2 = 0.$$

$\therefore$ The plane (i) will cut the cone (ii) in perpendicular generators if

$$(b + c) (al)^2 + (c + a) (bm)^2 + (a + b) (cn)^2 = 0 \quad \textbf{(Note)}$$

$\therefore$ The locus of the line $x/l = y/m = z/n$ is

$(b + c) a^2x^2 + (c + a) b^2y^2 + (a + b)c^2z^2 = 0.$ **Hence proved.**

Example 80:

Find the condition that the lines of section of the plane $lx + my + nz = 0$ and cones $fyz + gzx + hxy = 0$, $ax^2 + by^2 + cz^2 = 0$ should be coincident.

Solution:

Let $x/\lambda = y/\mu = z/\nu$ be one the lines in which the given plane $lx + my + nz = 0$ cuts the cone $fyz + gzx + hxy = 0$.

then we have $l\lambda + m\mu + n\nu = 0$ ...(i)

and $f\mu\nu + g\nu\lambda + h\lambda\mu = 0$ ...(ii)

Eliminating ν between (i) and (ii), we get

$$f\mu\,[-\,l\lambda + m\mu)/n] + g\lambda\,[-l\lambda + m\mu)/n] + h\lambda\mu + 0$$

$$\Rightarrow \quad -\,lf\,\lambda\mu + mf\mu^2 - gl\lambda^2 - gm\lambda\mu + nh\lambda\mu = 0$$

$$\Rightarrow \quad gl\lambda^2 + (lf + gm - nh)\,\lambda\mu + mf\mu^2 = 0$$

$$\Rightarrow \quad gl\,(\lambda/\mu)^2 + (lf + gm - nh)\,(\lambda/\mu) + mf = 0. \quad ...(iii)$$

If $x/\lambda_1 = y/\mu_1 = z/\nu_1$ and $x/\lambda_2 = y/\mu_2 = z/\nu_2$ be two lines of section then

from (iii) we get $\frac{\lambda_1}{\mu_1}\cdot\frac{\lambda_2}{\mu_2}$ = product of roots = $\frac{mf}{gl}$

$$\Rightarrow \quad \frac{\lambda_1\lambda_2}{f/l} = \frac{\mu_1\mu_2}{g/m} = \frac{\nu_1\nu_2}{h/n}, \text{ by symmetry} \quad ...(iv)$$

Again if $x/p = y/q = z/r$ be one of the lines in which the plane $lx + my + nz = 0$ cuts the cone $ax^2 + by^2 + cz^2 = 0$

then we have $lp + mq + nr = 0$ ...(v)

and $ap^2 + bq^2 + cr^2 = 0$ ...(vi)

Eliminating r between (v) and (vi), we get

$$ap^2 + bq^2 + c\,[-\,(lp - mq)/n]^2 = 0$$

$$\Rightarrow \quad (an^2 + cl^2)\,p^2 - 2clm\,pq + (bn^2 + cm^2)\,q^2 = 0$$

$$\Rightarrow \quad (an^2 + cl^2)\,(p/q)^2 - 2clm\,(p/q) + (bn^2 + cm^2) = 0 \quad ...(vii)$$

If $x/p_1 = y/q_1 = z/r_1$ and $x/p_2 = y/q_2 = z/r_2$ be the two lines of section

then from (vii) we get $\frac{p_1}{q_1}\cdot\frac{p_2}{q_2}$ = product of the roots = $\frac{bn^2 + cm^2}{an^2 + cl^2}$

$$\frac{p_1p_2}{bn^2 + cm^2} = \frac{q_1q_2}{cl^2 + an^2} = \frac{r_1r_2}{am^2 + bl^2}. \text{ by symmetry}$$

Now if the lines of section of the plane $lx + my + nz = 0$

and cones $fyz + gzx + hxy = 0$ and $ax^2 + by^2 + cz^2 = 0$ are coincident

then we must have $\frac{\lambda_1\lambda_2}{p_1p_2} = \frac{\mu_1\mu_2}{q_1q_2} = \frac{\nu_1\nu_2}{r_1r_2}$

$$\Rightarrow \quad \frac{f/l}{bn^2 + cm^2} = \frac{g/m}{cl^2 + am^2} = \frac{h/n}{am^2 + bl^2}$$

$$\Rightarrow \quad \frac{fmn}{bn^2 + cm^2} = \frac{gln}{cl^2 + an^2} = \frac{hlm}{am^2 + bl^2} \text{ are the required conditions.}$$

Example 81:

Prove that the angle between the lines in which the plane $x + y + z = 0$ cuts the cone $ayz + bzx + cxy = 0$ will be $\pi/2$ if $a + b + c = 0$.

Solution:

From the equation of the cone we observe that the sum of the coefficients of x^2, y^2 and z^2 is zero. *i.e.,*, the given curve has three mutually perpendicular generators. If the two lines in which the plane $x + y + z = 0$ cuts the given cone are at right angles, then the third generator for the cone will be the normal to the given plane through the vertex *i.e.,* $x/1 = y/1 = z/1$ and as such its direction cosines will satisfy the equation of the cone and so we have

$a . 1 . 1 + b . 1 . 1 + c . 1. 1. = 0 \Rightarrow a + b + c = 0$. **Hence proved.**

CYLINDER

Definition : *The surface generated by a variable straight line moving parallel to a fixed straight line and satisfying one more condition (intersecting a given curve or touching a given surface) is called a* ***cylinder****.*

If the generating line is always at a constant distance from the fixed straight line then the cylinder so generated is called a *right circular cylinder* whose radius is this constant distance and axis is the fixed straight line.

EQUATION OF A CYLINDER

Let the generators of the cylinder be parallel to the line

$$x/l = y/m = z/n \qquad \text{...(i)}$$

and the equation of the conic be

$$ax^2 + 2hxy + by^2 + 2gx + 2fy + c = 0,\ z = 0 \qquad \text{...(ii)}$$

Let P (x_1, y_1, z_1) be any point on the cylinder then the equation of generator through P with the help of (i) are given by

$$\frac{x - x_1}{l} = \frac{y - y_1}{m} = \frac{z - z_1}{n} \qquad \text{...(iii)}$$

This generator meets the plane $z = 0$ in the point given by

$$\frac{x - x_1}{l} = \frac{y - y_1}{m} = \frac{0 - z_1}{n}$$

i.e.,
$$\left(x_1 = \frac{l z_1}{n},\ y_1 - \frac{m z_1}{n},\ 0\right)$$

$\therefore$ The generator (iii) meets the conic (ii) if

$$a\left(x_1 = \frac{l z_1}{n}\right) + 2h\left(x_1 - \frac{l z_1}{n}\right)\left(y_1 - \frac{m z_1}{n}\right) + b\left(y_1 - \frac{m z_1}{n}\right)^2$$

$$+ 2g\left(x_1 - \frac{l z_1}{n}\right) + 2f\left(y_1 - \frac{m z_1}{n}\right) + c = 0$$

$$\Rightarrow \quad a\,(nx_1 - lz_1) + 2h\,(nx_1 - lz_1)\,(ny_1 - mz_1) + b\,(ny_1 - mz_1)^2$$
$$+ 2gn\,(nx_1 - lz_1) + 2fn\,(ny_1 - mz_1) + cn^2 = 0$$

$\therefore$ The required equation of the cylinder on the locus of P (x_1, y_1, z_1) is given by

$$a\,(nx - lz)^2 + 2h\,(nx - lz)\,(ny - mz) + b\,(ny_1 + mz)^2$$
$$+ 2gn\,(nx - lz) \pm 2fn\,(ny - mz) + cn^2 = 0$$

Corollary 1 : If the generator of the cylinder is parallel to z-axis whose d.c.'s are 0, 0, 1 *i.e.*, $l = 0 = m$, then the above equation of the cylinder reduces to

$$ax^2 + 2hxy + by^2 + 2gx + 2fy + c = 0 \qquad \textbf{(Note)}$$

In general every equation of the form f (x, y) = 0 represents a cylinder passing through the curve f (x, y) = 0, z = 0 and whose generators are parallel to the z-axis. **(Remember)**

Similarly, the equation f (y, z) = 0 represents a cylinder through the f_1 curve f (y, z) = 0, x = 0 and generators parallel to x-axis and the equation f (z, x) – 0 represents a cylinder through the curve f (z, x) = 0 and generators parallel to y-axis.

Corollary 2 : The equation of the cylinder which intersects the curve (x, y, z) = 0, f_2 (x, y, z) = 0 and whose generators are parallel to x-axis is obtained by eliminating x between f_1 (x, y, z) = 0 and f_2 (x, y, z) = 0.

Similarly by eliminating y (or z) between these equations, the equation of under passing through the curve given by these equations and generators any (or z) axis can be obtained.

EQUATION OF A RIGHT CIRCULAR CYLINDER

To find the equation of the right circular cylinder whose axis is the line

$$\frac{x-\alpha}{l} = \frac{y-\beta}{m} = \frac{z-\gamma}{n} \quad \textit{and whose radius is r.}$$

Let (x, y, z) be any point on the cylinder. Then the perpendicular distance 'p' of the point from the given axis is given by

$$p^2 = (x-\alpha)^2 + (y-\beta)^2 + (z-\gamma)^2 - \frac{[l(x-\alpha) + m(y-\beta) + n(z-\gamma)]^2}{(l^2+m^2+n^2)}$$

Since the radius of the given cylinder is r, so by definition of a right circular cylinder we have p = r *i.e.*, $r^2 = p^2$.

Hence the equation of the required right circular cylinder is

$$(x-\alpha)^2 + (y-\beta)^2 + (z-\gamma)^2 - \frac{[l(x-\alpha) + m(y-\beta) + n(z-\gamma)]^2}{(l^2+m^2+n^2)} = r^2.$$

Example 1:

Find the equation of the right circular cylinder whose axis is x – 2 = z, y = 0 and passes through the point (3, 0, 0).

Solution:

The equations of the axis of the cylinder are given as

$$\frac{x-2}{1} = \frac{y}{0} = \frac{z}{1} \qquad \text{...(i)}$$

Also (radius of the cylinder)^2

= [length of perpendicular from (3, 0, 0) to the line (i)]^2.

$$= \frac{1}{(1^2+0^2+1^2)}\,[(0.1-0.0)^2 + \{0.1-1.(3-2)\}^2 + \{0.(3-2)-1.0\}^2]$$

= (1/2) [1] = 1/2 *i.e.*, radius of the cylinder = $1/\sqrt{2}$

Now let P (x_1, y_1, z_1) be any point on the cylinder.

Then the length of the perpendicular from P (x_1, y_1, z_1) to the axis (i) must be equal to radius $1/\sqrt{2}$ of the cylinder *i.e.*,

$$\left(\frac{1}{2}\right)[1^2 + 0 + 1^2] = \{1.y_1 - 0.z_1\}^2 + \{1.z_1 - 2)\}^2 + \{0.(x_1-2) - 1.y_1\}^2$$

$$\Rightarrow \quad 1 = y_1^2 + (z_1 - x_1 + 2)^2 + y_1^2$$

$$\Rightarrow \quad x_1^2 + 2y_1^2 + z_1^2 - 2z_1x_1 - 4x_1 + 2z_1 + 3 = 0$$

∴ The locus of P (x_1, y_1, z_1) or the required equation of the cylinder is

$$x^2 + 2y^2 + z^2 - 2zx - 4x + 2z + 3 = 0.$$ **Ans.**

Example 2:

Find the equation of the cylinder with generators parallel to Oy, which passes through the curve of intersection of the surfaces $x^2 + y^2 + 2z^2 = 12$, $x - y + z = 1$.

Solution:

Since the generators are parallel to Oy, so we shall eliminate y between the given two equations.

Substituting $y = x + z - 1$ in $x^2 + y^2 + 2z^2 = 12$, we get

$$x^2 + (x + z - 1)^2 + 2z^2 = 12.$$

Hence $2x^2 + 3z^2 + 2zx - 2x - 2z - 11 = 0$ is the equation of the required cylinder.

Example 3:

Find the equation of the right circular cylinder of radius 2 whose axis is the line $(x - 1)/2 = (y - 2) = (z - 3)/2$.

Solutio:

The required equation of the right circular cylinder is

$$(x - \alpha)^2 + (y - \beta)^2 + (z - \gamma)^2 - \frac{[l\,(x-\alpha) + m\,(y-\beta) + (z-\gamma)^2}{l^2 + m^2 + n^2} = r^2$$

$$\Rightarrow \quad (x - 1)^2 + (y - 2)^2 + (z - 3)^2$$

$$\frac{1}{(4+1+4)}\,[2\,(x - 1) + 1\,(y - 2) + 2\,(z - 3)]^2 = (2)^2.$$

$$\Rightarrow \quad 9\,(x^2 + y^2 + z^2 - 2x - 4y - 6z + 14) - (2x + y + 2z - 10)^2 = 36.$$

On simplification, we obtain

$$5x^2 + 8y^2 + 5z^2 - 4xy - 4yz - 8zx + 22x - 16y - 14z - 10 = 0.$$

Example 4:

Find the equation of the circular cylinder whose guiding circle is $x^2 + y^2 + z^2 - 9 = 0$, $x - y + z = 3$.

Solution:

The axis of the circular cylinder is perpendicular to the plane $x - y + z = 3$ of the given circle. So direction ratios of the axis are 1, – 1, 1.

The generator through (α, β, γ) and parallel to the axis has equations

$$\frac{x-\alpha}{1} = \frac{y-\beta}{-1} = \frac{x-\gamma}{1} = r.$$

Any point on this line $(r + \alpha, \beta - r, r + \gamma)$ lies on the given circle, if

$$(r + \alpha)^2 + (\beta - r)^2 + (r + \gamma)^2 - 9 = 0, \qquad ...(1)$$

and $\quad r + \alpha - \beta + r + r + \gamma = 3. \Rightarrow \quad 3r = 3 - (\alpha - \beta + \gamma). \quad ...(2)$

From (1), $3r^2 + 2r(\alpha - \beta + \gamma) + \alpha^2 + \beta^2 + \gamma^2 = 9. \qquad ...(3)$

Multiplying both sides of (3) by 3 and using (2), we obtain

$$\{3 - (\alpha - \beta + \gamma)\}^2 + 2\{3 - (\alpha - \beta + \gamma)\}(\alpha - \beta + \gamma) + 3(\alpha^2 + \beta^2 + \gamma^2) = 27$$

$$\Rightarrow \quad 3(\alpha^2 + \beta^2 + \gamma^2) - (\alpha - \beta + \gamma)^2 = 18$$

$$\Rightarrow \quad \alpha^2 + \beta^2 + \gamma^2 + \beta\gamma - \gamma\alpha + \alpha\beta = 9.$$

Hence the locus of (a, b, g) is

$$x^2 + y^2 + z^2 + yz - zx + xy = 9.$$

Example 5:

Obtain the equation of the right circular cylinder described on the circle through the three points (1, 0, 0), (0, 1, 0). (0, 0, 1) as guiding circle.

Solution:

The equation of the circle through the three given points is

$$x^2 + y^2 + z^2 - x - y - z = 0, \; x + y + z = 1. \qquad ...(1)$$

[Take $a = b = c = 1$]

The axis of the given cylinder is perpendicular to the plane $x + y + z = 1$. So direction ratios of the axis are 1, 1, 1. The generator through (α, β, γ) and parallel to the axis has equations

$$\frac{x-\alpha}{1} = \frac{y-\beta}{1} = \frac{z-\gamma}{1} = r.$$

Any point on this line $(r + \alpha, r + \beta, r + \gamma)$ lies on the circle (1), if

$$r + \alpha + r + \beta + r + \gamma = 1 \Rightarrow 3r = 1 - (\alpha + \beta + \gamma), \qquad ...(2)$$

and $\quad (r + \alpha)^2 + (r + \beta)^2 + (r + \gamma)^2 - (r + \alpha + r + \beta + r + \gamma) = 0$

$$\Rightarrow \quad 3r^2 + 2r(\alpha + \beta + \gamma) + \alpha^2 + \beta^2 + \gamma^2 - 1 = 0, \text{ using (2)} \qquad ...(3)$$

Multiplying (3) by 3 and using (2), we obtain

$$[1 - (\alpha + \beta + \gamma)]^2 \; 2(\alpha + \beta + \gamma)[1 - (\alpha + \beta + \gamma)] + 3(\alpha^2 + \beta^2 + \gamma^2 - 1) = 0$$

$$\Rightarrow \quad 3(\alpha^2 + \beta^2 + \gamma^2 - 1) - (\alpha + \beta + \gamma)^2 + 1 = 0$$

$$\Rightarrow \quad \alpha^2 + \beta^2 + \gamma^2 - \alpha\beta - \beta\gamma - \gamma\alpha = 1.$$

Hence the locus of (α, β, γ) is

$$x^2 + y^2 + z^2 - xy - yz - zx = 1,$$

which is the equation of the required cylinder.

Example 6:

Find the equation of the cylinder whose generators are parallel to the line $x = -y/2 = z/3$ and whose guiding curve is the ellipse $x^2 + 2y^2 = 1$, $z = 0$.

Solution:

Let P (α, β, γ) be any point the cylinder. Any generator through the point P and parallel to $x = -y/2 = z/3$ is

$$\frac{x-\alpha}{1} = \frac{y-\beta}{-2} = \frac{z-\gamma}{3}.$$

This line meets the plane $z = 0$ in the point

$$\left(\alpha - \frac{\gamma}{3}, \beta + \frac{2\gamma}{3}, 0\right).$$

This points lies on the ellipse $x^2 + 2y^2 = 1$.

$$\therefore \qquad \left(\alpha - \frac{\gamma}{3}\right)^2 + \left(\beta + \frac{2\gamma}{3}\right)^2 = 1$$

$$\Rightarrow \qquad (3\alpha - \gamma)^2 + 2\,(3\beta + 3\gamma)^2 = 9$$

$$\Rightarrow \qquad 3\alpha^2 + 6\beta^2 + 3\gamma^2 + 8\beta\gamma - 2\alpha\gamma - 3 = 0.$$

Hence the locus of (α, β, γ) is

$$3x^2 + 6y^2 + 3z^2 + 8yz - 2xz - 3 = 0,$$

which is the required equation of the cylinder.

Example 7:

Find the equation of the cylinder whose generators are parallel to the line $x = -\frac{1}{2}, y = \frac{1}{2}z$ and whose guiding curve is the ellipse $x^2 + 2y^2 = 1$, $z = 3$.

Solution:

Let P (α, β, γ) be any point on the cylinder. Any generator through the point P and parallel to $x = -\frac{1}{2}\,y = \frac{1}{2}z$ is

$$\frac{x-\alpha}{1} = \frac{y-\beta}{-2} = \frac{z-\gamma}{3}.$$

This line meets the plane z = 3 in the point

$$\left(\alpha+\frac{1}{3}(3-\gamma),\ \beta-\frac{2}{3}(3-\gamma),\ 3\right).$$

The above point lies on $x^2 + 2y^2 = 1$, therefore

$$\left\{\alpha+\frac{1}{3}(3-\gamma)\right\}^2+2\left\{\beta-\frac{2}{3}(3-\gamma)\right\}^2=1$$

$$(3\alpha - \gamma + 3)^2 + 2\,(3\beta + 2\gamma - 6)^2 = 9$$

$$3(\alpha^2 + 2\beta^2 + \gamma^2) + 8\beta\gamma - 2\gamma\alpha + 6\alpha - 24\beta - 18\gamma + 24 = 0.$$

Hence the locus of (α, β, γ) is

$$3(x^2 + 2y^2 + z^2) + 8yz - 2zx + 6x - 24y - 18z + 24 = 0,$$

which is the required equation of the cylinder.

Example 8:

Find the equation of the circular cylinder whose generating lines have the direction cosines l, m, n and which passes through the fixed circle $x^2 + z^2 = 1$ in the zox-plane.

Solution:

If (α, β, γ) is any point on the cylinder, then any generator through P is

$$\frac{x-\alpha}{l}=\frac{y-\beta}{m}=\frac{z-\gamma}{n}.$$

This line meets the zox-plane (y = 0) in the point

$$\left(\alpha-\frac{l}{m}\beta,\ 0,\ \gamma-\frac{n}{m}\beta\right).$$

The line intersects $x^2 + z^2 = 1$ if

$$\left(\alpha-\frac{l}{m}\beta\right)^2+\left(\gamma-\frac{n}{m}\beta\right)^2=1$$

$$\Rightarrow \qquad (m\alpha - l\beta)^2 + (m\gamma - n\gamma)^2 = m^2.$$

Hence the locus of (α, β, γ) is

$$(mx - ly)^2 + (mz - ny)^2 = m^2.$$

Example 9:

Find the equation of the cylinder whose generators intersect the curve $ax^2 + by^2 = 2z$, $lx + my + nz = p$ and are parallel to the z-axis.

Solution:

Let P (α, β, γ) be any point on the cylinder. Any generator through P and parallel to the z-axis is

$$\frac{x-\alpha}{0}=\frac{y-\beta}{0}=\frac{z-\gamma}{1}=1.$$

Any point on this line $(\alpha, \beta, r+\gamma)$ lies on the given conic if

$$a\alpha^2 + b\beta^2 = 2(r+\gamma), \qquad ...(1)$$

and $\quad l\alpha + m\beta + n(r+\gamma) = p.$

$$\Rightarrow \qquad (r+\gamma) = \frac{1}{n}(p - l\alpha - m\beta). \qquad ...(2)$$

Substituting (2) in (1), we obtain

$$n(a\alpha^2 + b\beta^2) = 2(p - l\alpha - m\beta).$$

Hence the locus of (α, β, γ) is

$$n(ax^2 + by^2) + 2lx + 2my - 2p = 0.$$

Note : We can also obtain the locus by eliminating z between the given equations of the curve.

Example 10:

Find the equation of the cylinder with generators parallel to x-axis and passing through the curve $ax^2 + by^2 + cz^2 - 1$, $lx + my \mid nz - p$.

Solution:

The given curve is

$$ax^2 + by^2 + cz^2 = 1, \qquad ...(1)$$

$$lx + my + nz = p. \qquad ...(2)$$

Since the generators of the cylinder are parallel to x-axis, so the equation of the cylinder will not contain terms of x. Thus, the equation of the cylinder will be obtained by eliminating x between (1) and (2). From (2), we get

$$x = \frac{1}{l}(p - my - nz).$$

Putting in (1), we obtain

$$\frac{a}{l^2}(p - my - nz)^2 + by^2 + c^2 = 1.$$

$$\Rightarrow \qquad a(p = my = nz)^2 + bl^2y^2 + cl^2z^2 = l^2.$$

Hence $(am^2 + bl^2)\,y^2 + (an^2 + cl^2)\,z^2 + 2amnyz - 2ampy$

$$- 2ampz + (ap^2 - l^2) = 0.$$

Example 11:

Find the equation of the cylinder whose generators are parallel to the line and $x/1 - y/2 = z/3$ passes through the curve

$$x^2 + y^2 = 16 \text{ and } z = 0.$$

Solution:

Let P (x_1, y_1, z_1) be any point on the cylinder, then the equations of the generator through P are given by $\frac{x-x_1}{1} = \frac{y-y_1}{2} = \frac{z-z_1}{3}$.

This generator meets the plane z = 0 in the point given by

$$\frac{x-x_1}{1} = \frac{y-y_1}{2} = \frac{0-z_1}{3}$$

i.e., in the point $\left[x_1 - \frac{1}{3} z_1, y_1 - (2/3)\, z_1, 0\right]$.

∴ The generator intersects the given curve if

$$\left(x_1 - \frac{1}{3} z_1\right)^2 + [y_1 - 2/3\, z_1]^2 = 16$$

∴ The locus of P (x_1, y_1, z_1) or the required equation of the cylinder is given by $\left(x - \frac{1}{3} z\right)^2 + [y - (2/3)\, z]^2 = 16$

⇒ $$x^2 + (1/9)\, z^2 - (2/3)\, xz + y^2 + (4/9)\, z^2 - (4/3)\, yz = 16$$

⇒ $$9x^2 + 9y^2 + 5z^2 - 12yz - 6zx - 144 = 0$$ **Ans.**

EQUATION OF TANGENT PLANE TO A CYLINDER

To find the equation of the tangent plane to the cylinder $ax^2 + 2hxy + by^2 + 2gx + 2fy + c = 0$ at the point (x_1, y_1, z_1) and to show that this tangent plane touches the cylinder along a generator.

The equation of the given cylinder is given by

$$ax^2 + 2hxy + by + 2gx + 2fy + c = 0 \qquad ...(i)$$

Let P be the point (x_1, y_1, z_1) which lies on the cylinder (i).

Therefore $ax_1^2 + 2hx_1y_1 + by_1^2 + 2gx_1 + 2fy_1 + c = 0$ ...(ii)

Equations of any line through P (x_1, y_1, z_1) are given as

$$\frac{x-x_1}{l}=\frac{y-y_1}{m}=\frac{z-z_1}{n} \quad ...(iii)$$

Any point on this line is $(x_1 + lr, y_1 + mr, z_1 + nr)$.

If this point lies on the cylinder (i), then we get

$$a(x_1 + lr)^2 + 2h(x_1 + lr)(y_1 + mr) + b(y_1 + mr)^2 + 2g(x_1 + lr) + 2f(y_1 + mr) + c = 0$$

$$\Rightarrow \quad r^2(al^2 + 2hlm + bm)^2 + 2r[l.(ax_1 + hy_1 + g) + m(hx_1 + by_1 + f)] + (ax_1^2 + 2hx_1.y_1 + by_1^2 + 2gx_1 + 2fy_1 + c) = 0$$

$$\Rightarrow \quad r^2(al^2 + 2hlm + bm^2) + 2r[l(ax_1 + hy_1 + g) + m(hx_1 + by_1 + f)] = 0, \text{ from (ii)} \quad ...(iv)$$

This is a quadratic in r and one value of r is zero. Now, if the line (iii) is a tangent line to the cylinder (i), then this line touches the cylinder *i.e.,* this line meets the cylinder only in one point (rather two coincident points and so both the values of r given by (iv) must be zero.

And from (iv) we find the if the other value of r given by (iv) is zero then

$$l(ax_1 + hy_1 + g) + m(hx_1 + by_1 + f) = 0 \quad ...(v)$$

$\therefore$ If the line (iii) is a tangent line to the cylinder (i) at P (x_1, y_1, z_1) the d.c.'s of the line (iii) must satisfy the relation (v). Also the locus of this tangent line to the cylinder (i) at P (x_1, y_1, z_1) is the required tangent plane.

$\therefore$ Eliminating l, m, n between (iii) and (v), the required equation of the tangent plane to the cylinder (i) at P (x_1, y_1, z_1) is given by

$$(x - x_1)(ax_1 + hy_1 + g) + (y - y_1)(hx_1 + by_1 + f) = 0$$

$$\Rightarrow \quad x(ax_1 + by_1 + g) + y(ax_1 + by_1 + f) = ax_1^2 + 2hx_1y_1 + by_1^2 + gx_1 + fy_1$$

Adding $(gx_1 + fy_1 + c)$ to both sides, we get

$$x(ax_1 + hy_1 + g) + (hx_1 + by_1 + f) + (gx_1 + fy_1 + c)$$

$$= ax_1^2 + 2hx_1y_1 + by_1^2 + 2gx_1 + 2fy_1 + c.$$

$$= 0, \text{ from (ii)}.$$

$$\Rightarrow axx_1 + h(xy_1 + x_1y) + byy_1 + g(x + x_1) + f(y_1 + y_1) + c = 0. \quad ...(vi)$$

Again we observe from (i) there is no term containing z in it, hence, the axis of the cylinder and consequently the generators of the cylinder are parallel to z-axis whose d.c.'s are 0, 0, 1. **(Note)**

∴ The equations of the generator through P (x_1, y_1, z_1) of the cylinder (i) are

$$\frac{x-x_1}{0}=\frac{y-y_1}{0}=\frac{z-z_1}{1} \qquad ...(vii)$$

Any point on this generator is $(x_1, y_1, r + z_1)$.

∴ Equation of tangent plane to (i) at this point $(x_1, y_1, z_1 + r)$ is

$$axx_1 + h(xy_1 + x_1y) + byy_1 + g(x + x_1) + f(y + y_1) + c = 0. \qquad ...(viii)$$

obtained by replacing x_1, y_1, z_1 by $x_1, y_1, z_1 + r$ respectively in (vi).

Now we observe that (vi) and (viii) are the same and also (viii) being free from r is the same for all values of r.

i.e., the equations of the tangent plane at every point of (vii) is the same. Hence the tangent plane to (i) at P (x_1, y_1, z_1) touches the cylinder (i) along the generator (vii) through the point P. **Hence proved.**

ENVELOPING CYLINDER

Definition : *The locus of the tangents to a surface drawn in a given direction is called the enveloping cylinder of the surface i.e., the enveloping cylinder is that cylinder whose generators touch a given surface and are parallel to a given straight line.*

Equation of the enveloping cylinder of sphere $x^2 + y^2 + z^2 = a^2$ and whose generators are parallel to the line $x/l = y/m = z/n$.

Proof:

Let P (α, β, γ) be a point on the cylinder. Then equations of the generator of the cylinder through P drawn parallel to the given line are

$$\frac{x-\alpha}{l}=\frac{y-\beta}{m}=\frac{z-\gamma}{n}=r \text{ (say)} \qquad ...(i)$$

Any point on this generator is $(\alpha + lr, \beta + mr, \gamma + nr)$. If this point lies on the given sphere, we have $(\alpha + lr)^2 + (\beta + mr)^2 + (\gamma + nr) = a^2$

$$\Rightarrow \quad r^2(l^2 + m^2 + n^2) + 2r(l\alpha + m\beta + n\gamma) + (\alpha^2 + \beta^2 + \gamma^2 - a^2) = 0 \qquad ...(ii)$$

Since the generator (i) is a tangent line to the given sphere, so the two values of r given by (ii) must be equal and the condition for the same is

$$[2(l\alpha + m\beta - n\gamma)]^2 = 4(l^2 + m^2 + n^2)(\alpha^2 + \beta^2 + \gamma^2 - a^2)$$

∴ The equation of the enveloping cylinder or the locus of P (α, β, γ) is

$$(lx + my + nz)^2 = (l^2 + m^2 + n^2)(x^2 + y^2 + z^2 - a^2) \qquad ...(iii)$$

$$(\alpha x + \beta y + \gamma z - a^2)^2 = (x^2 + y^2 + z^2 - a^2)(\alpha^2 + \beta^2 + \gamma^2 - a^2)$$

If the vertex of the above cone be (lr, mr, nr), then above equation reduces to $r^2\left(lx+my+nz-\frac{a^2}{r}\right)^2 = (x^2 + y^2 + z^2 - a^2)$

$(l^2r^2 + m^2r^2 + n^2r^2 - a^2)$

$$\Rightarrow \quad \left(lx+my+nz-\frac{a^2}{r}\right)^2 = (x^2 + y^2 + z^2 - a^2)\left(l^2+m^2+n^2-\frac{a^2}{r^2}\right)$$

Now if $r \to \infty$ *i.e.*, if the vertex of the cone is at an infinite distance, then the above equation reduces to

$$(lx + my + nz)^2 = (x^2 + y^2 + z^2 - a^2)(l^2 + m^2 + n^2),$$

which is the same as (ii) above.

Hence, the enveloping cylinder of the sphere $x^2 + y^2 + z^2 = a^2$ can be taken as a limiting case of its (sphere's) enveloping cone whose vertex is at an infinite distance.

(b) Equation of the enveloping cylinder of the sphere $x^2 + y^2 + z^2 + 2ux + 2vy + 2wz + d = 0$ whose generators are parallel to the line

$$x/l = y/m = z/n.$$

Proof:

Let P (α, β, γ) be a point of the cylinder. Then the equations of the generator of the cylinder through P drawn parallel to the given line are

$$\frac{x-\alpha}{l} = \frac{y-\beta}{m} = \frac{z-\gamma}{n} = r \text{ (say)} \qquad \text{...(i)}$$

Any point on this generator is $(\alpha + lr, \beta + mr, \gamma + nr)$. If this point lies on the given sphere, we have

$$(\alpha + lr)^2 + (\beta + mr)^2 + (\gamma + nr)^2 + 2u(\alpha + lr) + 2v(\beta + mr) + 2w(\gamma + nr) + d = 0$$

$$\Rightarrow \quad r^2(l^2 + m^2 + n^2) + 2r(l\alpha + m\beta + n\gamma + ul + vm + wn) + (\alpha^2 + \beta^2 + \gamma^2 + 2u\alpha + 2v\beta + 2w\gamma + d) = 0. \qquad \text{...(ii)}$$

Since the generator (i) is a tangent line to the given sphere, so the two values of r given by (ii) must be equal and the condition for the same is

'$B^2 + 4AC$'

$$[2(l\alpha + m\beta + n\gamma + ul + vm + wn)]^2 = 4(l^2 + m^2 + n^2)(\alpha^2 + \beta^2 + \gamma^2 + 2u\alpha + 2v\beta + 2w\gamma + d)$$

$\therefore$ The equation of the enveloping cylinder or the locus of P (α, β, γ) is

$$(lx + my + nz + ul + vm + wn)^2$$

$$= (l^2 + m^2 + n^2)(x^2 + y^2 + z^2 + 2ux + 2vy + 2wz + d)$$

$$[l(x + u) + m(y + v) + n(z + w)]^2$$

$$= (l^2 + m^2 + n^2)(x^2 + y^2 + z^2 - 2ux + 2uy + 2wz + d).$$

Example 1:

Find the equation of the right circular cylinder through the circle of intersection of $x^2 + y^2 + z^2 = 1$ *and* $x + y + z = 1$.

Solution:

The d. ratios of the axis of cylinder, which is perpendicular to plane of circle given by $x + y + z = 1$ are 1, 1, 1.

Let one of the generators of the cylinder passing through any point P (α, β, γ) on the cylinder be $(x - \alpha)/1 = (y - \beta)/1 = (z - \gamma)/1$

Any point on this generator at a distance r from (α, β, γ) is $(\alpha + r, \beta + r, \gamma + r)$ and if it lies on the given circle, we have

$$(\alpha + r)^2 + (\beta + r)^2 + (\gamma + r)^2 = 1, (\alpha + r) + (\beta + r) + (\gamma + r) = 1$$

$$\Rightarrow (\alpha + r)^2 + (\beta + r)^2 + (\gamma + r)^2 = 1, 3r = 1 - (\alpha + \beta + \gamma)$$

Eliminating r, we get

$$\left[\alpha + \frac{1-(\alpha+\beta+\gamma)}{3}\right]^2 + \left[\beta + \frac{1-(\alpha+\beta+\gamma)}{3}\right]^2 + \left[\gamma + \frac{1-(\alpha+\beta+\gamma)}{3}\right]^2 = 1$$

$\Rightarrow \quad \alpha^2 + \beta^2 + \gamma^2 - \beta\gamma - \gamma\alpha - \alpha\beta = 1$, on simplifying

$\therefore$ Locus of P (α, β, γ) or the required equation of the right circular cylinder is

$$x^2 + y^2 + z^2 - yz - zx - xy = 1.$$ **Ans.**

Example 2:

Prove that the equation of the right circular cylinder whose one section is the circle $x^2 + y^2 + z^2 - x - y - z = 0$, $x + y + z = 1$ *is*

$$x^2 + y^2 + z^2 - yz - zx - xy = 1,$$

Solution:

Do yourself. **Ans.** $x^2 + y^2 + z^2 - yz - zx - xy = 1.$

Example 3:

Find the equation of the right circular cylinder which passes through the circle $x^2 + y^2 + z^2 = 9$, $x - y + z = 3$.

Solution:

The direction ratios of the axis of the cylinder, which is perpendicular to the plane of the circle given by $x - y + z = 3$ are $1, -1, 1$.

So let one of the generators of the cylinder passing through the any point (α, β, γ) on the cylinder be given as $\frac{x-\alpha}{1} = \frac{y-\beta}{-1} = \frac{z-\gamma}{1}$

Any point lies on the generator at a distance r from (α, β, γ) is

$$(\alpha + r, \beta - r, \gamma + r)$$

If this point lies on the given circle, then we have

$$(\alpha + r)^2 + (\beta - r)^2 + (\gamma + r)^2 = 9, (\alpha + r) - (\beta - r) + (\gamma + r) = 3$$

$$\Rightarrow \alpha^2 + \beta^2 + \gamma^2 + 2r (\alpha - \beta + \gamma) + 3r^2 = 9, \alpha - \beta + \gamma + 3r = 3.$$

Eliminating r we get

$$\alpha^2 + \beta^2 + \gamma^2 + 2 (\alpha - \beta + \gamma)$$

$$\left[\frac{1}{3}(3-\alpha+\beta-\gamma)\right]+3\left[\frac{1}{3}(3-\alpha+\beta-\gamma)\right]^2 = 9$$

$$\Rightarrow \quad 3 (\alpha^2 + \beta^2 + \gamma^2) + 2 (\alpha - \beta + \gamma) (3 - \alpha + \beta - \gamma) + (3 - \alpha + \beta - \gamma)^2=27$$

$$\Rightarrow \quad 3 (\alpha^2 + \beta^2 + \gamma^2 + (3 - \alpha + \beta - \gamma) (3 + \alpha - \beta + \gamma) = 27$$

$$\Rightarrow \quad 3 (\alpha^2 + \beta^2 + \gamma^2) + 9 - (\alpha - \beta + \gamma)^2 = 27 \qquad \textbf{(Note)}$$

$$\rightarrow \quad \alpha^2 + \beta^2 + \gamma^2 + \alpha\beta - \alpha\gamma + \beta\gamma - 9 = 0.$$

∴ The equation of the cylinder *i.e.*, the locus of P (x_1, y_1, z_1) is

$$x^2 + y^2 + z^2 + xy - xz + yz - 9 = 0.$$ **Ans.**

Example 4:

Find the equation of the enveloping cylinder of the conicoid $ax^2 + by^2 + cz^2 = 1$ whose generators are parallel to the line

$$x = y = z.$$

Solution:

Let P (α, β, γ) be any point on the enveloping cylinder.

Then the equations of the generator through P (α, β, γ) are

$$\frac{x-\alpha}{1} = \frac{y-\beta}{1} = \frac{z-\gamma}{1} = r \text{ say}$$

here the generators being parallel to t line $x/1 = y/1 = z/1$.

Any point $(\alpha + r, \beta + r, \gamma + r)$ on the given conicoid, then we have

$$2 (\alpha + r)^2 + b (\beta + r)^2 + c (x + 5)^2 = 1$$

$$\Rightarrow \quad r^2 (a + b + c) + 2r (a\alpha + b\beta + c\gamma) + (a\alpha^2 + b\beta^2 + c\gamma^2 - 1) = 0$$

Since this generator through $P (\alpha, \beta, \gamma)$ is a tangent to the given conicoid, so the two values of r obtained from (i) must be equal and the condition for the same is "$B^2 = 4AC$"

i.e., $$4 (a\alpha + b\beta + c\gamma)^2 = 4 (a + b + c) (a\alpha^2 + b\beta^2 + c\gamma^2 - 1)$$

∴ The locus of $P (\alpha, \beta, \gamma)$ or the required equation of the enveloping cylinder of the given ellipsoid is

$$(ax + by + cz)^2 = (a + b + c) (ax^2 + by^2 + cz^2 - 1)$$

$$\Rightarrow \quad (b + c) x^2 (c + a) y^2 + (a + b) z^2 - 2abxy - 2bcyz - 2cazx - (a + b + c) = 0$$ **Ans.**

Example 5:

Find the equation of a right circular cylinder which envelopes a sphere with centre (a, b, c) and radius r and has the generators parallel to the direction (l, m, n).

Solution:

The equation of the given sphere is given as

$$(x - a)^2 + (y - b)^2 + (z - c)^2 = r^2$$

$$\Rightarrow \quad x^2 + y^2 + z^2 - 2ax - 2by - 2cz + (a^2 + b^2 + c^2 - r^2) = 0 \quad ...(i)$$

Let $P (\alpha, \beta, \gamma)$ be any point on the enveloping cylinder.

Then the equations of the generator through $P (\alpha, \beta, \gamma)$ are given as

$$\frac{x-\alpha}{l} = \frac{y-\beta}{m} = \frac{z-\gamma}{n} = k \text{ (say)}.$$

The generators are given parallel to the direction (l, m, n)

Any point on this generator is $(\alpha + lk, \beta + mk\ \gamma + nk)$.

If this point lies on the sphere (i), then we get

$$(\alpha + lk)^2 + (\beta + mk)^2 + (\gamma + nk)^2 - 2a (\alpha + lk) - 2b (\beta + mk) - 2c (\gamma + nk) + (a^2 + b^2 + c^2 - r^2) = 0$$

$$\Rightarrow \quad (l^2 + m^2 + n^2) k^2 + 2k (l\alpha + m\beta + n\gamma - al - bm - cn) + (\alpha^2 + \beta^2 + \gamma^2 - 2a\alpha - 2b\beta - 2c\gamma + a^2 + b^2 + c^2 - r^2) = 0$$

$$\Rightarrow \quad k^2 (l^2 + m^2 + n^2) - 2k [l (a - \alpha) + m (b - \beta) + n (c - \gamma)] + [(a - \alpha)^2 + (b - \beta)^2 + (c - \gamma)^2 - r^2] = 0 \quad ...(ii)$$

Since this generator is a tangent to the given sphere (i), so the two values of k obtained from (ii) must be equal and the condition for the same is

$$[l\ (a-\alpha)+m\ (b-\beta)+n\ (c-\gamma)]^2=(l^2+m^2+n^2)\ [(a-\alpha)^2 + (b-\beta)^2+(c-\gamma)^2-r^2]$$

$\therefore$ The locus of P (α, β, γ) or the required equation of the cylinder is

$$[l\ (a-x)+m\ (b-y)+n\ (c-z)]^2 = (l^2+m^2+n^2)\ [(a-x)^2+(b-y)^2+(c-z)^2-r^2]$$ **Ans.**

Example 6:

Find the equation of the enveloping cylinder of the ellipsoid $\frac{x^2}{a^2}+\frac{y^2}{b^2}+\frac{z^2}{c^2}=1$*, whose axis is* $x=y=z$ *or whose generators are parallel to the line* $x=y=z$.

Solution:

Let P (α, β, γ) be any point on the enveloping cylinder.

Then the equations of the generator through P (α, β, γ) are

$$\frac{x-\alpha}{1}=\frac{y-\beta}{1}=\frac{z-\gamma}{1}=1\ \text{(say)}$$

[Note that the generators are parallel to $x/1 = y/1 = z/1$]

Any point on this generator is $(\alpha+r, \beta+r, \gamma+r)$.

If this point $(\alpha+r, \beta+r, \gamma+r)$ lies on the given ellipsoid, then we have

$$\frac{(\alpha+r)^2}{a^2}+\frac{(\beta+r)^2}{b^2}+\frac{(\gamma+r)^2}{c^2}=1$$

$$\Rightarrow \quad r^2\left(\frac{1}{a^2}+\frac{1}{b^2}+\frac{1}{c^2}\right)+2r\left(\frac{\alpha}{a^2}+\frac{\beta}{b^2}+\frac{\gamma}{c^2}\right)+\left(\frac{\alpha^2}{a^2}+\frac{\beta^2}{b^2}+\frac{\gamma^2}{c^2}-1\right)=0$$

Since this generator through (α, β, γ) is a tangent to the given ellipsoid, so the two values of r obtained from (i) must be equal and the condition for the same is "$B^2 = 4AC$".

i.e.,
$$4\left(\frac{\alpha}{a^2}+\frac{\beta}{b^2}+\frac{\gamma}{c^2}\right)=4\left(\frac{1}{a^2}+\frac{1}{b^2}+\frac{1}{c^2}\right)\left(\frac{\alpha^2}{a^2}+\frac{\beta^2}{b^2}+\frac{\gamma^2}{c^2}-1\right)$$

$\therefore$ The locus of P (α, β, γ) or the required equation of the enveloping cylinder of the given ellipsoid is

$$\left(\frac{x}{a^2}+\frac{y}{b^2}+\frac{z}{c^2}\right)^2 = 4\left(\frac{1}{a^2}+\frac{1}{b^2}+\frac{1}{c^2}\right)\left(\frac{x^2}{a^2}+\frac{y^2}{b^2}+\frac{z^2}{c^2}-1\right).$$

$$\Rightarrow \quad (xa^{-2} + yb^{-2} + zc^{-2})^2 = (a^{-2} + b^{-2} + c^{-2})(x^2a^{-2} + y^2b^{-2} + z^2c^{-2} - 1).$$

Example 7:

Find the equation of the enveloping cylinder of the sphere $x^2 + y^2 + z^2 - 2x + 4y = 1$ having its generators parallel to the line $x = y = z$.

Solution:

Let P (α, β, γ) be any point on the enveloping cylinder then the equations of the generator through P (α, β, γ) are

$$\frac{x-\alpha}{1} = \frac{y-\beta}{1} = \frac{z-\gamma}{1} = r \text{ (say)}$$

Any point on this generator is $(\alpha + r, \beta + r, \gamma + r)$

If this point lies on the given sphere, then we get

$$(\alpha + r)^2 + (\beta + r)^2 + (\gamma + r)^2 - 2(\alpha + r) + 4(\beta + r) = 1$$

$$\Rightarrow 3r^2 + 2r(\alpha + \beta + \gamma + 1) + (\alpha^2 + \beta^2 + \gamma^2 - 2\alpha + 4\beta - 1) = 0 \quad \text{...(i)}$$

Since this generator is a tangent to the given sphere, so the values of r obtained from (i) are equal and the condition for the same is

$$[2(\alpha + \beta + \gamma + 1)]^2 = 4^2.3(\alpha^2 + \beta^2 + \gamma^2 - 2\alpha + 4\beta - 1)$$

$$\Rightarrow \quad \alpha^2 + \beta^2 + \gamma^2 - \beta\gamma - \gamma\alpha - \alpha\beta - 4\alpha + 5\beta - \gamma - 2 = 0$$

$\therefore$ The locus of P (α, β, γ) *i.e.*, the required equation of the enveloping cylinder is $x^2 + y^2 + z^2 - yz - zx - xy - 4x + 5y - z - 2 = 0$. **Ans.**

Example 8:

Find the equation of the enveloping cone of the ellipsoid $\frac{x^2}{a^2}+\frac{y^2}{b^2}+\frac{z^2}{c^2} = 1$ and deduce from it the equation of the enveloping cylinder whose generators are parallel to the line $\frac{x}{l} = \frac{y}{m} = \frac{z}{n}$.

Solution:

We can find that the equation of the enveloping cone of ellipsoid $x^2/a^2 + y^2/b^2 + z^2/c^2 = 1$ with vertex at (x_1, y_1, z_1) is

$$\left(\frac{x^2}{a^2}+\frac{y^2}{b^2}+\frac{z^2}{c^2}\right)\left(\frac{x_1^2}{a^2}+\frac{y_1^2}{b^2}+\frac{z_1^2}{c^2}-1\right) = \left(\frac{xx_1}{a^2}+\frac{yy_1}{b^2}+\frac{zz_1}{c^2}-1\right)^2 \quad \text{...(i)}$$

Also any point on the line $\frac{x}{l} = \frac{y}{m} = \frac{z}{n}$ can be taken as (lr, mr, nr)

$\therefore$ If the vertex of the cone (i) is (lr, mr, nr), the above equation reduces to

$$\left(\frac{x^2}{a^2}+\frac{y^2}{b^2}+\frac{z^2}{c^2}-1\right)\left(\frac{l^2r^2}{a^2}+\frac{m^2r^2}{b^2}+\frac{n^2r^2}{c^2}-1\right)=\left(\frac{xlr}{a^2}+\frac{ymr}{b^2}+\frac{nzr}{c^2}-1\right)^2$$

$$\Rightarrow \left(\frac{x^2}{a^2}+\frac{y^2}{b^2}+\frac{z^2}{c^2}-1\right)\left(\frac{l^2}{a^2}+\frac{m^2}{b^2}+\frac{n^2}{c^2}-\frac{1}{r^2}\right)=\left(\frac{xl}{a^2}+\frac{ym}{b^2}+\frac{nz}{c^2}-\frac{1}{r}\right)^2,$$

dividing both sides by r^2. **(Note)**

Now if $r \to \infty$ *i.e.,* the vertex of cone is at an infinite distance then the above equation reduces to

$$\left(\frac{x^2}{a^2}+\frac{y^2}{b^2}+\frac{z^2}{c^2}-1\right)\left(\frac{l^2}{a^2}+\frac{m^2}{b^2}+\frac{n^2}{c^2}\right)=\left(\frac{xl}{a^2}+\frac{my}{b^2}+\frac{nz}{c^2}\right)^2$$

which is the required equation of the enveloping cylinder. **Ans.**

Example 9:

Show that the enveloping cylinder of the conicoid $ax^2 + by^2 + cz^2 = 1$ with generators perpendicular to x-axis meets the plane $z = 0$ in parabolas.

Solution:

The d.c.'s of the z-axis are 0, 0, 1.

$\therefore$ The d.r.'s of the line perpendicular to z-axis are l, m, 0.

Let P (α, β, γ) be a point on the enveloping cylinder.

Then the equations of the generator through P (α, β, γ) are

$$\frac{x-\alpha}{l} = \frac{y-\beta}{m} = \frac{z-\gamma}{0} = r \text{ (say)}$$

Any point on it is $(\alpha + lr, \beta + mr, \gamma)$

If this point lies on the given conicoid, we get

$$a(\alpha + lr)^2 + b(\beta + mr)^2 + c(\gamma)^2 = 1$$

$$\Rightarrow \quad r^2 (al^2 + bm)^2 + 2r(a\alpha l + b\beta m) + (a\alpha^2 + b\beta^2 + c\gamma^2 - 1) = 0 \quad ...(i)$$

Since this generator is tangent to the given conicoid so the two values of r obtained from (i) must be equal and the condition for the same is

$$(a\alpha l + b\beta m)^2 = (al^2 + bm^2)(a\alpha^2 + b\beta^2 + c\gamma^2 - 1)$$

$\therefore$ The equation of the enveloping cylinder of the given concoid ***i.e.***, the locus of P(α, β, γ) is $(alx + bmy)^2 = (al^2 + bm^2)(ax^2 + by^2 + cz^2 - 1)$

Its section by the plane z = 0 is

$$(alx + bmy)^2 = (al^2 + bm^2)(ax^2 + by^2 - 1),\ z = 0$$

$$\Rightarrow \quad a^2l^2x^2 + b^2m^2y^2 + 2ablmxy = a^2l^2x^2 + abl^2y^2 - al^2 + abm^2x^2 + b^2m^2y^2 - bm^2;\ z = 0$$

$$\Rightarrow \quad ab\,(m^2x^2 + l^2y^2 - 2lmxy) = al^2 + bm^2,\ z = 0$$

$$\Rightarrow \quad ab\,(mx - ly)^2 = al^2 + bm^2,\ z = 0$$

which represents a parabola as the second degree terms form a perfect square.

Example 10:

Prove that the enveloping cylinder of the ellipsoid $(x^2/a^2) + (y^2/b^2) + (z^2/c^2) = 1$ whose generators are parallel to the line

$$\frac{x}{0} = \frac{y}{\pm\sqrt{(a^2 - b^2)}} = \frac{z}{c}$$

meet the plane z = 0 in circles.

Solution:

Let P (α, β, γ) be any point on the enveloping cylinder. Then the equations of the generator through P (α, β, γ) are

$$\frac{x}{0} = \frac{y - \beta}{\pm\sqrt{(a^2 - b^2)}} = \frac{z - \gamma}{c} = r \quad \text{(say)}$$

Any point on this generator is $(\alpha, \beta, \pm\sqrt{(a^2 - b^2)}, \gamma + cr)$

If this point lies on the given ellipsoid, then we get

$$\frac{\alpha^2}{a^2} + \frac{\left\{\beta \pm r\sqrt{(a^2 + b^2)}\right\}}{b^2} + \frac{(\gamma + cr)^2}{c^2} = 1$$

$$\Rightarrow r^2\left[\frac{a^2 - b^2}{b^2} + \frac{c^2}{c^2}\right] + 2r\left[\frac{c\gamma}{c^2} + \frac{\beta\sqrt{(a^2 + b^2)}}{b^2}\right] + \left(\frac{\alpha^2}{a^2} + \frac{\beta^2}{b^2} + \frac{\gamma^2}{c^2} - 1\right) = 0 \quad \text{...(i)}$$

Since this generator is a tangent to the given ellipsoid, so the two value of r obrained from (i) must be equal and the condition for the same is "$b^2 = 4ac$" *i.e.*,

$$\left[\frac{\gamma}{c} \pm \frac{\beta\sqrt{(a^2 - b^2)}}{b^2}\right]^2 = \left[\frac{a^2 - b^2}{b^2} + \frac{c^2}{c^2}\right]\left[\frac{\alpha^2}{a^2} + \frac{\beta^2}{b^2} + \frac{\gamma^2}{c^2} - 1\right]$$

$$\Rightarrow \quad \frac{\gamma^2}{c^2} \pm \frac{\beta\left(a^2-b^2\right)}{b^4} + \frac{2\beta\gamma\sqrt{\left(a^2-b^2\right)}}{b^2 c} = \left(\frac{a^2}{b^2}\right)\left[\frac{\alpha^2}{a^2}+\frac{\beta^2}{b^2}+\frac{\gamma^2}{c^2}-1\right]$$

∴ The locus of P (α, β, γ) or the equation of the cylinder is

$$\frac{z^2}{c^2} + \frac{b^2\left(a^2+b^2\right)}{b^4} \pm \frac{2yz\sqrt{\left(a^2+b^2\right)}}{b^2 c} = \frac{a^2}{b^2}\left[\frac{x^2}{a^2}+\frac{y^2}{b^2}+\frac{z^2}{c^2}-1\right]$$

This meets the plane z = 0 in the curve

$$\frac{y^2\left(a^2-b^2\right)}{b^4} = \frac{a^2}{b^2}\left(\frac{x^2}{a^2}+\frac{y^2}{b^2}-1\right), \quad z = 0 \qquad \textbf{(Note)}$$

i.e., $$\frac{x^2}{b^2}+\frac{y^2}{b^2}=\frac{a^2}{b^2}, \quad z = 0$$

i.e., $x^2 + y^2 = a^2, \quad z = 0$

which is a circle of radius a on the plane z = 0. **Hence proved.**

Example 11:

Find the equation of the enveloping cylinder of the conicoid $x^2/a^2 + y^2/b^2 + z^2/c^2 = 1$ *whose generators are parallel to the line*

$$x/l = y/m = z/n.$$

Solution:

Let P (α, β, γ) be any point on the enveloping cylinder.

Then the equations of the generator through P (α, β, γ) are given as

$$\frac{x-\alpha}{l}=\frac{y-\beta}{m}=\frac{z-\gamma}{n}=r \text{ (say)}$$

Any point on this generator is $(\alpha + lr, \beta + mr, \gamma + nr)$.

If this point lies on the given conicoid, then have

$$\frac{(\alpha+lr)}{a^2}+\frac{(\beta+mr)^2}{b^2}+\frac{(\gamma+nr)^2}{c^2}=1$$

$$\Rightarrow r^2\left(\frac{l^2}{a^2}+\frac{m^2}{b^2}+\frac{n^2}{c^2}\right)+2r\left(\frac{l\alpha}{a^2}+\frac{m\beta}{b^2}+\frac{n\gamma}{c^2}\right)+\left(\frac{\alpha^2}{a^2}+\frac{\beta^2}{b^2}+\frac{\gamma^2}{c^2}-1\right)=1 \quad \text{...(i)}$$

Since this generator is a tangent to the given conicoid, so the two values of r obtained from (i) must be equal and the condition for same is

$$\left(\frac{l\alpha}{a^2}+\frac{m\beta}{b^2}+\frac{n\gamma}{c^2}\right)^2=\left(\frac{l^2}{a^2}+\frac{m^2}{b^2}+\frac{n^2}{c^2}\right)\left(\frac{\alpha^2}{a^2}+\frac{\beta^2}{b^2}+\frac{\gamma^2}{c^2}-1\right)$$

∴ The locus of P (α, β, γ) or the required equation of the cylinder is given

$$\left(\frac{lx}{a^2}+\frac{my}{b^2}+\frac{nz}{c^2}\right)^2=\left(\frac{l^2}{a^2}+\frac{m^2}{b^2}+\frac{n^2}{c^2}\right)\left(\frac{x^2}{a^2}+\frac{y^2}{b^2}+\frac{z^2}{c^2}-1\right)$$ **Ans.**

Example 12:

Find the equation of the cylinder whose generators are parallel to the line $x/1 = y/(-2) = z/3$ and passing through the curve $x^2 + 2y^2 = 1$, $z = 0$.

Solution:

Let P (x_1, y_1, z_1) be any point on the cylinder, then the equations of the generator through P are given as

$$\frac{x-x_1}{1}=\frac{y-y_1}{-2}=\frac{z-z_1}{3} \quad ...(i)$$

This generator meets the plane z = 0 in the point given by

$$\frac{x-x_1}{1}=\frac{y-y_1}{-2}=\frac{0-z_1}{3}$$

i.e., in the point $\left[x_1-\frac{1}{3}z_1,\ y_1+\frac{2}{3}z_1,\ 0\right]$

∴ This generator (i) intersects the given conic if

$$\left(x_1-\frac{1}{3}z_1\right)^2+2\left[y_1+\frac{2}{3}z_1\right]^2=1$$

∴ The locus of P (x_1, y_1, z_1) or the required equation of the cylinder is

$$\left(x-\frac{1}{3}z\right)^2+2\left[y+\frac{2}{3}z\right]^2=1$$

⇒ $[x^2 + (1/9)\, z^2 - (2/3)\, xz] + 2\, [y^2 + (4/9)\, z^2\, (4/3)\, yz] = 1$

⇒ $9x^2 + z^2 - 6xz + 18y^2 + 8z^2 + 24yz - 9 = 0$

⇒ $3x^2 + 6y^2 + 3z^2 - 2xz + 8yz - 3 = 0.$ **Ans.**

Example 13:

Find the equations of the quadric cylinder which intersects the curve $ax^2 + by^2 + cz^2 = 1$, $lx + my + nz = p$ and whose generators are parallel to the axis of z.

Solution:

The guiding curve is given by the equations

$$ax^2 + by^2 + cz^2 = 1 \quad \text{...(i)}$$

$$lx + my + nz = p \quad \text{...(ii)}$$

Now as the generators of the cylinder are parallel to z-axis, so the equation of the cylinder will not contain terms of z. **(Note)**

Hence the required equations of the cylinder will be obtained by eliminating z between (i) and (ii).

From (ii) we get $z = (p - lx - my)/n$.

Substituting in (i) we get $ax^2 + by^2 + c\,[(p - lx - my)/n]^2 = 1$

$$\Rightarrow \quad an^2x^2 + bn^2y^2 + c\,(p^2 + l^2x^2 + m^2y^2 - 2plx - 2pmy + 2lmxy) - n^2 = 0$$

$$\Rightarrow \quad (an^2 + cl^2)\,x^2 + (bn^2 + cm^2)\,y^2 + 2lcmxy - 2cplx - 2cpmy + (cp^2 - n^2) = 0 \quad \textbf{Ans.}$$

Example 14:

Find the equation of the circular cylinder, whose generating lines have the direction cosines l, m, n and which passes through the fixed circle $x^2 + y^2 = a^2$ *in the zox-plane.*

Solution:

Let $P\,(x_1, y_1, z_1)$ be any point on the cylinder, then the equations of the generator through P are

$$\frac{x - x_1}{l} = \frac{y - y_1}{m} = \frac{z - z_1}{n} \quad \text{...(i)}$$

This generator meets the zox-plane *i.e.,* $y = 0$ in the point given by

$$\frac{x - x_1}{l} = \frac{0 - y_1}{m} = \frac{z - z_1}{n}$$

i.e., in the point $[x_1 - (ly_1/m), 0, z_1 - (ny/m)]$

$\therefore$ This generator (i) intersects the given circle if

$$[x_1 - (ly_1/m]^2 + [z_1 - (ny_1/m)]^2 = 1$$

$$(mx_1 - ly_1)^2 + (mz_1 - ny_1)^2 = m^2$$

$\therefore$ The locus of $P\,(x_1, y_1, z_1)$ or the required equation of the cylinder is

$$(mx - ly)^2 + (mz - ny)^2 = m^2 \quad \textbf{Ans.}$$

Example 15:

Find the equation of the surface generated by a straight line which is parallel to the line $y = mx$, $z = nx$ *and intersects the ellipse* $x^2/a^2 + y^2/b^2 = 1$, $z = 0$.

Solution:

Equations of the given line are $\frac{x}{l} = \frac{y}{m} = \frac{z}{n}$. ...(i)

Let P (x_1, y_1, z_1) be any point on the surface (which is a cylinder by definition), then the equations of the generator through P, which is parallel to (i) are given by

$$(x - x_1)/l = (y - y_1)/m = (z - z_1)/n. \quad ...(ii)$$

This generator meets the plane z = 0 in the point given as

$$\frac{x - x_1}{l} = \frac{y - y_1}{m} = \frac{0 - z_1}{n}$$

i.e., in the point $[x_1 - (z_1/n), y_1 - (mz_1/n), 0]$ **(Note)**

This point must lie on the given guiding curve and so we have

$$\frac{[x_1 - (z_1/n)]^2}{a^2} + \frac{[y_1 - (mz_1/n)]^2}{b^2} = 1,$$

from $\frac{x^2}{a^2} + \frac{y^2}{b^2} = 1$

$\Rightarrow$ $b^2 (nx_1 - z_1)^2 + a^2 (ny_1 - mz_1)^2 = a^2b^2n^2$

$\therefore$ The locus of P (x_1, y_1, z_1) or the equation of required surface is

$$b^2 (nx - z)^2 + a^2 (ny - mz)^2 = a^2b^2n^2.$$ **Ans.**

Example 16:

Find the equation of the cylinder with generators parallel to z-axis and passing through the curve $ax^2 + by^2 + 2cz$, $lx + my + nz = p$.

Solution:

The curve is given by $ax^2 + by^2 = 2cz$, ...(i)

$$lx + my + nz = p. \quad ...(ii)$$

Since generators are parallel to z-axis. so eliminating z between (i) and (ii) we get the required equation as

$$ax^2 + by^2 + 2c (p - lx - my)/n$$

$\Rightarrow$ $n (ax^2 + by^2) + 2c (lx + my) - 2pc = 0.$ **Ans.**

Example 17:

Find the equation of the cylinder with generators parallel to the axis of x and passing through the curve

$$ax^2 + by^2 + cz^2 = 1, \; lx + my + nz = p.$$

Solution:

The curve is given by

$$ax^2 + by^2 + cz^2 = 1 \quad ...(i)$$

and $\quad lx + my + nz = p \quad ...(ii)$

Since the generators are parallel to x-axis, so eliminating x between (i) and (ii) we get the required equations as a

$$[(p - my - nz)/l]^2 + by^2 + cz^2 = 1$$

$$\Rightarrow \quad a(p - my - nz)^2 + bl^2y^2 + cl^2z^2 = l^2$$

$$\Rightarrow \quad (am^2 + bl^2)\, y^2 + (an^2 + cl^2)\, z^2 + 2amnyz - 2ampy$$

$$- 2anpz + (ap^2 - l^2) = 0. \qquad \textbf{Ans.}$$

Example 18:

Find the equation to the cylinder whose generators are parallel to the line x/1 = y (– 2) = z/3 and the guiding curve is the ellipse

$$x^2 + 2y^2 = 1,\ z = 3.$$

Solution:

Let P (x_1, y_1, z_1) be any point on the cylinder, then the equations of the generator through P are given as $\dfrac{x - x_1}{1} = \dfrac{y - y_1}{-2} = \dfrac{z - z_1}{3}$. ...(i)

This generator meets the plane z = 3 in the point given by

$$\frac{x - x_1}{1} = \frac{y - y_1}{-2} = \frac{3 - z_1}{3}$$

i.e., in the point $\left[x_1 - \frac{1}{3} z_1 + 1,\ y_1 + (2/3)\, z_1 - 2,\ 3\right]$. **(Note)**

∴ This generator intersects the given conic if

$$[x_1 - (1/3)\, z_1 + 1]^2 + 2\,[y_1 + (2/3)\, z_1 - 2]^2 = 1.$$

∴ The locus of P (x_1, y_1, z_1) or the required equation of the cylinder is

$$\left(x - \frac{1}{3}z + 1\right)^2 + 2\,[y + (2/3)\, z - 2]^2 = 1$$

$$\Rightarrow \quad x^2 + (1/9)\, z^2 + 1 - (2/3)\, xz + 2x - (2/3)\, z + 2\,[y^2 + (4/9)\, z^2$$

$$+ 4 + (4/3\ yz - (8/3)\, z] = 1$$

$$\Rightarrow \quad x^2 + 2y^2 + z^2 - \left(\frac{2}{3}xz\right) + \left(\frac{8}{3}yz\right) + 2x - 8y - 6z + 8 = 0.$$

Example 19:

Find the equation of a right circular cylinder whose axis is

$$x/l = y/m = z/n.$$

Solution:

Let r be the radius of the cylinder and the equations of its axis are given by

$$\frac{x}{l} = \frac{y}{m} = \frac{z}{n}. \quad \text{...(i)}$$

If P (x_1, y_1, z_1) be any point on the cylinder, then the length of the perpendicular from P to (i) must be r and so from the Chapter on Straight Lines we have

$$r^2 (l^2 + m^2 + n^2) = \{ny_1 - mz_1\}^2 + \{lz_1 - nx_1\}^2 + mx_1 - ly_1\}^2$$

$\therefore$ The locus of P (x_1, y_1, z_1) or the required equation to the cylinder is given by $(ny - mz)^2 + (lz - nx)^2 + (mx - ly)^2 = r^2 (l^2 + m^2 + n^2)$. **Ans.**

Example 20:

Find the equation of right circular cylinder whose axis is x = 2y = – z and radius is 4.

Solution:

Let P (x_1, y_1, z_1) be any point on the cylinder. Then the length of pependicular from P to the given axis $\frac{x}{1} = \frac{y}{1/2} = \frac{z}{-1}$,

must be equal to the radius 4.

So we have $4 [(1)^2 + (1/2)^2 + (-1)^2]$

$= [y_1 (-1) - (1/2) z_1]^2 + [z_1 (1) - (-1) x_1]^2 + [x_1 (1/2) - y_1 (1)]^2.$

$\therefore$ Locus of P (x_1, y_1, z_1) or the required equation to the cylinder is given by

$$4^2 [1 + (1/4) + 1] = (1/4) (2y + z)^2 + (z + x)^2 + (1/4\text{\textbackslash} (x - 2y)^2$$

$$\Rightarrow \quad 144 = (2y + z)^2 + 4 (x + z)^2 + (x - 2y)^2$$

$$\Rightarrow \quad 5x^2 + 8y^2 + 5z^2 + 4yz + 8xz - 4xy = 144.$$ **Ans.**

Example 21:

Find the equation of the right circular cylinder of radius 2 and having as axis the line $\frac{1}{2}(x - 1) = (y - 2) = \frac{1}{2}(z - 3)$.

Solution:

Let P (x_1, y_1, z_1) be any point on the cylinder. Then length of the perpendicular from P (x_1, y_1, z_1) to the given line

$$\frac{x-1}{2}=\frac{y-2}{1}=\frac{z-3}{2}$$

must be equal to the radius 2. So from Straight Lines we get

$$2^2\,[2^2 + 1^2 + 2^2] = 2\,(y_1 - 2) - 1\,(z_1 - 3)]^2$$
$$+ \{2\,(z_1 - 3) - 2\,(x_1 - 1)\}^2 + \{1\,(x_1 - 1) - 2\,(y_1 - 2)\}^2$$

$$\Rightarrow \quad 4\,[9] = (2y_1 - z_1 - 1)^2 + (2z_1 - 2x_1 - 4)^2 + (x_1 - 2y_1 + 3)^2$$

$\therefore$ The required equation or the locus of P (x_1, y_1, z_1) is

$$(2y - z - 1)^2 + (2z - 2x - 4)^2 + (x - 2y + 3)^2 = 36$$

$$\Rightarrow \quad 5x^2 + 8y^2 + 5z^2 - 4yz - 8xz - 22x - 16y - 14z - 10 = 0.$$ **Ans.**

Example 22:

Find the equation of the right circular cylinder of radius 2 whose axis passes through (1, 2, 3) and has direction cosines proportional (2, – 3, 6).

Solution:

Let P (x_1, y_1, z_1) be any point on the cylinder. Then the length of the perpendicular from P (x_1, y_1, z_1) to the given axis

$$\frac{x-1}{2}=\frac{y-2}{-3}=\frac{z-3}{6}$$

must be equal to 2. So we have

$$2^2\,[2^2 + (-3)^2 + 6^2] = \{6\,(y_1 - 2) - (-3)\,(z_1 - 3)\}^2$$
$$+ \{2\,(z_1 - 3) - 6\,(x_1 - 1)\}^2 + \{(-3)\,(x_1 - 1) - 2\,(y_1 - 2)\}^2$$

$$\Rightarrow \quad 4\,(49) = 9\,(2y_1 + z_1 - 7)^2 + 4\,(z_1 - 3x_1)^2 + (3x_1 + 2y_1 - 7)^2$$

$$\Rightarrow \quad 45x_1^2 + 40y_1^2 + 13z_1^2 + 36y_1z_1 - 24z_1x_1 + 12x_1y_1 - 42x_1 - 280y_1$$
$$- 126z_1 + 294 = 0.$$

$\therefore$ The locus of P (x_1, y_1, z_1) or the required equation of the cylinder is

$$45x^2 + 40y^2 + 13z^2 + 36yz - 24zx + 12xy - 42x$$
$$- 280y - 126z + 294 = 0.$$ **Ans.**

Example 23:

Find the equation of the right circular cylinder of radius and axis $\frac{1}{2}(x-1)=\frac{1}{2}(y-3)=\frac{1}{7}(5-z)$.

Solution:

Let P (x_1, y_1, z_1) be any point on the cylinder. Then the length of perpendicular from p to the given line

$$\frac{x-1}{2} = \frac{y-3}{2} = \frac{z-5}{-7} \text{ must be equal to 3}$$

$$3^2 [2^2 + 2^2 + (-7)^2] = \{-7(y_1 - 3) - 2(z_1 - 5)\}^2$$

$$\Rightarrow \quad 513 = (7y_1 + 2z_1 - 31)^2 + (7x_1 + 2z_1 + 17)^2 + 4(x_1 - y_1 + 2)^2$$

$$= 49y_1^2 + 4z_1^2 + 961 + 28y_1z_1 - 434y_1 - 124z_1 + 49x_1^2 + 4z_1^2 + 289$$

$$+ 28x_1z_1 + 238x_1 + 68z_1 + 4x_1^2 + 4y_1^2 + 16 - 8x_1y_1 + 16x_1 - 16y_1$$

$$\Rightarrow \quad 53x_1^2 + 53y_1^2 + 8z_1^2 + 28y_1z_1 + 28z_1x_1 - 8x_1y_1$$

$$+ 254x_1 - 450y_1 - 56z_1 + 753 = 0.$$

$\therefore$ Locus of P (x_1, y_1, z_1) or the required equation of the curve is

$$53x^2 + 53y^2 + 8z^2 + 28yz + 28xz - 8xy + 254x$$

$$- 450y - 56z + 753 = 0. \text{ **Ans.**}$$

Example 24:

Find the equation of the right circular cylinder whose axis is $\frac{1}{2}(x-1) = y-2 = \frac{1}{2}(z-3)$ *and radius 5.*

Solution:

Let P (x_1, y_1, z_1) be any point on the cylinder. Then length of perpendicular from P (x_1, y_1, z_1) to the given line must be equal to the radius 5. So from straight lines, we get

$$5^2 [2^2 + 1^2 + 2^2] = \{2(y_1 - 2) - 1(z_1 - 3)\}^2$$

$$+ \{2(z_1 - 3) - 2(x_1 - 1)\}^2 + (1(x_1 - 1) - 2(y_1 - 2)\}^2$$

$$\Rightarrow \quad 25(9) = (2y_1 - z_1 - 1)^2 + (2z_1 - 2x_1 - 4)^2 + (x_1 - 2y_1 + 3)^2$$

$\therefore$ The required equation or the locus of P (x_1, y_1, z_1) is

$$25(9) = (2y - z - 1)^2 + (2z - 2x - 4)^2 + (x - 2y + 3)^2$$

$$\Rightarrow \quad 5x^2 + 8y^2 + 5z^2 - 4xy - 4yz - 8zx + 22x - 16y - 14z = 199.$$

Example 25:

Show that the equation of the right circular cylinder described on the circle through the three points A (1, 0, 0), B (0, 1, 0) and C (0, 0, 1) as the guiding curve is $x^2 + y^2 + z^2 - yz - zx - xy = 1$.

Solution:

The equation of the plane ABC is $x + y + z = 1$. ...(i)

Let the sphere through A (1, 0, 0), B (0, 1, 0) and C (0, 0, 1) pass through the origin O (0, 0, 0) also. Then the equation of the sphere OABC is

$$\begin{vmatrix} x^2+y^2+z^2 & x & y & z & 1 \\ 1 & 1 & 0 & 0 & 1 \\ 1 & 0 & 1 & 0 & 1 \\ 1 & 0 & 0 & 1 & 1 \\ 0 & 0 & 0 & 0 & 1 \end{vmatrix} = 0$$

$$x^2 + y^2 + z^2 - x - y - z = 0 \quad ...(ii)$$

The equations of guiding circle of the cylinder are

$$x^2 + y^2 + z^2 - x - y - z = 0;\ x + y + z = 1 \quad ...(iii)$$

Now the d.r.'s of the axis of the cylinder which is perpendicular to the plane of the circle given by $x + y + z = 1$ are 1, 1, 1.

So let one of the generators of the cylinder passing through any point P (α, β, γ) on the cylinder be $\frac{x-\alpha}{1} = \frac{y-\beta}{1} = \frac{z-\gamma}{1}$

Any point on the generator is $(\alpha + r, \beta + r, \gamma + r)$

If the point lies on the circle (iii), we have

$$(\alpha + r)^2 + (\beta + r)^2 + (\gamma + r)^2 - (\alpha + r) - (\beta + r) - (\gamma + r) = 0$$

and $$(\alpha + r) + (\beta + r) + (\gamma + r) = 1$$

$\Rightarrow$ $$(\alpha^2 + \beta^2 + \gamma^2 - \alpha - \beta - \gamma) + r(2\alpha + 2\beta + 2\gamma - 3) + 3r^2 = 0 \quad ...(iv)$$

and $$r = (1/3)(1 - \alpha - \beta - \gamma).$$

Eliminating r between (iv) and (v) we get

$$(\alpha^2 + \beta^2 + \gamma^2 - a - \beta - \gamma) + (1/3)(2\alpha + 2\beta - 3)(1 - \alpha - \beta - \gamma) + 3(1/9)(1 - \alpha - \beta - \gamma)^2 = 0$$

$$\Rightarrow 3\alpha^2 + 3\beta^2 + 3\gamma^2 - 3\alpha - 3\beta - 3\gamma + (2\alpha + 2\beta + 2\gamma - 3)(1 - \alpha - \beta - \gamma) + (1 - \alpha - \beta - \gamma)^2 = 0$$

$\Rightarrow$ $(\alpha^2 + \beta^2 + \gamma^2) - (\alpha\beta + \beta\gamma + \gamma\alpha) - 1 = 0$, on simplifying.

$\therefore$ The required equation of the cylinder or the locus of P (α, β, γ) is

$$x^2 + y^2 + z^2 - xy - yz - zx = 1.$$ **Hence proved.**

EXERCISES

1. Find the equation of the circular cone whose axis is the line $x = y = z$ and semi-vertical angle is 30°, the vertex being at the origin. **Ans.** $5(x^2 + y^2 + z^2) = 8(xy + yz + zx)$.
2. Obtain the equation to a right circular cone whose vertex is the origin, axis the y-axis the semi-vertical angle is a.

   Ans. $x^2 - y^2 \tan^2 a + z^2 = 0$
3. Find the equation of the right circular cone whose vertex is the origin, axis the y-axis and the semi-vertical angle is 30°.

   Ans. $3x^2 - y^2 + 3z^2 = 0$
4. Find the equation of the right circular cone whose vertex is the origin, axis is the axis of y and the semi-vertical angle is $\pi/3$.

   Ans. $x^2 + z^2 = 3y^2$
5. Find the equation of a right circular cone whose vertex is (2, –3, 5) and axis makes equal angles with the axes of co-ordinates and passes through (1, – 2, 3).
6. Find the condition that the section of the surface $ax^2 + by^2 + cz^2 = 1$ by the $lx + my + nz = 0$ may be (i) a parabola, (ii) an ellipse, (iii) a hyperbola.
7. A right circular cone has its vertex at (2, 3, – 5), its axis passes through (3, – 2, 6) and its semi-vertical angle is 30°. Find the equation of the cone.

Ans. $5x^2 + 5y^2 + 5z^2 + 8xy - 8yz + 8zx - 4x - 86y + 58z + 278 = 0$

8. Find the locus of a luminous point which moves so that the sphere $x^2 + y^2 + z^2 = 2az$ casts a parabolic shadow on the plane $z = 0$.

   Ans. $z^2(x^2 + y^2 + z^2 - 4az + 4a^2) = 2a(x^2 + y^2)$
9. Find the locus of luminous point of the ellipsoid $x^2/a^2 + y^2/b^2 + z^2/c^2 = 1$ which castes a circular shadow on the plane $z = 0$.
10. Find the locus of the points from which three mutually perpendicular tangent lines can be drawn to the surface $ax^2 + by^2 + cz^2 = 0$.

    Ans. $a(b + c)^2 x^2 + b(c + a) y^2 + c(a + b) z^2 = a + b + c$.
11. Find the equation of the cylinder whose generating line is parallel to z-axis and the guiding curve is $x^2 + y^2 = z$, $x + y + z = 0$.
12. Find the equation of the cylinder which intersects the curve $ax^2 + by^2 + cz^2 = 1$, $lx + my + nz = p$ and whose generators are parallel to the axis of x.

Ans. $(bl^2 + am^2)\, y^2 + (cl^2 + am^2)\, z^2 + 2amnyz - 2ampy - 2ampz + (ap^2 - l^2) = 0$

13. Find the equation of the circular cylinder whose guiding circle is $x^2 + y^2 + z^2 = 9$, $x - y + z = 0$.

Ans. $x^2 + y^2 + z^2 + xy + yz - zx - (27/2) = 0$

14. Find the equation of the enveloping cylinder to the surface $ax^2 + by^2 + cz^2 = 1$, whose generators are parallel to the line $x/l = y/m = z/n$.

Ans. $(alx + bmy + cnz)^2 = (al^2 + bm^2 + cn^2)(ax^2 + by^2 + cz^2 - 1)$

15. Prove that the line $\frac{1}{2}x = -y = \frac{1}{2}z$ is the generator of the cone $2x^2 + 4y^2 - 3x^2 = 0$.

16. Find the equation of the cone through the co-ordinate axes and the in which $lx + my + nz = 0$, cuts the cone $ax^2 + by^2 + cz^2 + 2fyz + 2gzx + 2hxy = 0$.

17. The plane $x - y - z = 1$ meets the co-ordinate axes in A, B, C. Prove that the equation of the cone generated by lines throgh O, the origin, to meet the circle ABC is $yz + zx + xy + 0$.

18. Find the equation of the generator of the cone $x^2 + y^2 - z^2 = 0$ through the point (3, 4, 5).

19. Find the equation to a cone which passes through the co-ordiantes axes and the two lines $x = 3y = 2z$ and $x = -3y = 3z$.

20. A variable plane is parallel to the given plane $(x/a) + (y/b) + (z/c) = 0$ and meets the axes in A, B, C respectively. Prove that the circle ABC lies on the curve

$$yz\left(\frac{b}{c}+\frac{c}{b}\right)+zx\left(\frac{a}{c}+\frac{c}{a}\right)+xy\left(\frac{a}{b}+\frac{b}{a}\right)=0.$$

21. Find the equation of the cone with vertex (0, 0, 1) and passing through the points of the circle $x^2 + y^2 = 1$, $z = 0$.

Ans. $x^2 + y^2 + z^2 + 2z - 1 = 0$

22. Find the equation of the cone with vertex at (1, 1, 1) which passes through the curve $x^2 + y^2 = 4$, $z = 6$.

Ans. $25x^2 + 25y^2 - 2z^2 + 10yz + 10zx - 60x - 16z + 68 = 0$

23. Find the angle between the generators in which the plane $x - 2y + z = 0$ cuts the cone $x^2 - 5y^2 + z^2 = 0$. **Ans.** $\cos^{-1}(5/6)$

24. Find the angle between the two lines in which the plane $6x - 10y - 7z = 0$ cuts the cone $108x^2 - 20y^2 + 7z^2 = 0$. **Ans.** $\cos^{-1}(16/21)$

25. Find the equations of the line along which the plane $x + y + z = 0$ cuts the cone $x^2 + 2y^2 + z^2 = 0$. Also find the acute angle between them.

26. Prove that the equation

$$x^2 - 2y^2 + 3z^2 - 4xy + 5yz - 6zx + 8x - 19y - 2z = 20$$

represents a cone. Show that the coordinates of its vertex are $(1, -2, 3)$.

27. Prove that the equation $2x^2 - 8xy - 4yz - 4x - 2y + 6z + 35 = 0$ represents a cone with its vertex at $(4, 3/2, -17/2)$.

28. Find the equation to the cone whose vertex is the point $(0, 0, 3)$ and guiding curve is the circle $x^2 + y^2 = 4$, $z = 0$.

29. Find the equation of the right circular **cone** whose vertex is the origin, axis the x-axis and the semi-vertical angle is $60°$.

Ans. $3x^2 - y^2 - z^2 = 0$.

30. Find the equation of the right circular cone whose vertex is origin, whose axis is the line $x = \frac{1}{2}y = \frac{1}{3}z$ and whose vertical angle is $60°$.

Ans. $19x^2 + 13y^2 + 3z^2 = 8xy + 24yz + 12zx$

31. The radius of a normal section of right circular cylinder is 2 units; the axis lies along the straight line $\frac{1}{2}(x - 1) = -(y + 3) = \frac{1}{5}(z - 2)$. Find the equation.

32. Show that the generating line of the cylinder given below is parallel to x-axis: $ax^2 + 2hyz + by^2 + 2gx + 2fy + k = 0$.

33. Prove that the equation of the right circular cylinder whose axis is $(x - 2)/2 = (y - 1)/1 = z/3$ and passes through the point $(0, 0, 3)$ is $10x^2 + 13y^2 + 5z^2 - 6yz - 12zx - 4xy - 36x - 18y + 30z - 135 = 0$.

34. Find the equation of the right circular cylinder whose axis is $\frac{1}{2}x = \frac{1}{3}y = \frac{1}{6}z$ and radius 5.

Ans. $45x^2 + 40y^2 + 13z^2 - 12xy - 36yz - 24zx - 1225 = 0$

35. Find the equation of a right circular cylinder with axis

$$\frac{x-1}{3}=\frac{y-2}{-1}=\frac{z-3}{2}$$ and radius 5.

Ans. $5x^2+13y^2+10z^2+4yz-12zx+6xy+14x-70t-56z-203 = 0$

36. Find the equation of the cylinder whose generators are parallel to the line $x = y/2 = z/2$ and whose guiding curve is the ellipse $x^2 + 2y^2 = 1, z = 0$.

Ans. $4x^2 + 8y^2 + 9z^2 - 4xz - 16yz - 4 = 0$

37. Find the **equation to the** cylinder whose generators are parallel to the **line** $x/1 = y/-2 = z/3$ and guiding curve is the ellipse $x^2 + 4y^2 = 1, z = 6$.

Ans. $9x^2 + 36y^2 + 25z^2 - 6zx + 48yz + 36x - 288y - 204z + 603 = 0$

38. Find the equation of the cylinder whose generators are parallel to the line $x = y/2 = z/3$ and whose guiding curve is the ellipse $x^2 + 2y^2 = 1, z = 0$.

Ans. $3x^2 + 6y^2 + 3z^2 - 2xz - 8yz - 3 = 0$

39. Find the equation of the right circular whose axis is $x/2 = y/3 = z/6$ and radius 4.

Ans. $45x^2 + 40y^2 + 13z^2 + 36yz + 24zx + 12xy - 441 = 0.$

40. Obtain the equation of the cylinder which passes through $y^2 = 4ax$, $z = 0$ and whose generators are parallel to the line $x = y - z$.

Ans. $(y - z)^2 = 4a (x - z)$

41. Find the equation of the cone which passes through three co-ordinate axes and the straight lines

$$x=-\frac{1}{2}y=\frac{1}{3}z;\ \frac{1}{3}x=\frac{1}{2}y=-z.$$

42. Show that the locus of the mid-points of chords of the cone $ax^2 + by^2 + cz^2 + 2fyz + 2gzx + 2hxy = 0$ drawn parallel to the line $x/l = y/m = z/n$ is the plane

$x (al + hm + gn) + y (hl + bm + fn) + z (gl + fm + cn) = 0.$

43. Prove that the equation of the right circular cone which passes through the line $x = 3y = - 5z$ and has $x = y = z$ as its axis also passes through the axes of co-ordinates.

Ans. Cone is $xy + yz + zx = 0$.

44. Show that the equation of pair of tangent to the cone $ax^2 + by^2 + cz^{+2} = 0$ passing through the line $x/l = y/m = z/n$ is

$(al^2 + bm^2 + cn^2) (ax^2 + by^2 + cz^2) = (alx + bmy + cnz)^2$

45. Obtain the general equation of cone passing through the coordinate axes.

46. A is a point on axis OX and B on axis OY, so that angle OAB constant and equal to α. Taking AB as diameter a circle is drawn whose plane is parallel to OZ. Prove that as AB varies, the circle generates the cone

$$2xy - z^2 \sin 2a = 0.$$

47. Show that f $(l, m, n) = 0$, where l, m, n are the direction cosines the generator of the cone f $(x, y, z) = 0$.

48. Find the locus of the vertices of enveloping cones of the surface $(x^2/a^2) + (y^2/b^2) + (z^2/c^2) = 1$, if sections of cones by the plane $z = 0$ are circles.

49. Cone $ax^2 + by^2 + cz^2 + 2fyz + 2gzx + 2hxy = 0$ has the pependicular generators if

(i) $(1/a) + (1/b) + (1/c) = 0$; (ii) $f + g + h = 0$;

(iii) $a + b + c = 0$; (iv) $a^2 + b^2 + c^2$

50. Write down the equation of the generator of the cylinder $x^2 + y^2 = a^2$ which passes through the point (a, 0, 0).

51. Show that the plane $z = a$ meets any enveloping cone of the sphere $x^2 + y^2 + z^2 = a^2$ in a conic which has a focus at the point (0, 0, 0).

52. Prove that the tangent planes to the cone $fyz + gzx + hxy = 0$ are perpendicular to the generators of the cone

$$f^2 x^2 + g^2y^2 + h^2z^2 - 2ghyz - 2fhzx - 2fgxy = 0.$$

53. Find the equations of the tangent planes to $2x^2 - 6y^2 + 3z^2 = 5$, which pass through the line $x + 9y - 3z = 0$, $3x - 3y + 6z = 5$.

54. Find the locus of the lines through the vertex of the cone $ax^2 + by^2 + cz^2 + 2fyz + 2gzx + 2gzx + 2hxy = 0$ perpendicular to its tangent planes.

55. Find the enveloping cone of the of the sphere $x^2 + y^2 + z^2 - 2x + 4z = 1$ with its vertex at (1, 1, 1)

Ans. $4x^2 + 3y^2 - 5z^2 - 6zy - 8x + 16z = 4$

56. Find the equation of the right circular cone, whose vertex is (3, 2, 1), axis is the line $\frac{1}{4}(x - 3) = y - 2 = \frac{1}{3}(z - 1)$ and semi-vertical angle is 30°.

Ans. $7x^2 + 37y^2 + 21z^2 - 16xy - 12yz - 48zx + 38x - 98y + 126z - 32 = 0.$

57. Find the equation to the right circular cone having the semi-vertical angle equal to α and the straight line $x/l = y/m = z/n$ as its axis, l, m, n being the direction cosines.

Ans. $(x^2 + y^2 + z^2) = (lx + my + nz)^2 \sec^2 \alpha$.

58. Find the equation of the right circular cylinder whose radius is 1 and axis is z-axis.

59. Find the equation of the right circular cylinder whose axis is $\frac{1}{2}(x - 1) = -\frac{1}{3}(y - 2) = \frac{1}{6}(z + 3)$ and radius 2.

60. Find the equation of the right circular cylinder whose axis is the line $(x - \alpha)/l = (y - \beta)/m = (z - \gamma)/n$ and radius 3.

61. Find the equation of the right circular cylinder whose axis $x = 2y = -z$ and radius is 4. Prove that area of section of this cylinder by plane $z = 0$ is 24π.

62. Find the equation of the cylinder whose axis are parallel to co-ordinate axes.

63. Find the equation of the right circular cylinder described on circle the points (1, 0, 0), (0, 1, 0), (0, 0, 1) as the guiding curve.

64. Show that the plane $x + y - z = 0$ cuts the conicoid $4x^2 + 2y^2 + z^2 + 3yz + zx - 1 = 0$ in a circle. What is the radius of the circle.

Ans. $1/\sqrt{3}$.

65. Find the equation to the cone whose vertex is (0, 0, 0) and which contains the curve given by $x^2 - y^2 + 4ax = 0$, $x + y + z = b$.

Ans. $(b + 4a)x^2 - by^2 + 4ay + + 4az = 0$.

66. Find the equation of the cone whose vertex is the origin and whose generators pass through the section of the sphere $x^2 + y^2 + z^2 + 2z + 2y + 2z + 5 = 0$ by the plane $x + y + z = 1$.

Ans. $4(x^2 + y^2 + z^2) + 7(xy + yz + zx) = 0$

67. Show that the line $x/l = y/m = z/n$, whose $l^2 + 2m^2 - 3n^2 = 0$, is a generator of the cone $x^2 + 2y^2 - 3z^2 = 0$.

68. Find the equation of the cone generated by straight lines drawn through the point (1, 2, 3) whose direction ratios satisfy the relation $2l^2 + 3m^2 - 4n^2 = 0$. **Ans.** $2x^2 + 3y^2 - 4z^2 - 4x - 12y + 24z = 22$.

69. Prove that the equation

$(x^2/a^2) + (y^2/b^2) + (z^2/c^2) + 2ux + 2vy + 2wz + d = 0$

represents a cone if $a^2u^2 + b^2v^2 + c^2w^2 = d$.

70. Find the equation to the cone whose vertex is the point P (a, b, c) and whose generating line intersects the conic $px^2 + qy^2 = 1, z = 0$.

Ans. $c^2 (px^2 + qy^2) + (a^2p + b^2q - 1)^2 - 2c(apzx + bqyz - z) = c^2$

71. Find the equation of the cone with vertex (5, 4, 3) and with $3x^2 + 2y^2 = 6, y = z = 0$ as base

Ans. $147x^2 + 87y^2 + 101z^2 + 90yz - 210(zx + xy)+84(y+z)=294$.

72. Prove that the equation $2x^2 + 2y^2 + 7z^2 - 10yz - 10zx + 2x + 2y + 26z - 17 = 0$ represents a cone whose vertex is at (2, 2, 1).

73. Find the cone whose vertex is (1, 1, 0) and whose guiding curve is $y = 0, x^2 + z^2 = 4$.

74. Find the equation of the cone whose vertex is (1, – 2, 3) and whose guiding curve is $x^2 + y^2 = 9, z = 0$.

Ans. $9x^2 + 9y^2 - 4z^2 + 12yz - 6zx + 54z - 81 = 0$

75. If $x = y/2 = z$ represents one of a set of three mutually perpendicular generators of the cone $11yz + 6zx - 14xy = 0$, find the equations of the other two.

[**Ans.** $x/2 = - y/3 = z/4, - x/11 = y/2 = z/7$]

76. If the plane $2x - y + cz = 0$ cuts the cone $yz + zx + xy = 0$ in perpendicular lines, find the value of c.

77. Show that the lines given by $x - y - z = 0$, $ayz + bzx + cxy = 0$ are at right angles if $a = b + c$.

[**Hint :** The line $x/1 = y/-1 = z/-1$, which is normal to the given plane, is a generator of the given cone.

$\therefore \quad a (-1) (-1)+ b (-1) (1) + c (1) (-1) = 0 \Rightarrow a = b + c$].